TO NANCY

THEORIES OF LIGHT
FROM DESCARTES TO NEWTON

A. I. SABRA
Harvard University

Paolo Mancosu
Yale University

CAMBRIDGE UNIVERSITY PRESS

Cambridge
London   New York   New Rochelle
Melbourne   Sydney

Published by the Press Syndicate of the University of Cambridge
The Pitt Building, Trumpington Street, Cambridge CB2 1RP
32 East 57th Street, New York, NY 10022, USA
296 Beaconsfield Parade, Middle Park, Melbourne 3206, Australia

First published by Oldbourne Book Co. Ltd 1967
First published by Cambridge University Press 1981

Printed in Canada

Library of Congress Cataloging in Publication Data
Sabra, A. I.
Theories of light, from Descartes to Newton.
Bibliography: p.
Includes index.
1. Light – History. I. Title.
QC401.S3 1981   535'.1   81-6108   AACR2
ISBN 0 521 24094 8 hard covers
ISBN 0 521 28436 8 paperback

Preface to New Edition

The first edition of this book was allowed to go out of print only a few years after its publication in 1967. Now, fourteen years later, it is being reprinted with only some minor corrections and an added bibliography. It did not take me long to decide that the present edition should remain substantially the same as the first. When I first turned my attention to the study of seventeenth-century optics, the questions that I had in mind were primarily philosophical or methodological in nature. Although during the writing of this book I had already begun to develop an interest in the historical approach to the study of science, my initial interests were bound to influence the mode of treatment as well as the choice of topics in the chosen area of my research. I was in effect asking philosophical and methodological questions in a historical context. Responses to the book on the part of both philosophers and historians indicate to me that my venture was well worth the effort. Here I must reassert my belief in the value of philosophical concepts and philosophical analysis for the historical investigation of science, and I confess that, by and large, my position on the issues discussed in the book is fundamentally the same now as it was when it was first published. Nevertheless, if I were to embark now on a study of the same subject, or extensively revise what I wrote earlier, the result would naturally be a different book – one, however, in which my debt to Sir Karl Popper and to Alexandre Koyré would remain undiminished.

Between 1955, when the *Theories* was first written, and 1967, when it first appeared, very few studies of seventeenth-century optics were published. Most important among these were the articles by Richard Westfall, which I was able to take notice of in the book and include in the bibliography. The situation has changed substantially in the last twelve years or so. Numerous studies on various aspects of seventeenth-century optics have appeared in print. The larger number of these are concerned with Newton, and they form part of the current surge in Newtonian studies. Most of them contain new and valuable information and insights. Frequently the authors of these studies have had occasion to comment on or disagree with views and interpretations expressed in my book. It would of course be impossible even to begin to do justice to these comments and criticisms within the limits of a preface, but I hope that the appended bibliography will at least make it easy for the reader to locate the results of more recent research.

Cambridge, Massachusetts   A. I. Sabra
January 1981

Contents

Preface		9
Introduction		11
Abbreviations		16

I   *Descartes' Theory of Explanation and the Founda-*
     tions of his Theory of Light   **17**

1. The position expressed in the *Dioptric* and the *Discourse:* arguing from suppositions, p. 17; discussion with Morin, p. 20; limitation of proof in physics, p. 22. 2. Metaphysical foundations of physics, p. 24. 3. The role of analogies: in the *Dioptric* and *Le Monde ou Traité de la Lumière*, p. 27, in the *Regulae*, p. 29; Cartesian analysis and Baconian induction, p. 30. 4. The role of natural history: Descartes and Bacon, p. 33. 5. The deductive procedure, p. 37; '*venir au-devant des causes par les effets*': the role of experiment, p. 38; conjectural explanation; absolute and moral certainties, p. 41. 6. Conclusion, p. 44

II   *Descartes' Doctrine of the Instantaneous Propaga-*
     tion of Light and his Explanation of the Rainbow
     and of Colours   **46**

1. The doctrine of the instantaneous propagation of light before Descartes, p. 46; place of the doctrine of instantaneous propagation in Descartes' system of physics, p. 48; the doctrine as expounded in the *Traité de la Lumière*: the cosmological fable, p. 50; the laws of nature and their cosmological application, p. 51; instantaneous propagation and the Cartesian concept of matter, p. 55; Descartes' argument from lunar eclipses, p. 57; the crossing of light rays, p. 59. 2. Theory of colours and the rainbow: birth of the *Meteors*, p. 60; experimental character of Descartes' investigations: explanation of the rainbow, p. 62; mechanical explanation of colours, p. 65

III   *Descartes' Explanation of Reflection. Fermat's*
     Objections   **69**

1. Explanation of reflection before Descartes: in antiquity, p. 69; by Ibn al-Haytham, p. 72; by Roger Bacon,

p. 76, and Kepler, p. 78. **2.** Descartes' explanation in the *Dioptric*: the ball analogy, p. 78; denial of *quies media* and of elasticity, p. 80; kinematical derivation of the law of reflection, p. 82. **3.** The controversy with Fermat, p. 85; Fermat's objection against Descartes' use of the parallelogram method, p. 87; Descartes' reply, p. 89

IV *Descartes' Explanation of Refraction. Fermat's 'Refutations'*   93

1. Explanation of refraction before Descartes: Ptolemy, p. 93; Ibn al-Haytham, p. 93; Roger Bacon, Witelo and Kepler, p. 98. **2.** Discovery of the law of refraction: Descartes and Snell, p. 99. **3.** Descartes' treatment of refraction: in the *Cogitationes privatae* (1619–21), p. 105; in the *Dioptric*, p. 107. **4.** Analysis of the Cartesian treatment: derivation of the refraction law, p. 110; physical interpretation of Descartes' assumptions, p. 112. **5.** The controversy with Fermat over refraction: Fermat's statement of 1664, p. 116; Fermat and Descartes' distinction between 'force' and 'determination' of motion, p. 117; Fermat's 'refutation' of Descartes' proof of the refraction law, p. 121. **6.** Fermat and Clerselier: the mathematical problem, p. 127; Clerselier's collision model, p. 133

V *Fermat's Principle of Least Time*   136

1. Introductory, p. 136. **2.** La Chambre, Fermat and the principle of economy, p. 137; Fermat's interpretation of the principle as a principle of least time, p. 139. **3.** Fermat's method of maxima and minima, p. 144; his *Analysis for Refractions*, p. 145; Fermat, Leibniz and Descartes, p. 147. **4.** Fermat's *Synthesis for Refractions*, p. 150. **5.** Conclusion, p. 154

VI *Huygens' Cartesianism and his Theory of Conjectural Explanation*   159

1. Mechanical explanation as the aim of physical science: Huygens' dispute with Roberval over the cause of gravity, p. 159; Huygens' Cartesianism, p. 163; Huygens and Newton, p. 164. **2.** Conjectures *versus* induction: Huygens and Bacon, p. 170; Baconian induction, p. 175; Huygens' position in the *Traité de la Lumière*, p. 181

VII   *Two Precursors of Huygens' Wave Theory: Hooke and Pardies*   185

1. Introductory: the formation of Huygens' theory and preceding achievements in optics, p. 185. 2. Hooke's investigations in the *Micrographia*, p. 187; his explanation of the refraction of waves, p. 192. 3. Pardies' wave hypothesis and Ango's *L'optique*, p. 195

VIII   *Huygens' Wave Theory*   198

1. Huygens and the 'difficulties' in Descartes' theory of light: *Projet du Contenu de la Dioptrique* (1673), p. 198, Huygens and Roemer, p. 202, the crossing of light rays, p. 207. 2. Mathematization of the Cartesian picture: Huygens' principle of secondary waves and his explanation of rectilinear propagation, p. 212; E. Mach and the origin of Huygens' principle, p. 215. 3. Huygens' construction for ordinary refraction, p. 216; derivation of Fermat's principle, p. 218; total and partial reflections, p. 219. 4. Explanation of double refraction: Huygens and Bartholinus, p. 221; Huygens' researches, p. 223; Newton and double refraction, p. 226. 5. Conclusion, p. 229

IX   *Newton's Theory of Light and Colours, 1672*   231

1. Introductory, p. 231. 2. Analysis of Newton's 1672 paper to the Royal Society: the geometrical problem of dispersion, p. 234; the *experimentum crucis* and the doctrine of white light, p. 239. 3. The historicity of the 1672 account, p. 244; Newton's inductive argument, p. 248

X   *Three Critics of Newton's Theory: Hooke, Pardies, Huygens*   251

1. Hooke's 'Considerations': his general attitude to Newton's theory, p. 251; his explanation of the generation of colours by refraction in the *Micrographia*, p. 254; his idea that white light might be represented by the 'coalescence' of vibrations or waves, p. 259; his insistence (in 1672) on his dualistic theory of colours, p. 261. 2. Pardies' suggestions: his difficulties in understanding Newton's experiments, and his final acceptance of the unequal refractions of colours, p. 264. 3. Huygens' reservations: his demand for a mechanical explanation of colours, and his final acceptance of their differential refrangibility, p. 268

XI *Newton's Dogmatism and the Representation of White Light*　273

1. Newton's general strategy with his critics, p. 273. **2.** White light as a heterogeneous mixture—Newton and Hooke, p. 276. **3.** Newton's empirical dogmatism and the use of hypotheses, p. 284; his *a priori* conception of rays as discrete entities, p. 287. **4.** Colours as qualities of the rays and Huygens' demand for a mechanical explanation, p. 290. **5.** Newton's atomism and his interpretation of the *experimentum crucis*, p. 294

XII *The Two Levels of Explanation: Newton's Theory of Refraction*　298

Introductory, p. 298. **1.** Dynamical explanation of refraction—Newton and Descartes, p. 299; Newton's demonstration of the refraction law in the *Principia*, p. 302, in the *Opticks*, p. 304. **2.** Hypothetical explanation of refraction in terms of impulsion, p. 308, in terms of attraction, p. 311; logical status of Newton's dynamical explanation, p. 313; the role of Foucault's experiment, p. 315

XIII *The Two Levels of Explanation: Newton's Theory of the Colours of Thin Plates*　319

Introductory: the problem of partial reflection, p. 319. **1.** Hooke and the colours of thin plates: Boyle and Hooke, p. 321; Hooke's researches in the *Micrographia*, p. 322; Hooke and Newton, p. 327. **2.** Newton's quantitative approach, p. 331; his theory of 'fits', p. 334; his explanation of partial reflection, p. 336; logical status of the theory, p. 337; alternative hypothetical explanations of the 'fits', p. 338

Bibliography　343

Corrections　356

Index　357

Preface

The following chapters contain the substance, revised in parts, of a thesis which earned the Ph.D. degree from the University of London in 1955. My first acknowledgement must be to the University of Alexandria for a generous scholarship which allowed me to devote much time to this work. My debt to Professor Sir Karl Popper, who supervised my research at the London School of Economics, is basic and manifold. It was he who kindled my interest in the growth of scientific knowledge, and from his writings and lectures I came to see in the history of science the history of man's most imaginative and most rational enterprise. The suggestions and criticisms which I received from him in the course of numerous discussions have influenced every aspect of my work. In common with other historians of science of my generation I am gratefully aware of the influence of the work of the late Alexandre Koyré whose *Études galiléennes* appeared to me, almost from the start, as a model of exploring the historical development of scientific ideas. To him I also owe the idea, which he suggested to me in a private conversation in 1952, that Newton's belief in atomism is the key to his interpretation of the *experimentum crucis*. So far as I know, Koyré did not develop this idea in his published work, and the responsibility for the way it is here presented and argued is entirely my own.

Since the original version of this book was written, a great deal of fresh information regarding Newton's optical manuscripts has come to light, and I have now used as much of this new material as is relevant to my argument and indicated such use wherever it is made. I must add, however, that although this recent research has yielded important results that were not known to me before 1955, these results have not obliged me to alter the argument itself; in fact they seem to me to support it at more than one point.

I wish to thank the many scholars and friends who, through conversations, criticisms and encouragement, have helped me in various ways. I am particularly grateful to J. O. Wisdom for his constant encouragement since my undergraduate years in Alexandria, to L. J. Russell for some very useful suggestions, and to Thomas S. Kuhn for an extensive and valuable commentary especially concerned with the chapters on Descartes and Huygens. My thanks are due to Michael Hoskin, editor of the Oldbourne History of Science Library, for his helpful suggestions and reading the proofs, but above all for his patience and good humour. It is a pleasure also to thank the staff of the Oldbourne Press for their helpfulness throughout.

In preparing the revision I made use of some of my time as Senior Research Fellow at the Warburg Institute (1962–4) and had the benefit of obtaining the advice of several of its members on various problems. While at Princeton University as Visiting Associate Professor for 1964–5 I was able to make further improvements in the light of discussions with members of its Program in the History and Philosophy of Science; moreover, the Program generously defrayed the expenses of typing a considerable part of the final manuscript.

To the library of the British Museum, where most of this book was written, I am very grateful. The courteous and unfailing help of its staff has been invaluable.

A word about translation. In general I have quoted already existing English translations of French and Latin texts, sometimes with minor alterations which are indicated as they occur. In my thesis I had the privilege of using an English translation made by Czeslaw Lejewski of the Latin letters exchanged between Newton and Pardies. My debt to Professor Lejewski is not diminished by the fact that I here quote the English translation printed in the *Philosophical Transactions* (abridged edition, 1809) and now made widely available in I. Bernard Cohen's edition of *Isaac Newton's Papers and Letters on Natural Philosophy* (1958). Unless otherwise noted, all other translations are my own.

A. I. Sabra

The Warburg Institute
University of London

Introduction

This book is not a survey of optics in the seventeenth century, nor does it claim to offer a complete account of the optical researches of any one investigator in that period—perhaps with the exception of Fermat. It is a study of problems and controversies which have appeared to me to be particularly important in the development of seventeenth-century theories about the nature of light and its properties.

The method I have followed is to compare actual practice, in so far as it can be historically determined, with the interpretations placed upon it by the practitioners themselves. The two were closely linked. When Descartes first published his views on light and colours in the *Dioptric* and in the *Meteors,* he presented them as fruits of 'the method' outlined in the *Discourse*; and he declared those views to have been conceived in accordance with a new meta-physically founded system of physics. Fermat's successful derivation of the law of refraction from his principle of least time was the culminating point of a long and tortuous controversy about what constituted a true law of nature and about the applicability of certain mathematical techniques to physical problems. Huygens in the preface to his *Treatise on Light* introduced a methodological theory that was obviously designed to support the type of mechanical explanations he offered for those properties of light with which he dealt. And, Newton, in his first paper to the Royal Society, con-joined his new theory of light and colours with sharply defined views on the proper way of conducting scientific inquiry. The stand on hypotheses which he took during the subsequent dispute with Hooke, Pardies and Huygens can be seen clearly reflected in the mode of exposition which he adopted in the *Opticks*, published some thirty years later.

None of these facts warrants the postulation of any simple and direct relationship between the substantive theories and the second-order speculations which accompanied them. In particular, they do not justify the assumption that the achievements of seventeenth-century optics were due to the application of previously conceived *règles de la méthode* or *regulae philosophandi*. Indeed, the following pages provide more than one illustration of the notorious fact that what scientists do is often quite different from what they say they do. It is, however, my belief that a study of the endeavours of seventeenth-century scientists, in which actual practice and actual results are confronted with their attendant evaluations and theories of method, should yield a fuller understanding of the accomplishments of those early masters of modern science, and a better appreciation of their aims, convictions and limitations.

There are two main reasons for beginning this study with Descartes. He was the first to publish the correct law of refraction without which no substantial progress in optics was forthcoming; and he incorporated this law into a physico-mathematical theory which, despite, or rather because of, its many defects, constituted the starting point for the investigations of Fermat, Hooke, Huygens and even Newton. Thus Descartes not only gave optical research a new impetus, but also a new set of problems, and hence a new direction. To be sure, many of the elements of his theory, both physical and mathematical, are to be found in the writings of his predecessors, such as Ibn al-Haytham, Witelo and Kepler; this we shall have to emphasize. Nevertheless, the cumulative aspect of the development of science should never blur the emergence of ideas, patterns and programmes that are, in an important sense, new. By clearly redefining the boundaries for subsequent discussions about light, Descartes in fact put the science of optics on a new path.

Descartes' commitment to a conception of a strictly full universe obliged him to conceive of the transmission of light as something more like the transfer of energy than the transport of body. His representation of the rays of light as nothing but the lines of direction of the static pressure exerted by the luminous object upon the surrounding matter, was very much in keeping with his plan to reduce the whole of physics to geometry. But when he came to investigate specific properties of light, such as refraction, he failed to mathematize this picture. Instead, he availed himself of a model consisting of a moving sphere which could be imagined to change direction and speed on meeting a refracting surface. The model may be called a 'substitute model' since it did not represent what, in Descartes' view, actually took place in reality, but was employed rather as a substitute to which a handy mathematical device—the parallelogram of velocities—could be directly applied. Descartes did manage, however, to formulate two mathematical assumptions which correctly yielded the law of sines.

Newton did not fail to notice the importance of this mathematical achievement of the Cartesian theory. His own theory of refraction, published in the *Principia* and in the *Opticks,* incorporated Descartes' two assumptions together with the conclusion, which was to be generally accepted until the middle of the nineteenth century, that the velocity of light was greater in denser media. But the mathematical assertions, which Descartes had failed to explain mechanically, received in the Newtonian treatment a plausible physical interpretation which regarded refraction as a special case of particle dynamics. Simultaneously, the model which Descartes had incongruously applied to his continuous picture now coincided with what it was taken to represent in Newton's theory; the sphere shrank into the light corpuscle. What finally remains in Newton's account of refraction is simply the dynamical situation for which the parallelogram method had originally been introduced.

But those who, like Pardies and Huygens, preferred a Cartesian-type physics, became chiefly concerned with repairing the obvious defects in Descartes' mechanical considerations. At the hands of Huygens, Cartesian matter ceased to be rigid and continuous. But the original picture of light as motion transmitted by contact remained, though the transmission was no longer instantaneous as it had been with Descartes. It was Huygens' merit to have been the first to mathematize this picture successfully. This he achieved by imagining a truly representative model of contiguous and elastic spheres to which he applied a new mathematical technique embodying what we now refer to as Huygens' principle.

For Descartes, matter had to be incompressible and space full because of a primary conception equating space and matter. He believed in fact that a significant part of physical science could be deductively developed from such primary conceptions; but he recognized the impossibility of extending this deductive process throughout the whole of physics. At one point or another in the process hypotheses have to be introduced for the purpose of explaining particular phenomena. These hypotheses acquire some of their plausibility from the fact that they explain a large number of phenomena. They must, however, satisfy the further condition of being formulated in terms already defined by the first principles of physics. In other words, the Cartesian physicist must never lose sight of the programme which has been delineated for him by an *a priori* decision, even when he is no longer moving within a system of *a priori* deductions. With the help of these distinctions we may now define Huygens' position in the following way: he rejected Descartes' apriorism, but fully subscribed to the Cartesian programme; thus he presented the whole of his theory of light as nothing more than a system of hypotheses, but hypotheses conceived in accordance with Cartesian mechanism.

The theories of Descartes, Fermat and Huygens were primarily concerned with refraction; the only important exceptions were Descartes' mechanical but qualitative account of colours, and his quantitative explanation of the rainbow. With Hooke and Newton the interest shifts to the problem of colour. Hooke's contribution to the investigation of this problem involved a series of pioneering and important observations on the colours of thin plates, which Newton later raised to an astonishingly high level of experimental sophistication. And yet Newton's fundamental theory of prismatic colours, his chief contribution to optics as the first satisfactory explanation of colour phenomena, must be understood in relation to the prevailing doctrine of refraction. For the problem immediately posed by his famous prism experiment is one of refraction, not of colours as such; it was the shape of the spectrum, not the colours in it, that formed the basis of his arguments. Moreover, what may be regarded as the firmly established part of his theory, the assertion that each colour is always connected with a constant degree of refrangibility, is the part directly related to this geometrical aspect of his experiment. Newton further believed that he was in possession of an *experimentum crucis* which positively established the composite nature of white light. He always insisted, however, on interpreting this compositeness in a narrow sense which, contrary to what is generally assumed by historians of optics, the experiment certainly did not prove. In the chapters dealing with this problem Newton's experiment is viewed against the background of a particular belief in atomism towards which Newton had inclined from the beginning of his intellectual career. This belief was to the effect that properties cannot be *created* but only made apparent, by separating the elements which have always possessed them. Without this belief Newton's insistence on his representation of white light would be left without explanation.

Newton's first published paper on light and colours is regarded here more as an argument devised to convince his audience of his new theory than as an exact autobiographical account of how he had in fact arrived at it. The argument is inductive and was expressly levelled against 'the Philosophers universall Topick', the hypo-thetico-deductive procedure. Newton continued to use Baconian inductivism as a stick with which to beat his Cartesian opponents, but the effect of this on Newton himself was unfortunate: it seri-ously impaired his insight into the structure of his own great achievements.

Abbreviations

B *The Works of Francis Bacon*, collected and edited by James Spedding, R. L. Ellis and D. D. Heath, 14 vols, London, 1857–74

D *Œuvres de Descartes*, edited by Charles Adam and Paul Tannery, 12 vols, Paris, 1897–1913

F *Œuvres de Fermat*, edited by Paul Tannery and Charles Henry, 4 vols, Paris, 1891–1912

H *Œuvres complètes de Christiaan Huygens*, published by the Société Hollandaise des Sciences, 22 vols, La Haye, 1888–1950

HR *The Philosophical Works of Descartes*, translated by Elizabeth Haldane and G. R. T. Ross, 2 vols, 2nd ed., Cambridge, 1931

'Huygens, *Treatise*' refers to the English translation by Silvanus P. Thompson, London, 1912
'Newton, *Correspondence*' refers to H. W. Turnbull's edition in three volumes, Cambridge, 1959, 1960, 1961 (continuing)
Unless otherwise indicated, references to Newton's *Opticks* are to the New York edition of 1952, and references to Newton's *Principia* are to the edition of Florian Cajori, Berkeley, California, 1934

Chapter One

DESCARTES' THEORY OF EXPLANATION AND THE FOUNDATIONS OF HIS THEORY OF LIGHT

1. The *Dioptric*, in which Descartes first published his views on light, appeared in 1637 as one of three treatises claiming to contain results which the author had arrived at through the application of a new method for discovering scientific truths.[1] The method itself was briefly outlined in the *Discourse* which formed a kind of preface to the three treatises. Yet a reader who has gone through the *Discourse*, and who now turns to the *Dioptric* expecting to find a concrete illustration of the four general *rules of method* that are set out in the former work, will be disappointed. Nowhere will he find any of these rules called upon to clear up any particular problem;[2] in fact he will meet with no reference to *the method*.[3] What will be perhaps more puzzling to the reader is the fact that he is being presented instead with another method, an old one which, according to Descartes himself, had been practised by astronomers since antiquity. Descartes in fact declares in the beginning of the *Dioptric* that since

---

[1] The title for the whole volume described the *Dioptric* together with the other two treatises, the *Meteors* and *Geometry*, as '*essais de cette Méthode*'.

[2] As Descartes later explained to Vatier, 22 February 1638, his aim in publishing the 1637 volume was neither to expound the method in the *Discourse* nor to illustrate its application in the three treatises that followed: 'mon dessein n'a point esté d'enseigner toute ma Methode dans le discours où ie la propose, mais seulement d'en dire assez pour faire iuger que les nouuelles opinions, qui se verroient dans la Dioptrique et dans les Meteores, n'estoient point conceuës à la legere, & qu'elles valoient peut être la peine d'estre examinées. Ie n'ay pû aussi monstrer l'vsage de cette methode dans les trois traittez que i'ay donnez, à cause qu'elle prescrit vn ordre pour chercher les choses qui est assez different de celuy dont i'ay crû deuoir user pour les expliquer. I'en ay toutesfois monstré quelque échantillon en décriuant l'arc-en-ciel . . .' (D, I, p. 559). See similar remarks in Descartes to Mersenne, March 1637, D, I, p. 349; and to*** 27 April, 1637?, ibid., p. 370

[3] The only reference to 'the method' in the three treatises occurs in the *Meteors* in connection with Descartes' explanation of the rainbow (D, VI, p. 325). See preceding note; and below, p. 61.

his aim in talking about light in this treatise is restricted to explaining how its rays enter the eye, and how they are made to change direction by the various bodies they encounter,

there will be no need that I undertake to say what in truth the nature [of light] is, and I believe it will be sufficient that I make use of two or three comparisons that help to conceive of it in the manner that seems to me most convenient, in order to explain all those of its properties that are known to us from experience, and to deduce afterwards all the others which cannot be so easily observed; imitating in this the Astronomers who, although their suppositions are almost all false or uncertain, nevertheless succeed in deriving from them many consequences that are very true and very assured, because of their agreement with the various observations they have made.[4]

In 1637 Descartes had reasons for not revealing the foundations of his physics. He had in fact almost completed, before July 1633, a comprehensive work, *Le Monde ou Traité de la Lumière*, which contained, in his own words, '*tout le cors de ma Physique*'.[5] But when the news of the condemnation of Galileo reached him (in November 1633), he decided to defer publication of this work for in it he had committed himself to the Copernican view which had brought Galileo to the notice of the Inquisition. Since in 1637 he was still unwilling to express disagreement with the doctrines favoured by the Church, the object of the volume published in that year was to give some results of the new method in less controversial matters without expressly stating the 'principles' from which these results were obtained.[6] Nor would he indicate those principles briefly lest they should be misunderstood and distorted by 'those who imagine that in one day they may discover all that another has arrived at in twenty years of work, so soon as he has merely spoken to them two or three words on the subject'.[7] Thus, in spite

---

[4] D, VI, p. 83.

[5] Letter to Vatier, 22 February 1638, D, I, p. 562.

[6] *Discourse*, Pt. VI, HR, I, p. 128: 'And I thought that it was easy for me to select certain matters which would not be the occasion for many controversies, nor yet oblige me to propound more of my principles than I wish, and which yet would suffice to allow a pretty clear manifestation of what I can do and what I cannot do in the sciences.'

[7] Ibid., p. 129.

of what Descartes says in the beginning of the *Dioptric*, the suppositions from which he starts in that treatise (and in the *Meteors*) are
not really of the same kind as those of 'the Astronomers'. For
whereas their suppositions may be false or uncertain (even though
the consequences agree with observation), his are founded on
principles which, Descartes maintained, were ultimately deducible
from certain primary truths:

And I have not named them [the matters of which he speaks at the
beginning of the *Dioptric* and *Meteors*] hypotheses [*suppositions*] with any
other object than that it may be known that while I consider myself able
to deduce them from the primary truths which I explained above, yet I
particularly desired not to do so, in order that certain persons may not
for this reason take occasion to build up some extravagant philosophic
system on what they take to be my principles, and thus cause the blame
to be put on me.[8]

Since the suppositions of the *Dioptric* and of the *Meteors* are
deducible from certain 'primary truths', they must themselves be
necessarily true. They are not conceived simply as hypotheses devised
only to explain the phenomena that are dealt with in these two
treatises, although the reader is invited to look on them as such for
the time being. If the reader insists on having some justification of
those assumptions, he may, for the present, find satisfaction in the
following argument:

If some of the matters of which I spoke in the beginning of the *Dioptrics*
and *Meteors* should at first sight give offence because I call them hypotheses
[*suppositions*] and do not appear to care about their proof, let them have
the patience to read these in entirety, and I hope that they will find
themselves satisfied. For it appears to me that the reasonings are so
mutually interwoven, that as the later ones are demonstrated by the
earlier, which are their causes, the earlier are reciprocally demonstrated
by the later which are their effects. And it must not be imagined that in
this I commit the fallacy which logicians name arguing in a circle, for,
since experience renders the greater part of these effects very certain, the
causes from which I deduce them do not so much serve to prove their

---

[8] Ibid., p. 129.

existence as to explain them; on the other hand, the causes are (proved) by the effects.[9]

The astronomer Morin was among the first to raise objections against this method of demonstration, ignoring, it seems, Descartes' intentions. He protested[10] that experience alone cannot establish the truth of a supposition. The apparent celestial movements, he argued, could be equally derived from one or the other of the two suppositions assuming the stability of the earth or its motion; experience was therefore not sufficient to decide which of these two 'causes' was the true one. Further, unconvinced by the argument in the *Discourse* (just quoted), Morin insisted that it was surely arguing in a circle to prove the effects by some causes and then to prove the causes by the same effects. Finally, he maintained that Descartes' procedure was artificial, since, in Morin's view, nothing was easier than to 'adjust' some causes to given effects.

Descartes' reply to these objections exhibits the same attitude already expressed in the *Discourse*:[11] while indicating that he believes his suppositions to be obtainable from higher principles which he has not yet divulged, he also defends his right to argue for the truth of those suppositions on purely empirical grounds. This, of course, is consistent with his intentions: for if we grant that the experiments bear out the proposed suppositions, then, Descartes hopes, we will be better prepared to accept his principles when they come to light. Thus, on the one hand, he readily agrees with Morin that the apparent celestial movements are deducible from either of the two suppositions mentioned, and adds: 'and I have desired that what I have written in the *Dioptric* about the nature of light should be received in the same way, so that the force of mathematical demonstrations which I have there attempted would not depend on any physical opinion.'[12] And later on in the same letter he admits that one is not obliged to believe any of the views expressed in the *Dioptric*, but

---

[9] Ibid. The English translation erroneously reads 'explained' for 'proved'. Cf. D, VI, p. 76.
[10] Morin to Descartes, 22 February 1638, D, I, pp. 538–9.
[11] Descartes to Morin, 13 July 1638, D, II, pp. 197ff. Cf. *Principles*, III, 4, D, VIII, pp. 81–2, IX, pp. 104–5.
[12] D, II, p. 197.

that he wants his readers to judge from his results 'that I must have some knowledge of the general causes on which they depend, and that I could not have discovered them otherwise'.[13]

On the other hand, Descartes would not accept Morin's objections that the demonstrations in the *Dioptric* are circular or that the proposed explanations are artificial. He grants that 'to prove some effects by a certain cause, then to prove this cause by the same effects', is arguing in a circle; but he would not admit that it is circular to explain some effects by a cause, and then to prove that cause by the same effects, 'for there is a great difference between *proving* and *explaining*'.[14] Descartes points out that he used the word 'demonstration' (in the passage quoted above from the *Discourse*) to mean either one or the other 'in accordance with common usage, and not in the particular sense given to it by Philosophers'.[15] Then he adds: 'it is not a circle to prove a cause by several effects which are known otherwise, then reciprocally to prove some other effects by this cause.'[16] As to Morin's last objection, Descartes writes: 'you say lastly that "nothing is so easy as to adjust some cause to a given effect". But although there exist in fact several effects to which it is easy to adjust diverse causes, one to each, it is however not so easy to adjust one and the same cause to several different effects, unless it be the true one from which they proceed; indeed, there are often effects such that one has sufficiently [*assez*] proved what their true cause is, if one has assigned to them one cause from which they can be clearly deduced; and I claim that those of which I have spoken [in the *Dioptric*] belong to this category.'[17]

The basic supposition in Descartes' *Dioptric* is that light is a certain action or movement that is transmitted to all distances through an all-pervading medium. This is the 'one cause' by which Descartes wants to explain the various effects of light. Some of these effects are 'known otherwise', that is, independently of any knowledge of the supposed cause; such, for example, is the rectilinear propagation of

---

[13] Ibid., p. 201.
[14] Ibid., pp. 197–8. See passage from the *Principles* quoted below on p. 36, n. 56.
[15] Ibid., p. 198.
[16] Ibid.
[17] Ibid., p. 199.

light, the fact that light rays are not interrupted when they cross one another, and the equality of the angles of incidence and reflection. Descartes' contention is that his supposition is 'sufficiently proved' by clearly deducing all these different effects from it. Moreover, he claims to have derived from the same supposition 'some other effects', notably the law of refraction which he presents as his own discovery; this provides him with independent evidence which saves his demonstration from circularity.

In this argument no reference is made to higher principles from which the suppositions in question could be derived. One may therefore ask whether Descartes believed, at the time of writing his letter to Morin, that *proving the cause by the effects*—which he called *a posteriori* proof[18]—constituted a conclusive demonstration establishing the truth of his suppositions. In other words, did Descartes here understand by 'proof' a form of argument that would allow the transmission of truth from the consequences to the supposed assumptions, in the same way that mathematical deduction allows truth to be transmitted from the premises to the conclusions? We are not, I think, entitled to answer this question in the affirmative; 'proof' in this context is not to be understood in a strictly logical sense. Descartes does not say that the suppositions are entailed by the experimentally verified consequences; they are only 'sufficiently proved' by them, that is to say 'proved' in a sense that he finds appropriate to the subject matter dealt with. It will be seen later that, as opposed to Francis Bacon's inductive-deductive procedure, the Cartesian conception of method is purely deductivist. This does not mean, however, that Descartes' method can dispense with experiment; on the contrary, deduction is understood in such a way

---

[18] 'As to what I have supposed at the beginning of the *Meteors*, I could not demonstrate it *a priori* without giving the whole of my Physics; but the experiments which I have necessarily deduced from it, and which cannot be deduced in the same way from any other principles, seem to me to demonstrate it sufficiently [*assez*] *a posteriori*' (Descartes to Vatier, 22 February 1638, D, I, p. 563). Later on in the same letter (pp. 563–4), after repeating his claim of being able to deduce his suppositions from 'the first principles of my Metaphysics', he compares his performance in the *Dioptric* to that of 'Thales, or whoever it was who first said that the Moon receives her light from the Sun, providing for this no other proof indeed except that by supposing it one explains the various phases of her light: which has been sufficient, ever since, for this opinion to have been accepted by everybody without contradiction'.

as to make experiment indispensable.[19] Moreover, the limitation of proof in physics was clearly recognized by Descartes in a letter to Mersenne, written less than two months before his reply to Morin:

You ask whether I consider what I have written about refraction [in the *Dioptric*] to be a demonstration; and I believe it is, at least in so far as it is possible to give a demonstration in this matter without having previously demonstrated the principles of Physics by Metaphysics (which I hope to do one day, but which hitherto has not been done), and in so far as any question of Mechanics, or of Optics, or of Astronomy, or any other matter that is not purely Geometrical or Arithmetical, has ever been demonstrated. But to require of me Geometrical demonstrations in a matter which depends on Physics· is to demand that I achieve impossible things. And if one wishes to call demonstrations only the proofs of Geometers, one should then say that Archimedes never demonstrated anything in Mechanics, nor Vitellion in Optics, nor Ptolemy in Astronomy, etc., which however one does not say. For one is satisfied in such matters if the Authors, presupposing certain things that are not manifestly contrary to experience, have subsequently spoken consistently and without Paralogism, even though their suppositions might not be exactly true. For I could demonstrate that the definition of the centre of gravity, which Archimedes has given, is false, and that such a centre does not exist; and the other things which he supposes elsewhere are not exactly true either. As for Ptolemy and Vitellion, they have suppositions that are much less certain, and nevertheless one should not for this reason reject the demonstrations they have deduced from them.[20]

No interpretation of Descartes' theory of explanation would be satisfactory without taking this passage, and the problems which it raises, into full consideration. Together with the claim repeatedly made about the deducibility of physical principles from metaphysics, we have a clear and obviously sincere admission that conclusive demonstrations in physics are impossible; the two are expressed almost in the same breath. Thus the limitations set upon physical proofs are not due simply to a temporary deficiency which will be remedied once the metaphysical foundations are brought to light;

---

[19] See below, pp. 24ff.   [20] Letter to Mersenne, 27 May 1638, D, II, 141–2.

these limitations are imposed by the subject matter of physics itself, and are therefore inevitable. But note Descartes' ambiguous reference to the works of Archimedes, Witelo and Ptolemy: he abides by what he takes to be the common view that their arguments are entitled to be called 'demonstrations', and to that extent one might get the impression that he does not hope to improve upon them in this respect. But we cannot forget his contention in the *Discourse* that he was able to derive his suppositions from self-evident truths; while those of his predecessors are here asserted to be either not exactly true or simply false. The problem raised by the preceding passage can be expressed thus: if, as Descartes says, one is satisfied with explanations in physics so long as they are free from logical errors and the suppositions on which they rest are not contradicted by experiment, then what is the function of the principles from which these suppositions are deducible? On the other hand, if the suppositions are ultimately demonstrable by metaphysics, then why is it impossible to achieve strict demonstrations in physics?[21] We are thus led to consider Descartes' contention regarding the relationship between his physics and his metaphysics. How should we understand this contention? In particular, what are its implications with respect to the use, if any, of experiments; and how far does it allow, if at all, the use of conjectural hypotheses?

2. Shortly before Descartes published his *Meditations* in 1641, he confided in a letter to Mersenne that 'these six Meditations contain

---

[21] In fact Descartes later maintained more than once that he admits only mathematical proofs in his physics, indeed that the whole of his physics is nothing but geometry. Thus writing to Mersenne on 27 July 1638, he declared that he had resolved to abandon 'abstract' geometry for another kind of geometry whose aim was the explanation of natural phenomena. For, he went on, if one considered what he had written (in the *Meteors*) about the salt, the snow and the rainbow, one would find that 'toute ma Physique n'est autre chose que Geometrie' (D, II, p. 268). Again, the résumé of the last section, Section 64, in the second part of the *Principles* runs (in the French version) as follows: 'That I do not admit principles in Physics, which are not also admitted in Mathematics, in order to prove by demonstration what I will deduce from them; and that these principles suffice, inasmuch as all the Phenomena of nature can be explained by their means.' (D, IX, p. 101, Latin version, D, VIII, p. 78.) It is to be noted, however, that these claims occur at the end of Part II of the *Principles*, where the more general propositions of physics are explained; his remarks at the end of Part IV are more cautious. See below, p. 43.

all the foundations of my Physics'.[22] The arguments directly bearing on our questions are to be found in the last two. In the fifth Meditation Descartes asks whether anything certain can be known about material things. In accordance with the rule to proceed in order, the (sensible) 'ideas' of material things are first examined; meanwhile, the question whether there exist external objects corresponding to these ideas is being postponed. The criteria to be applied in this examination are those of clarity and distinctness.

Now, according to Descartes, all that can be clearly and distinctly conceived in a physical object is that it has length, breadth and depth. From this he concludes that we can have certain knowledge of material things only in so far as they constitute the subject matter of geometry. It remains for the sixth Meditation to prove their existence. Here, the criteria of clarity and distinctness are not sufficient; the clear and distinct idea of a triangle, for example, does not imply that such a figure really exists, and the same is true of the ideas of all material things. Our belief in the existence of these is derived from a 'feeling' or 'inclination' such as we are conscious of in sense-perception. This inclination can be relied upon only if guaranteed by a truthful God. We feel that the ideas of sensible things are produced in us by something other than ourselves. Moreover, we have a strong inclination to believe that they are conveyed to us by corporeal things. Since God could not have created this strong inclination simply to deceive us, we may safely believe that corporeal things exist. 'However, they are perhaps not exactly what we perceive by the senses, since this comprehension by the senses is in many instances very obscure and confused; but we must at least admit that all things which I perceive in them clearly and distinctly, that is to say, all things which, speaking generally, are comprehended in the object of pure mathematics, are truly to be recognised as external objects.'[23]

---

[22] Letter to Mersenne, 28 January 1641: '. . . I will tell you, between you and me, that these six Meditations contain all the foundations of my Physics. But this should not be made public, if you please, for it would probably make it more difficult for those who favour Aristotle to accept them; and I am hoping that those who will read them, will insensibly get used to my principles, and recognize their truth before they learn that they destroy those of Aristotle' (D, III, pp. 297–8).

[23] Meditation VI, HR, I, p. 191.

In physics we are concerned with what is truly existing. The veracity of God guarantees the truth of our judgements only when we ascribe existence to what we can clearly and distinctly conceive in external objects. As physicists, therefore, we are not to consider in external objects anything other than their extension, shape and movement—to the exclusion of all sensible and other qualities not reducible to clear and distinct geometrical properties. That is the lesson we learn from the *Meditations* regarding the metaphysical foundations of physics.[24] According to it, metaphysics lays down, in *a priori* fashion, the general framework within which all physical explanations should be sought; and to that extent metaphysics has, with regard to physics, only a programmatic function.

But that is not all. Descartes also believed that he could base on this foundation, again in *a priori* manner, a number of general propositions of far-reaching implications.[25] For example, that the nature of body wholly consists in pure extension; that matter is infinitely divisible and therefore atoms do not exist; that absolute void is impossible; that all motions are circular; that the quantity of motion in the universe is constant; and the laws of impact governing the communication of motion when moving bodies encounter one another. These are the 'principles' which are developed not only independently of experiment but, sometimes, against the verdict of experiment.[26] Since their truth is established *a priori* they are beyond doubt and revision.

Thus not only has the aim of the Cartesian physicist been laid down for him once and for all; quite a considerable and funda-mental part of his work has already been done for him, never to be undone.

---

[24] Cf. J. Laporte, *Le rationalisme de Descartes*, Paris, 1945, pp. 204–6.

[25] Cf. *Principles*, Part II.

[26] The seven laws of impact proposed by Descartes in the second part of the *Principles* are almost all experimentally false. Suspecting this, his belief in the correctness of their demonstra-tion was hardly shaken: 'we should have more faith in our reason than in our senses' was his comforting comment (*Principles*, II, 52, D, IX, p. 93). But why should we trust these laws when they appear to be contradicted by experiment? Answer: Because they relate to an abstract situation whose conditions (absolute hardness of the bodies considered, and tota absence of friction with any surrounding matter) are not possible to realize. Cf. R. Dugas *Histoire de la mécanique* (1950), pp. 155–6; H. Bouasse, *Introduction à l'étude des théories de la mécanique* (1895), Chap. X.

Looking now at the suppositions in the *Dioptric* and *Meteors*, we find in fact that, in agreement with Descartes' contention, they either belong with those principles, or they are more or less directly obtainable from them. In the *Meteors*, for example, Descartes 'supposes' such bodies as water, earth, and air to be composed of many small parts of various shapes and magnitudes.[27] This is not moving far from the divisibility thesis and the precept that we are to consider only geometrical properties such as shapes and magnitudes. He also 'supposes' in the same work that the gaps existing between the small parts of bodies are filled with a very subtle and imperceptible matter; which is another way of denying the existence of absolute void. In the *Dioptric* the fundamental supposition takes light to be a certain power or action that follows the same laws as local motion.[28] This is simply a consequence of the metaphysically determined principle that motion is the only power that can be rationally (that is, clearly and distinctly) asserted to exist in nature.

We may thus, in accordance with Descartes' statements, conceive of the basic part of his physical theory as a deductive system whose premises are rooted in metaphysics. By the basic part I mean the principles and the deductions made from them up to, and including, the so-called suppositions in the *Dioptric* and *Meteors*. Furthermore, some of these principles may have consequences which penetrate quite far into the theory. For example, Descartes' doctrine that light is propagated instantaneously follows from the only conception he allowed of the medium through which light is transmitted.[29]

3. Can we say that, for Descartes, *all* propositions in physics could be obtained in the same way?

The law of optical refraction is immediately derived in the *Dioptric*, not from the fundamental supposition that light is an action that follows the laws of motion, but from assumptions made with reference to certain mechanical 'comparisons' or analogies. In one

---

[27] Cf. *Meteors*, I, D, VI, p. 233.
[28] Cf. *Dioptric*, D, VI, pp. 84 and 100; Letter to Mersenne [27 May 1638], D, II, p. 143.
[29] See below, pp. 48ff.

case the comparison is with a moving ball that receives a hit from a tennis racket, thus acquiring added speed at one point of its journey; in another case the ball is supposed to strike a frail cloth which it breaks, and in so doing loses part of its speed. These analogies, and the assumptions they involve, may be said to be determined by the fundamental supposition inasmuch as they relate to the only power, i.e. motion, whose laws are taken to govern the action of light. But they cannot be said to be uniquely determined by that supposition in the sense of being deductive consequences of it. Nor does Descartes attempt to establish such a deductive relationship in the *Dioptric*. There is thus a logical gap separating the particular assumptions entailing the experimental law of refraction from the fundamental supposition in accordance with which these assumptions, as assumptions about motions, have been conceived.[30]

Now some of Descartes' statements may give one to understand that he believed his use of comparisons in the *Dioptric* to be purely accidental and that therefore they do not represent a significant feature of what, in his view, constituted a perfected physical explanation. In 1638, for example, he wrote to Vatier that since he had amply explained the properties of light in the treatise *De la Lumière*, he only attempted in the *Dioptric* to give 'some idea' of his views '*par des comparaisons & des ombrages*'.[31] Yet in *De la Lumière* we find that the author uses comparisons and calls them by that name.[32] In fact Descartes gives no other proof of the laws of optical reflection and refraction than the one in which he compares these phenomena to the mechanical reflection and refraction of a ball. Instead of providing better proofs in the *Traité de la Lumière*, as we might be led to expect, he simply refers to the proofs, by comparisons, that are given in the *Dioptric*. True, the *Traité* goes further by presenting the framework within which the comparisons are designed. Yet it

---

[30] Representing optical refraction, as Descartes does, as an impact phenomenon makes it in fact impossible to account for the *increase* of velocity that is required by his assumptions when the refraction is from a rare into a dense medium. The racket used in the comparison which illustrates this case thus represents a force that must be ruled out by the 'principles', let alone being derived from them. See below, pp. 14f, 124, 134f.

[31] Letter to Vatier, 22 February 1638, D, I, p. 562.

[32] *Le Monde*, D, XI, p. 102.

is important to note that, even after the physical principles have been stated, comparisons still have a function to perform in the explanations of the properties of light. These comparisons may be looked at as manifest mechanical models that must be used, at one stage or another, for ascertaining the hidden mechanism of light. As will be seen later, Descartes recognized that this mode of explanation does not attain the certainty of the principles under whose auspices it is performed.[33]

The role of analogy in a purely deductive scheme of explanation is clearly pointed out by Descartes in the exposition of Rule VIII of his *Regulae*.[34] Rules I–VII prescribe the order to be followed in the investigation of any given problem. As required by Rules V and VI one must first decompose the difficulty into simpler and simpler components until the simplest elements in it are reached. Then, starting from an intuition of the simplest and most absolute propositions, that is, propositions whose apprehension does not logically depend on the knowledge of any other elements in the problem, we gradually advance to the more and more complex propositions by a deductive process in which every step is an intuitive cognition of the necessary connection between the preceding link and the one that follows. But, for Descartes, this is only an ideal which it may not be always possible to realise. The aim of Rule VIII is, precisely, to advise the investigator against insisting on applying the preceding Rules when 'the nature of the problem itself, or the fact that he is human, prevents him'.[35] Descartes also recognizes that it may not be always possible to advance in knowledge through a series of rigorously connected intuitions. Fortunately for our purpose, he illustrates this Rule by the investigation of the problem that is in fact tackled in the *Dioptric* and *Geometry*.[36]

Descartes imagines[37] a mathematician trying to find the shape of the refracting surface (the so-called *anaclastic*) that would collect

---

[33] See below, pp. 42ff.

[34] Descartes wrote the *Regulae* in or about 1628 (cf. D, X, pp. 486–8). It was first published posthumously in 1701.

[35] *Regulae*, HR, I, p. 23.

[36] Cf. *Dioptric*, VIII; *Geometry*, II.

[37] *Regulae*, HR, I, pp. 23–24.

parallel rays into one focus.[38] He will first easily realise, by application of the Rules of analysis and synthesis, that the solution of the problem depends upon knowledge of the exact proportion observed by all the parallel rays on their passage through the surface. But this proportion he cannot determine by mathematical means alone. All he would be able to do as a mathematician would be to conjecture some proportion or other in accordance with which he would then determine the shape of the surface. But this would not lead to the discovery of what is wanted, but only to what is in agreement with his assumption. Further, Rule III prevents him from trying to learn the true proportion from 'the Philosophers or to gather it from experience'. Our mathematician, if he is unable to reason beyond the limits of his particular science, should stop short at this point, since he is faced with a physical, not a mathematical problem.

But suppose we consider instead 'a man who does not confine his studies to Mathematics, but, in accordance with the first rule, tries to discover the truth on all points'. Meeting with the same difficulty, he will find 'that this ratio between the angles of incidence and of refraction depends upon changes in their relation produced by varying the medium'. Next he will realize that 'these changes depend upon the manner in which the ray of light traverses the whole transparent body'. Again, knowledge of this manner of propagation 'presupposes a knowledge of the nature of the action of light; to understand which finally we must know what a natural potency is in general, this last being the most absolute term in the whole series in question'.

In this analytic procedure experience plays a certain role; for it is only from experience that the investigator learns, for example, that variation in the relationship between the angles of incidence and refraction depends upon varying the media. Yet analysis here should not be confused with Baconian induction. Bacon had con-

---

[38] Descartes was introduced to the problem of the anaclastic through the work of Kepler, whom he considered to have been 'my first teacher in Optics' (Letter to Mersenne, 31 March 1638, D, II, p. 86). The *Dioptrice*, in which Kepler formulated the problem, was published in 1611; see *Gesammelte Werke*, IV, pp. 371–2.

ceived of induction as a method of passing gradually from experimental information to knowledge of first causes through a sequence of 'axioms' or general propositions. In this picture experience is the only source of information, and the higher-level generalizations are obtained from those directly below them. Alternatively, knowledge of the lower-level generalizations is required for arriving at the axioms immediately above [39]. It was Newton who later identified induction in this Baconian sense with 'analysis'. As he wrote in Query 31 of the *Opticks*: 'This Analysis consists in making Experiments and Observations, and in drawing general Conclusions from them by Induction. . . .'[40] For Descartes, however, analysis of the problem he is here considering does not itself yield the general conclusions which will later serve as premises in the reverse process of synthesis or composition. Rather, it is a means of discovering and ordering the *conditions* for resolving the problem. The final result of the analysis is a statement of the most fundamental among those conditions, namely *that* 'we must know what a natural potency is in general'. But it does not assert *what* natural potency is. A knowledge of this, which is required for the knowledge of all other terms in the problem, must now come through another operation, i.e. intuition. As opposed to the ascending axioms of Bacon, the conditions ascertained by Descartes' method of analysis are such that the last among them in the order of discovery is the first in the order of knowledge.

Having clearly comprehended, 'by a mental intuition' (*per intuitum mentis*), what a natural power is in general, the investigator must now proceed backwards through every step of the series in the reverse order. In the ideal case (as in pure mathematics) the movement of the mind in this synthesis should consist in a straightforward deduction necessarily leading from one step to the next. But there might be a hitch: 'And if when he comes to the second step he is unable straight way to determine the nature of light, he will, in accordance with the seventh rule, enumerate all the other natural potencies, in order that the knowledge of some other of them may

---

[39] *Novum Organum*, I, 103–5, B, IV, pp. 96–98.   [40] *Opticks*, Query 31, p. 404.

help him, at least by analogy [*per imitationem*] ..., to understand this.'[41]

It is clearly admitted here by Descartes that one may not be able directly to deduce the nature of light from the absolute intuition of what constitutes a natural power in general. This even seems to be imposed on the investigator by the fact that the action of light, being a *particular* form of action, cannot be uniquely derived from the *general* concept of natural power. Descartes therefore suggests that we must first enumerate all forces in nature so that we may be able to understand what light is by its analogy with some other force. Enumeration, in this sense, involves an appeal to experience as a prerequisite for any further advance in the deductive process.

It therefore appears that the analogies or comparisons used in the *Dioptric* are not there by accident, or merely to facilitate the understanding of the explanations offered. These analogies indicate the kind of argument which, according to Descartes, one has to rely upon in order to establish the second deductive step after an intellectual intuition of what constitutes a natural force in general has been grasped.[42]

Descartes recognizes one and only one fundamental force or power in nature, namely, motion. The effects of heat, sound, magnetism, gravity, and all the other 'powers' of nature, must be produced by

---

[41] *Regulae*, HR, I, p. 24. Cf. D, X, p. 395.

[42] In a passage of the *Cogitationes privatae* (1619–21) Descartes expresses the idea that philosophy (i.e. natural philosophy) consists in drawing successful analogies from the objects of sense experience: '*Cognitio hominis de rebus naturalibus, tantum per similitudinem eorum quae sub sensum cadunt: & quidem eum verius philosophatum arbitramur, qui res quaesitas felicius assimilare poterit sensu cognitis*' (D, X, pp. 218–9). The idea is illustrated in the *Regulae*, Rule IX: 'For example if I wish to examine whether it is possible for a natural force to pass at one and the same moment to a spot at a distance and yet to traverse the whole space in between, I shall not begin to study the force of magnetism or the influence of the stars, not even the speed of light, in order to discover whether actions such as these occur instantaneously; for the solution of this question would be more difficult than the problem proposed. I should rather bethink myself of the spatial motions of bodies, because nothing in the sphere of motion can be found more obvious to sense than this' (HR, I, p. 29). The analogy here is that of a stick whose parts all move simultaneously when one end is moved.

Descartes' procedure may thus be described as, first, to explain the known by the unknown – the manifest properties of gravity, light, magnetism and so forth by the action of imperceptible particles in motion; and then to explain the unknown by the known—the hidden mechanism by the manifest motions of sensible bodies. The outcome is to explain the known by the better known.

various motions or combinations of motions. On these assumptions the question about the nature of light and its physical properties reduces to the following: what kind of motion is the action of light, and how is it propagated, reflected, refracted, etc.? The *Dioptric* answers this question by comparing the action of light with the action which simultaneously moves one end of a stick when applied to the other end, and the reflection and refraction of light with the mechanical reflection and refraction of a ball. That the answer is given by analogies and not by rigorous deduction is, we have seen, in perfect accordance with Descartes' rules of method. What Descartes omits to mention in the *Dioptric* are the principles demanding that light, and all the other forces of nature, *must be a form of movement*. That is to say, the *Dioptric* takes up the deductive argument at the second step of synthesis where analogies are not only permissible but, rather, required by the nature of the problem.[43] We have already indicated how the first step is determined.

4. We have seen that the Cartesian method of explanation, though based on *a priori* grounds, establishes contact with experience at various points. Apart from suggesting the physical problem to be investigated, experience assists in the operation of analysis, in the process of *enumeratio sive inductio* (as in the surveying of natural powers mentioned above), and in the construction of mechanical analogies or models. Also let us not forget that the objects of intuition itself, the pure and simple natures, may be revealed through experience.[44] Rule V of the *Regulae* seems to have been formulated

---

[43] That the method of the *Dioptric* is wholly synthetic seems to be clear enough from the fact that it begins with suppositions from which the various properties of light are deduced. For a different view, claiming that this work illustrates the method of analysis, see L. J. Beck, *The Method of Descartes* (1952), p. 167. Beck also remarks (ibid., p. 263) that the *Dioptric* 'may ... be profitably considered as an application of the method to a problem of physics'. This is contradicted by Descartes' own statement that the method would have required a different order from that which he adopted in the *Dioptric*. See above, p. 17, n. 2. All we can say (and all Descartes wants us to believe) is that the *Dioptric* contains results which he has discovered by his method without showing how this has in fact been done.
[44] *Regulae*, Rule VI, D, X, p. 383: '*Notandum 2. paucas esse duntaxat naturas puras & simplices, quas primo & per se, non dependenter ab alijs vllis, sed vel in ipsis experimentis, vel lumine quodam in nobis insito, licet intueri. . . .*'

with mainly empirical problems in mind; and it emphasizes the importance of experiment. An investigator who does not comply with this rule would be, according to Descartes, 'like a man who should attempt to leap with one bound from the base to the summit of a house, either making no account of the ladders provided for his ascent or not noticing them'.[45] This metaphor of the ladder is one that occupied a fundamental and conspicuous place in the *Novum Organum*; and 'leaping' from the data of a problem to the principles of its solution is precisely what Bacon had called *anticipatio*, the false method to which he opposed his own true method of *interpretatio*. Significantly enough, examples of those who sin against this rule (unlike most other examples in the *Regulae*) are all drawn from the empirical field: the astrologers who pronounce on the influence of the stars without having determined the nature of the heavens or even observed the movements of heavenly bodies; the instrument devisers who rashly pursue their profession with no knowledge of physics; and 'those Philosophers who, neglecting experience, imagine that truth will spring from their brain like Pallas from the head of Zeus'.[46]

All this reminds one rather strongly of Bacon. Indeed a case has been repeatedly made for a Baconian influence on Descartes.[47] Yet this influence was not of a fundamental character:[48] while Descartes recommended Bacon's view regarding the making of experiments, and even sometimes followed his example in the collection of natural histories,[49] he completely ignored induction *as a method of deriving general propositions from experiential data;*[50] more-

---

[45] *Regulae*, HR, I, p. 14.

[46] Ibid., p. 15.

[47] Cf. A. Lalande, 'Quelques textes de Bacon et de Descartes', *Revue de métaphysique et de morale*, XIX (1911), pp. 296–311; G. Milhaud, *Descartes savant* (1921), Chap. X; S. V. Keeling, *Descartes* (1934), pp. 47–50; and a comparison between the Baconian and Cartesian methods in L. Roth, *Descartes' Discourse on Method* (1948), pp. 52–71.

[48] It is plausible that Descartes did not read Bacon until after the crucial idea of philosophical reasoning as something akin, though inferior, to poetic imagination occurred to him in the winter of 1619–20. Cf. G. Milhaud, *Descartes savant*, pp. 216, 222–3, and Chap. II; Descartes, *Cogit. privat.*, D, X, p. 217.

[49] As, for instance, in his *Excerpta anatomica* (D, XI, pp. 549ff), and *De partibus inferiori ventre contentis* (D, XI, pp. 587ff). Cf. E. Gilson, *Commentaire*, pp. 451–2.

[50] Descartes sometimes speaks in the *Regulae* of 'deducing' (Rule XII, HR, I, p. 47) or 'inferring' (Rule XIII, ibid., pp. 49f), say, the nature of the magnet from the observations

over, his view of the role of experiment and observation was diametrically opposed to that of Bacon. This is not surprising in view of the fact that they start from opposite points: Bacon, from the observations and experiments themselves; and Descartes, from *a priori* considerations which first give him the general principles of all physical explanations.[51]

Let us first look at Descartes' references to Bacon; they all come from a period during which he was engaged in writing *Le Monde*. He wrote to Mersenne in January 1630:

I thank you for the [list of] qualities which you have drawn from Aristotle; I have already made a longer list of them, partly derived from Verulam, partly from memory, and this is one of the first things which I shall attempt to explain; and that will not be as difficult as might be thought; for the foundations being laid down, these qualities follow of themselves.[52]

As has been remarked by G. Milhaud,[53] one feels that Descartes had Bacon's writings within easy reach. But he had something more at his disposal: the foundations with reference to which the 'qualities' derived from Bacon would be readily explained. Far from attempting to arrive at the causes of these qualities through an examination free from all preconceived ideas, as Bacon's method would have

---

that have previously been made regarding it. But even here the function of natural history seems to be the same as that indicated in the *Principles* (below, p. 36, n.56) and elsewhere (below, pp. 36ff). In Rule XII, for instance, Descartes is concerned to warn against the misleading tendency of always looking for 'some new kind of fact' (ibid., p. 47) whenever one faces the difficulty of explaining a new phenomenon. To him all explanations should rather be sought in the perfectly known simple natures which form the objects of intuition, and in the connections which we intuitively perceive to hold between them. Thus the question about the nature of the magnet reduces to the following question: what *particular combination* of simple natures is necessary to produce all those effects of the magnet which we have gathered from experiments? Deducing or inferring the nature of the magnet would thus seem to mean no other than choosing one among the many possible combinations of simple natures in the light of the data to be explained.

The distinction, briefly stated in the *Regulae*, between seeking to know the causes from the effects and the effects from the causes (D, X, pp. 428, 433, 434) is not one between induction and deduction, but rather between theoretical and applied physics (cf. ibid., pp. 471–2).

For Descartes' meaning of 'venir au-devant des causes par les effects', see below, pp. 38f.

[51] This crucial difference was noted by Gassendi who was also aware of similarities between Bacon and Descartes. Cf. L. J. Beck, *The Method of Descartes*, p. 163, n. 1.

[52] Letter to Mersenne, January 1630, D, I, p. 109.

[53] G. Milhaud, *Descartes savant*, p. 213.

required him to do, he was already convinced of what the general nature of these causes must be.

That a knowledge of the general causes of physical things must precede the making of experiments, at least the more particular ones, was later stated by Descartes himself—again in connection with a reference to Bacon. To Mersenne, who desired to know 'a way of making useful experiments', Descartes replied:

> To this I have nothing to say, after what Verulam has written about it, except that without being too curious to seek after all the small particularities touching a certain matter, one should chiefly make general surveys of all things which are more common, and which are very certain and can be known without expense.... As for the more particular [experiments], it is impossible not to make many of them that are superfluous, and even false, if one does not know the truth of things [*la vérité des choses*] before making them.[54]

The reservation expressed in the latter part of this passage is extremely important; it amounts in fact to the proposal of a method that is the exact opposite of Bacon's induction. For the whole point of induction, of *interpretatio* as opposed to *anticipatio*, is that 'the truth of things' will only emerge at the very end when experiments of all kinds have been made and the axioms duly derived from them. Therefore, when, two years later, Descartes expressed the wish that someone would prepare for him 'the history of celestial appearances, according to the method of Verulam . . . without introducing in it any reasons or hypotheses',[55] he was not planning to embark upon a project similar to that of the *Novum Organum*.[56] The implication of his demand was rather this: he suspected the opinions which the prospective collector of history might mingle with the results of pure observation; but he did not distrust the physical principles which had flowered in his own mind from the seeds of knowledge which he believed to have been planted there by nature. Thus in

---

[54] Letter to Mersenne, 23 December 1630, D, I, pp. 195–6.

[55] Letter to Mersenne, 10 May 1632, D, I, p. 251.

[56] In the *Principles*, III, 4 (On the use of phenomena or experiments), Descartes explains the role of natural history as follows: '. . . je feray icy vne briéve description [Latin: brevem historiam] des principaux Phainomenes, dont je pretens rechercher les causes, *non point afin d'en*

the hands of Descartes the *historia naturalis* was made to serve an essentially deductive method.[57]

5. It is in the light of the ideas expressed in these references to Bacon that we should, I think, approach the well-known passage in the sixth part of the *Discourse* where Descartes explains the order of his procedure. As we have seen, 'seeking after all the small particularities touching a certain matter' was a feature of Bacon's *historia* about which Descartes had certain reservations. The passage in question opens with a sentence which indicates a further reservation regarding Bacon's method. Among Prerogative Instances, i.e. instances of particular help to the understanding in the process of *interpretatio*, the *Novum Organum* mentions the so-called Deviating Instances: 'For we have to make a collection or particular natural history of all prodigies and monstrous births of nature; of everything in short that is in nature new, rare, and unusual.'[58] Opposing this recommendation, it would seem, Descartes maintains that in the beginning of an inquiry 'it is better to make use simply of those [experiments] which present themselves spontaneously to our senses . . . rather than to seek out those which are more rare and recondite; the reason for this is that those which are more rare often mislead us as long as we do not know the causes of the more common. . . .'[59] What then was the order which he followed?

I have first tried to discover generally the principles or first causes of everything that is or that can be in the world, without considering anything that might accomplish this end but God Himself who has created the world, or deriving them from any source excepting from certain germs of truth which are naturally existent in our souls.[60]

---

tirer des raisons qui seruent à prouuer ce que j'ay à dire cy-apres: car j'ay dessein d'expliquer les effets par leurs causes, et non les causes par leurs effets; *mais afin que nous puissions choisir, entre vne infinité d'effets qui peuuent estre deduits des mesme causes, ceux que nous deuons principalement tascher d'en déduire'* (D, IX, p. 105, italics added).

[57] One might describe Descartes' position by saying that for him the investigation of a *particular* physical problem begins with observation and experiment, while *natural science* must begin with *a priori* principles.

[58] *Novum Organum*, I, 29, B, IV, p. 169.

[59] *Discourse*, VI, HR, I, p. 120.

[60] Ibid., p. 121.

The first step, yielding the 'principles or first causes', is thus clearly deductive. It may be noted here that the idea of seeds of truth from which the principles are derived goes back to the important period of 1619–21, and it recurs in the *Regulae* (1628).[61] The second step, Descartes tells us, was also deductive; from these causes he attempted to derive their 'primary and most ordinary effects'; and in this way he found (*j'ai trouvé*), as if by accident, 'the heavens, the stars, an earth, and even on the earth, water, air, fire, the minerals and some other such things, which are the most common and simple of any that exist, and consequently the easiest to know'.[62] A problem then presented itself when he wanted to descend to the more particular effects; for the particular objects of experience were of many various kinds, and the problem was how to 'distinguish the forms or species of bodies which are on the earth from an infinitude of others which might have been so if it had been the will of God to place them there. . . .'[63] This distinction, Descartes believed, was not possible 'if it were not that we arrive at the causes by the effects [*si ce n'est qu'on vienne au-devant des causes par les effets*], and avail ourselves of many particular experiments'.[64]

In spite of the words 'we arrive at the causes by the effects' the solution here suggested does not, in my opinion, constitute a breakdown of the deductive process and a recourse to something like induction.[65] For the 'first causes' have already been established. Yet they are such that they admit of an infinitude of possible particular effects, any of which could have been realised. Since the

---

[61] *Cogitationes privatae* (1619–21), D, X, p. 217: 'sunt in nobis semina scientiae, vt in silice, quae per rationem à philosophis educuntur, per imaginationem à poetis excutiuntur magisque elucent.' See also *Regulae*, Rule IV, D, X, pp. 373 and 376; HR, I, pp. 10 and 12.

[62] *Discourse*, VI, HR, I, p. 121.

[63] Ibid.

[64] Ibid., p. 121; D, VI, p. 64.

[65] That the knowledge of the cause must precede that of the effect is clearly stated by Descartes in the following sentence from the *Regulae*, Rule VI (HR, I, p. 16): 'For though Philosophers make cause and effect correlative, we find that here even, if we ask what the effect is, we must first know the cause and not conversely.' Bacon regarded the highest and most general axioms (i.e. the first causes) as 'prior and better known in the order of nature' (*Novum Organum*, I, 22, B, I, p. 160; IV, p. 50). But the investigator reaches them only at the end of the inductive procedure. As Bacon says in the Preface to the *Great Instauration*: 'Now my plan is to proceed regularly and gradually from one axiom to another, so that the most general are not reached till the last. . . .' (B, IV, p. 25).

aim of the deduction is to explain those effects which actually exist, recourse to particular experiments is necessary to ascertain the *existence* of what is to be explained. All this is, of course, in agreement with Descartes' doctrine that belief in the existence of material things is based on an empirical feeling (whose reliability is guaranteed by the truthfulness of God), rather than on a necessary deduction from clear and distinct ideas. Similarly, in explaining a particular effect there is need for empirical information presenting the investigator with the particular *explanandum*, i.e. the existing fact to be explained by deduction from the already determined principles or causes. Thus for Descartes the problem of formulating a physical explanation is one in which two terms are given: the one term, given in experience, represents the phenomena to be explained; the other, given *a priori*, includes the principles from which the explanation must start. The explanation is fulfilled when a way has been discovered of deducing the former from the latter. 'Arriving at the causes by the effects' simply means determining the *real* effects so that the deduction from the given causes of 'everything that is or that can be in the world' would not remain in the realm of mere possibilities.

Descartes now goes on to point out another problem to which he is led by the nature of the principles from which he started. He first states that there was nothing in the sphere of his experience which he could not 'easily explain' by those principles.

But I must also confess that the power of nature is so ample and so vast, and these principles are so simple and general, that I observed hardly any particular effect as to which I could not at once recognise that it might be deduced from the principles in many different ways; and my greatest difficulty is usually to discover in which of these ways the effect does depend upon them.[66]

This difficulty brings out a distinction between mathematical deduction, which has to do with necessary connections, and the kind of deduction which Descartes here had in mind. In a mathematical

---

[66] HR, I, p. 121.

system the same conclusion might be deducible from either of two different groups of axioms belonging to the same system. Also, starting from identical or different groups of axioms belonging to the same system, deduction might lead to the same conclusion through different groups of theorems. In either case we may say that the conclusion can be deduced from the axioms in different ways. But the question 'in which one of these ways does the conclusion depend upon the axioms?' does not arise. For, since the alternative deductions are all valid, there is no question of *choosing* one of these alternatives to the exclusion of all the others, although preference might be made on aesthetic or practical grounds. Therefore, the different ways in which the effect may depend upon the principles cannot—in Descartes' problem—be represented by different chains of logically valid deductions. If such were the representation meant by Descartes, his problem would be difficult to understand. For we should be satisfied with *any* of the ways in which the effect is deduced from the principles, provided that the deduction is logically necessary. It would therefore seem that the alternative ways of deduction here envisaged by Descartes are in fact only *possible* explanations of the effect in terms of the given principles: that is to say they are hypothetical constructions which are equally compatible with the 'simple and general' principles without following from them of necessity. Thus again the problem is for Descartes that of bridging the gap between possibility and actuality, between possible causes and *the* true cause. And again the solution lies in recourse to experiments: by increasing more and more the empirical content of the *explanandum*, the number of possible explanations could be reduced further and further. For example, what should we do in the simplified case where we have only two possible methods or ways of explaining the same effect, and how should we decide between them? 'As to that, I do not know any other plan but again to try to find experiments of such a nature that their result is not the same if it has to be explained by one of the methods, as it would be if explained by the other.'[67]

---

[67] Ibid.

Nothing in Descartes' preceding remarks suggests that experimental results may serve as premises from which general conclusions are drawn. Experiments are in his system assigned a different but essential function: they propose problems by pointing out the *existent* effects which it is the business of the physicist to explain; and they provide crucial tests between competing *possible* explanations. These are the only ways in which experiments may be said to help in ascertaining or 'arriving at' the causes. The terms in which the problems would be solved and the 'principles or first causes' to which all explanations must ultimately be reduced are themselves determined beforehand.

In the passage with which we have been concerned Descartes does not explicitly say whether he considers the possible ways of explaining the same effect to be always finite in number. If they are, then it is always possible, at least in principle, to establish *the true* explanation by a finite number of crucial tests which eliminate *all* the false ones. Crucial experiments would thus have a role similar to that of indirect proof in mathematics, which is the role assigned to them by Bacon.[68]

But what if the number of possible explanations is infinite?

Descartes faces this possibility in the *Principles*. There, in fact, this possibility becomes the actual truth by replacing the power of nature, which is 'so ample and so vast' but not necessarily infinite, by the power of God, which cannot be finite. Having illustrated at length (in Parts III and IV of the *Principles*) how in his view all natural phenomena can be explained by various motions attributed to small and imperceptible material particles of various shapes and magnitudes, he then[69] explains how he came to his knowledge of the properties of those particles. First, says Descartes, he considered the clear and distinct notions which we naturally have of material things. These are the notions of shape, magnitude and movement, together with the laws that govern them, namely the principles of geometry and mechanics. Being convinced that all knowledge of nature must be derived from these notions alone, he then considered

---

[68] See below, pp. 178f.   [69] *Principles*, IV, 203.

what sensible effects could be produced by particles of various shapes, magnitudes and movements in various combinations. Observing similar effects in nature he inferred that these might have been produced in the ways he envisaged. Descartes says that in all this he was helped by considering the example of machines: a clock-maker can ordinarily tell[70] by looking at some parts of a clock which he has not made what the other parts are; similarly, by considering the sensible effects of natural bodies he has attempted to discover the hidden mechanisms by which they have been produced.

But here it may be said that although I have shown how all natural things can be formed, we have no right to conclude on this account that they were produced by these causes. For just as there may be two clocks made by the same workman, which though they indicate the time equally well and are externally in all respects similar, yet in nowise resemble one another in the composition of their wheels, so doubtless there is an infinity of different ways in which all things that we see could be formed by the great Artificer [without it being possible for the mind of man to be aware of which of these means he has chosen to employ]. This I most freely admit; and I believe that I have done all that is required of me if the causes I have assigned are such that they correspond to all the phenomena manifested by nature [without inquiring whether it is by their means or by others that they are produced].[71]

Accordingly Descartes distinguishes two kinds of certainty. The one, certainty proper, or absolute certainty, holds 'when we think that the thing cannot be other than as we conceive it'.[72] Founded on the metaphysical principle that our Creator cannot lead us into error as long as we follow what we perceive clearly and distinctly, this certainty is to be found in mathematical truths (as, for instance, in the assertion that two and three cannot together be less than five), in our knowledge—based on the truthfulness of God—that material things exist, and in whatever can be demonstrated about them 'by

---

[70] Latin: 'facile cojiciunt', French: 'peut ordinairement juger'. D, IX, p. 322 ; D, VIII, p. 326.
[71] *Principles*, IV, 204, HR, I, p. 300. Brackets as in the English translation; they indicate additions made by Descartes in the French version.
[72] *Principles*, IV, 206, D, IX, p. 324.

the principles of Mathematics or by other [principles] equally evident and certain'.[73] Now Descartes urges further that this kind of certainty extends to the matters which he has dealt with in the *Principles*, but the French version adds the significant qualification '*at least the principal and more general*'[74] among those matters. The same qualification occurs a little later in the same context, this time both in the Latin and French versions. Descartes first maintains that the particular explanations presented in the *Principles* rest on the fundamental assertion that the heavens are fluid, i.e. that they are composed of small parts constantly moving in relation to one another. Then he adds: 'So that this sole point being recognized as sufficiently demonstrated by all the effects of light, and by all the things which I have explained as following from it, I believe that one must also recognize that I have proved by Mathematical demonstrations all the doctrines that I have proposed, *at least the more general ones* which concern the fabric of the heaven and of the earth. . . .'[75]

It is, therefore, the more particular matters that Descartes had in mind when he spoke of the other kind of certainty which he called 'moral certainty'.[76] This is the certainty which we have regarding the things we ordinarily do not question even though they might in principle be false. For instance, those who have not been to Rome do not doubt its existence, although it is possible that they might have been deceived by all reports which have reached them. It is also the certainty of the cryptogram reader that he has conjectured the correct interpretation of individual letters when he can make sense of a large number of them. Descartes believes that the number of natural phenomena which he can successfully explain is so large as to make it *unlikely* that his explanations should be false. But, just as in the case of cipher reading one cannot be absolutely sure that one has discovered the meaning intended by the author (for there is always the possibility of obtaining another meaning by interpreting the individual letters differently), so in the explanation of particular natural phenomena one cannot be absolutely certain

---

[73] Ibid., p. 324.   [74] Ibid.   [75] Ibid., p. 325, *italics* added. See also: D, VIII, pp. 328–9.
[76] *Principles*, IV, 205, D, IX, p. 323.

that the effects have been produced in the manner which has been supposed. This appears in fact inevitable when we consider that 'there is an infinity of different ways in which all things that we see could be formed by the great Artificer. . . .' Thus, one might say, while the truthfulness of God guarantees the certainty of our knowledge of the principles of physics and of general matters which follow from them, His unlimited power necessitates, on the other hand, that our explanations of the more particular phenomena must remain hypothetical. This position Descartes himself described to Mersenne, four years before the publication of the Latin edition of the *Principles*, in the following words: 'I do believe that one can explain one and the same particular effect in many different ways which are possible; but I believe that one cannot explain the possibility of things in general except in one way only, which is the true one.'[77]

6. To sum up: The Cartesian system of physics consists of two parts, one on a higher-level than the other and both of which are deductive. The first, higher and metaphysically determined part consists of the 'primary truths' and their logical consequences, the so-called 'principles' of physics. This is the domain of *a priori* truth, of strict demonstration and of absolute certainty. As repeatedly asserted by Descartes, the 'suppositions' at the beginning of the *Dioptric* belong to this part, even though they are claimed in that work to be defendable by a kind of *a posteriori* proof which Descartes identified with the method of 'the Astronomers'. The second, lower-level part may be described as a hypothetico-deductive system. Its premises are not deductive consequences of the first part standing above it, but conjectures devised to explain particular phenomena. These conjectures are not free from all limitations; they must not contradict the higher principles, and the terms in which they are formulated are already defined by these principles. In this latter part strict demonstration is—on the whole—impossible;

---

[77] Letter to Mersenne, 28 October 1640, D, III, p. 212.

only a greater or less degree of 'moral' or practical certainty is all that one can reasonably hope for. The qualification 'on the whole' is required to indicate the fact that some of the assertions which we might regard as belonging to this part can be logically determined by the higher-level principles. An example is the doctrine of the instantaneous propagation of light which is dealt with in the next chapter.

Chapter Two

DESCARTES' DOCTRINE OF THE INSTANTANEOUS PROPAGATION OF LIGHT AND HIS EXPLANATION OF THE RAINBOW AND OF COLOURS

1. Aristotle censured Empedocles for having spoken of light as travelling, that is, taking time to go from one place to another. Light for Aristotle was not, as it was for Empedocles, a material effluence which streamed from the luminous object with finite speed; nor was it a successive modification of the transparent medium which, according to him, was necessary for its transmission. Rather than being a process or movement, light for Aristotle was a state or quality which the medium acquired all at once from the luminous object, just as water may conceivably freeze at all parts simultaneously.[1] So far as the speed of propagation is concerned, it was the Aristotelian view which prevailed for many centuries afterwards. From the second to the twelfth century it was adopted by such influential writers as Galen,[2] Philoponus,[3] al-Kindi,[4] Avicenna[5] and Averroës[6]. A notable exception was Avicenna's contemporary

---

[1] Aristotle, *De anima*, II, 7; *De sensu*, VI.

[2] Cf. John I. Beare, *Greek Theories of Elementary Cognition*, Oxford, 1906, p. 59, n. 1.

[3] Cf. S. Sambursky, 'Philoponus' interpretation of Aristotle's theory of light', *Osiris*, XIII (1958), p. 119.

[4] Chapter 15 of al-Kindi's *De aspectibus* presents a 'demonstration' of the instantaneous propagation of light (cf. *Alkindi, Tideus und pseudo-Euklid*, ed. A. A. Björnbo and S. Vogl, Leipzig, 1912, pp. 25–27). Al-Kindi's argument is essentially the same as that produced previously by Aristotle against Empedocles (*De anima*, II, 7) and later by Witelo (*Optica*, II, 2, p. 63; III, 55, p. 110) and by Descartes (see below, pp. 55ff): if indeed light moved with a finite though very great speed, an appreciable interval of time would be perceived when the distance traversed is very great; this, however, is not the case; therefore, etc.

[5] Avicenna, *De anima*, III, 8, in *Opera*, Venetiis, 1508, fol. 16ʳb: 'Verum est autem quod simulacrum visi redditur mediante translucente membro receptibili apto, leni illuminato, ita ut non recipiat illud substantia translucentis aliquo modo secundum quod est ipsa forma; sed cadit in illud, secundum oppositionem non in tempore.'

[6] Cf. *Aristotelis De anima cum* Averrois *Commentariis*, Venetiis, apud Iuntas, 1562 (facsimile reproduction, Frankfurt am Main, 1962, Suppl. II), fol. 86ᵛ. Averroës here simply endorses

46

Ibn al-Haytham (d. *c.* 1039), known to the Latin medieval writers by the name Alhazen, who asserted that the movement of light required a finite, though imperceptible, interval of time.[7] Roger Bacon produced new arguments to defend Ibn al-Haytham's view;[8] but many writers in Europe up to the seventeenth century held to the doctrine of instantaneous transmission, and it was shared by Grosseteste,[9] Witelo[10] and Kepler.[11] Descartes subscribed to the

---

Aristotle's criticism of Empedocles. In his *Epitome* of Aristotle's *De anima*, not translated into Latin, he says that 'luminosity is one of the perfections that are not divisible by division of the body [of which they are perfections] and are not realized in time' (*Kitāb al-nafs*, p. 28, in *Rasā'il Ibn Rushd*, Hyderabad, 1947).

[7] Alhazeni *Optica*, II, 21 in *Opticae thesaurus*, ed. F. Risner, Basel, 1572, p. 37: 'Et dicemus, quod color in eo, quod est color, et lux in eo, quod est lux, non comprehendetur a visu, nisi in tempore, scilicet, quod instans, apud quod erit comprehensio coloris in eo, quod est color, et comprehensio lucis in eo, quod est lux, est diversum ab instanti, quod est primum instans, in quo contingit superficiem visus aer deferens formam. Quoniam color in eo, quod est color, et lux in eo, quod est lux, non comprehenduntur a sentiente, nisi post perventum formae in corpore sensibili, et non comprehenduntur ab ultimo sentiente, nisi post perventum formae ad concavum nervi communis, et perventus formae ad concavum nervi communis, est sicut perventus lucis a foraminibus, per quae intrat lux ad corpora opposita illis foraminibus, perventus igitur lucis a foramine ad corpus oppositum foramini, non erit, nisi in tempore, quamvis lateat sensum.' The discussion is continued in the same section. Ibn al-Haytham begins here by asserting that perception (*comprehensio*) of light as such and of colour as such takes place at a later instant than that at which the light makes contact with the surface of the eye. It is the act of perception, therefore, that requires an interval of time. He ends, however, by comparing the extension of light through the common nerve to its extension in space from an opening to a body placed opposite the opening, and asserts that the latter extension also takes place in time. M. Naẓif has remarked that Ibn al-Haytham's argument in support of this statement does not establish the sucessive movement of light in the direction of propagation (*al-Ḥasan ibn al-Haytham* . . . I, Cairo, 1942, pp. 118–20). But, as the same author also showed, the failure of the argument is no indication of the nature of the desired conclusion, namely the finite speed of light which Ibn al-Haytham affirms in many passages. It may be remarked here that as distinguished from Descartes who (as we shall see) could only speak of the 'force' or 'ease' with which light passed from one medium into another, but never of its speed, Ibn al-Haytham readily and frequently spoke of light itself as being quicker or slower in different media. It may also be added that Ibn al-Haytham's belief in the non-instantaneous transmission of light was connected with another opinion which he held—viz. that no absolutely transparent body exists, and that therefore all transparent bodies, including the body of the heavens, must offer some resistance to light. See below, pp. 93ff.

[8] Cf. *Opus majus*, V (*De scientia perspectiva*), pars I, distinctio IX, capa. 3 & 4, ed. Bridges, II, Oxford, 1897, pp. 68–74. Bacon's arguments are directed against Aristotle and al-Kindi; see above, p. 46, n. 4.

[9] *De luce*, in *Die philosophischen Werke*, ed. Baur, p. 51: 'Lux enim per se in omnem partem se ipsam diffundit, ita ut a puncto lucis sphaera lucis quamvis magna subito generetur, nisi obsistat umbrosum.' Cf. A. C. Crombie, *Grosseteste*, pp. 106–7.

[10] Cf. Vitellonis *Optica*, II, 2, in *Opticae thesaurus*, ed. Risner, Basel, 1572, p. 63; ibid. III, 55, p. 110. See Crombie, *Grosseteste*, p. 217, n. 4.

[11] *Paralipomena*, Francofurti, 1604, p. 36; cf. Crombie, *Grosseteste*, p. 284, n. 4.

doctrine of instantaneous propagation, but with him something new emerged: for his was the first uncompromisingly mechanical theory that asserted the instantaneous propagation of light in a material medium; a theory that had no use for the 'forms' of Ibn al-Haytham and Witelo, or the *species* of Grosseteste, Roger Bacon and Kepler.[12] Indeed, mechanical analogies had been used to explain optical phenomena long before Descartes, but the Cartesian theory was the first clearly to assert that light itself was nothing but a mechanical property of the luminous object and of the transmitting medium. It is for this reason that we may regard Descartes' theory of light as the legitimate starting point of modern physical optics.[13]

The doctrine of instantaneous transmission, far from embarrassing Descartes' system of physics, in fact formed an inseparable part of it. That this was Descartes' own view is clearly indicated in a letter which he wrote in 1634 to an unknown person. First he wrote:

I said to you lately, when we were together, not in fact as you write that light moves in an instant, but (which you take to be the same thing) that it reaches our eyes from the luminous object in an instant; and I even added that for me this was so certain, that if it could be proved false, I should be ready to confess that I know absolutely nothing in philosophy.[14]

---

[12] Descartes, *Dioptric*, I, D, VI, p. 85: 'En suite de quoy vous aurés occasion de iuger, qu'il n'est pas besoin de supposer qu'il passe quelque chose de materiel depuis les obiects iusques a nos yeux, pour nous faire voir les couleurs & la lumiere, ny mesme qu'il y ait rien en ces obiects, qui soit sembable aux idées ou aux sentimens que nous en auons: tout de mesme qu'il ne sort rien des corps, que sent vn aueugle, qui doiue passer le long de son baston iusques a sa main, & que la resistence ou le mouuement de ces corps, qui est la seule cause des sentimens qu'il en a, n'est rien de semblable aux idées qu'il en conçoit. Et par ce moyen vostre esprit sera deliuré de toutes ces petites images voltigeantes par l'air, nommées des *especes inten-tionelles*, qui trauaillent tant l'imagination des Philosophes.'

[13] This statement may give rise to misunderstanding, and I therefore wish to make it clear that I share the view that in the history of science there are no absolute beginnings. As will be shown in the two following chapters, some of the important elements of the Cartesian theories of reflection and refraction had existed for a long time before Descartes made use of them. Nevertheless, what Descartes did was more than simply to continue the efforts of his predecessors. By embodying those elements in a new system of ideas he gave them a new meaning and was able to suggest a new set of problems.

[14] Descartes to [Beeckman], 22 August 1634, D, I, pp. 307–8. 'Dixi nuper, cum vna essemus, lumen in instanti non quidem moueri, vt scribis, sed (quod pro eodem habes) à corpore luminoso ad oculum peruenire, addidique etiam hoc mihi esse tam certum, vt si falsitatis argui posset, nil me prorsus scire in Philosophia confiteri paratus sim.' The letter was first published in a French translation from the Latin original in Clerselier's edition of *Lettres de Mr. Descartes*, II, Paris, 1659, pp. 139–45. The edition did not mention the name of Descartes' correspondent,

To this he significantly added that if the finite velocity of light could be proved experimentally (as his correspondent had claimed) he would be prepared to admit that 'the whole of my Philosophy [i.e. physics] was shaken to its foundations'.[15]

This readiness on Descartes' part to stake the whole of his physics on one experiment implies a recognition of the decisive role of experimental evidence. It also reveals, however, a certain rigidity in his system: he is not just willing to give up those parts of his physics which would be directly related to the problem at issue, but finds himself compelled, should the experiments prove him wrong in this case, to renounce the whole of his scientific edifice. The reason is that the doctrine of instantaneous propagation of light happens to be one of these doctrines which Descartes asserts on *a priori* grounds. Consequently, the danger which the Cartesian physics would have to face if that doctrine were to be proved false, is the kind of danger that necessarily threatens a system of ideas which has been conceived *en bloc*.

Descartes distinguishes, in the passage just quoted, between

---

who, in the Adam and Tannery edition of Descartes' works, was conjecturally identified as Beeckman. The same editors later doubted their conjecture when Beeckman's *Journal* had been discovered and nothing in it was found to suggest that the author visited Descartes at any time in 1634 (cf. D, X, pp. 551–4). (The assumption made by J. F. Scott in *The Scientific Work of René Descartes*, London, 1952, pp. 40–41 that Descartes' correspondent was Mersenne is gratuitous.) It is, however, certain from Beeckman's *Journal* that he had adopted the doctrine of finite velocity of light since 1615 or 1616, and that he had imagined various experiments to verify his doctrine. Cf. Cornelis de Waard, *L'expérience barométrique*, Thouars, 1936, p. 86, n. 4; Isaac Beeckman, *Journal*, ed. C. de Waard, I (1939), p. 99, III (1945), pp. 49, 108–9, 112. No experiment in the *Journal* is identical with that which Descartes attributed to his correspondent. (See also Descartes, *Correspondance*, ed. C. Adam and G. Milhaud, I (1936), pp. 267–74, where the conjecture is made that Descartes' letter may have been addressed to Hortensius.)

[15] D, I, p. 308; *Letters*, ed. Clerselier, II, p. 140. Thus it is not an accident that Descartes entitled the treatise which, in his own words, 'contains the whole of my physics': *De la Lumière*; cf. letter to Vatier, 22 February 1638, D, I, p. 562. In the brief account he gives of this treatise at the beginning of Part V of the *Discourse*, the principal components of the cosmos are differentiated according to their optical functions: the sun and fixed stars (first element) emit light, the heavens (second element) transmit it, and the planets (third element) reflect it. Similarly, earthly bodies are distinguished by their properties of luminosity, transparency and opacity; man, as the spectator of all, is a receptor of light (HR, I, p. 107). Adopting a current medieval terminology deriving from Avicenna's *De anima*, III, 1, Descartes called the action of light in the luminous object *lux*, and the action of the subtle matter constituting the second element *lumen* (cf. Letter to Morin, 13 July 1638, D, II, p. 205). Cf. *Principles*, III, 48–52.

instantaneous *movement*, a doctrine which he does not hold, and instantaneous *communication* of movement, which he obviously takes to be a conceivable form of physical action.[16] For, according to him, light does not consist in an actual motion, but rather in a 'tendency to motion' which is transmitted to the eye through a medium. What, then, is this tendency; and what reasons did Descartes have for believing that by its action the light reached the eye in an instant?

In the same letter to X he gives an experimental argument for the instantaneous propagation of light, which is based on observations of the eclipses of the moon. This argument will be discussed later, but first in order to appreciate the extent to which his conception of light was determined by his general physics we have to turn to the *Traité de la Lumière* where the matter is explained in detail.

Descartes presents us in that treatise with a cosmological 'Fable' depicting how the world *might* have come to be as we see it. He claims that he does not propose to talk about the existing world, but to feign another at pleasure which would nevertheless coincide in appearance with our world.[17] There is no doubt that by adopting this mode of exposition he was only hoping to avoid opposition to his cosmological doctrines which were in agreement with the Copernican system.[18]

---

[16] Descartes to Morin, 13 July 1638, D, II, p. 215: 'Et tout ce que vous disputez en suite fait pour moy, excepté seulement ce que vous semblez vouloir dire à la fin, que *si la Lumiere est vn mouuement, elle ne se peut donc transmettre en vn instant*. A quoy ie répons que, bien qu'il soit certain qu'aucun mouument ne se peut faire en vn instant, on peut dire toutefois qu'il se transmet en vn instant, lors que chacune de ses parties est aussi-tost en vn lieu qu'en l'autre, comme lors que les deux bouts d'un baston se meuuent ensemble.'

[17] *Le Monde ou Traité de la Lumière*, D, XI, p. 36: 'Et mon dessein n'est pas d'expliquer comme eux [i.e. 'les Philosophes'] les choses qui sont en effet dans le vray monde; mais seulement d'en feindre vn à plaisir, dans lequel il n'y ait rien que les plus grossiers esprits ne soient capables de concevoir, & qui puissent toutefois estre crée tout de mesme qui je l'auray feint. Cf. ibid., p. 31. The description of *Le Monde* as a 'fable' appears in Descartes' correspondence for the first time in a letter to Mersenne, 25 November 1630 (D, I, p. 179), but must go back to an earlier time (D, XI, p. 699).

[18] In Part V of the *Discourse* Descartes explains that in order to express himself freely in *Le Monde* he decided to leave the existing world to the disputes of the learned 'and to speak only of what would happen in a new world if God now created, somewhere in an imaginary space, matter sufficient wherewith to form it, and if He agitated it in diverse ways, and without any order, the diverse portions of this matter, so that there resulted a chaos as confused as the poets ever feigned, and concluded His work by merely lending His concurrence to Nature in the usual way, leaving her to act in accordance with the laws which He had established' (HR,

We are invited to imagine[19] an indefinite extension of matter in the nature of which there is nothing that cannot be known 'as perfectly as possible';[20] a matter, that is, which has none of the scholastic or even sensible qualities, and whose essence is completely comprised in pure extension. Such a matter, conceived as a perfectly solid body, can be divided into parts of various shapes which can be moved in various ways. Solidity here means nothing but the being at rest of these parts relative to one another together with the fact that there is no void between one part and another.[21] Since extension can be conceived without motion, it is supposed that God, who has created the matter of this new world in the first place, has also actually divided it into parts of diverse shapes and magnitudes (which does not mean creating gaps between them) and set them in motion. Everything that subsequently happens in this imagined world is a necessary consequence of this initial action, in accordance with certain laws also imposed by the Creator.

What are these laws? They are called by Descartes Laws of Nature in as much as they govern changes in the material world which cannot be ascribed to the immediate action of God, since He does not change. They are based, however, on the fact that God continues to conserve matter 'in the same way [façon] as he created it'.[22] The first law states 'that every part of matter in particular continues always to be in the same state as long as it is not forced to change it by coming into contact with other parts'.[23] Among the 'states' of matter which are thus conserved Descartes counts motion as well as figure, magnitude and rest. He is concerned to distinguish his conception of motion from that of 'the Philosophers', that is, the Aristotelians. Referring to the Aristotelian doctrine that

---

I, p. 107). In the Preface (signed 'D.R.') to the first edition of Le Monde, published posthumously in 1664, we read: 'Il [Descartes] savoit que, si quelque part on defendoit de parler du Systeme de Copernic comme d'une verité, ou encore comme d'une hypothese, on ne deffendoit pas d'en parler comme d'une Fable. Mais c'est une Fable qui, non plus que les autres Apologues ou Profanes ou Sacrés, ne repugne pas aux choses, qui sont par effet' (D, XI, p. ix).

[19] Le Monde, VI, D, XI, pp. 31–36.
[20] Ibid., p. 33.
[21] Cf. Le Monde, III and IV; Principles, II, 54–63.
[22] Le Monde, VII, 'Des Loix de la Nature de ce Nouveau Monde', D, XI, p. 37.
[23] Ibid., p. 38.

natural rectilinear motion comes to an end when the moving body reaches its natural place, Descartes remarks that this gives motion an exceptional status as the only mode of being that would seek its own destruction.[24] Furthermore, he finds the resulting definition of motion, as *actus entis in potentia, prout in potentia est*, completely incomprehensible. The only motion he understands is that assumed by Geometers in their definition of lines and surfaces:

the nature of the motion about which I intend to speak here is so easy to understand that even the Geometers who are, of all men, the most careful to conceive very distinctly the objects of their considerations, have found it more simple and more intelligible than their surfaces and lines: as this appears from the fact that they have explained the line by the motion of a point, and the surface by the motion of a line.[25]

That is to say, the only motion which Descartes recognizes in his world is that by which a body passes from one place to another, successively occupying all intermediate positions.[26]

The second law asserts that when two bodies meet, the one loses as much motion as is acquired by the other.[27] These two laws, remarks Descartes, are in agreement with experiment;[28] but they

---

[24] Cf. also *Principles*, II, 37.   [25] *Le Monde*, D, XI, p. 39.

[26] Alexandre Koyré has commented on this passage as follows: 'Le mouvement cartésien, ce mouvement qui est la chose la plus claire et la plus aisée à connaître, n'est pas, Descartes nous l'a dit, le mouvement des philosophes. Mais ce n'est pas, non plus, le mouvement des physiciens. Ni même des corps physiques. C'est le mouvement des géomètres. Et des êtres géométriques: le mouvement du point qui trace une ligne droite, le mouvement d'une droite qui décrit un cercle. . . . Mais ces mouvements-là, à l'encontre des mouvements physiques, n'ont pas de vitesse, et ne se font pas dans le temps. La géométrisation à outrance—ce péché originel de la pensée cartésienne—aboutit à l'intemporel: elle garde l'espace, elle élimine le temps. Elle dissout l'être réel dans le géométrique' (*Études galiléennes*, II, 49). The diagnosis that Cartesian physics is ill with *géométrisation à outrance* is penetrating, and I intend to make use of it later, but it seems to me that the interpretation of Descartes' conception of *movement* as something independent of time (or not performed in time) cannot be justified. Time is indeed irrelevant in the definition of line as the movement of a point, etc.; but Descartes could not have intended to exclude time from the movement which he characterized by *successive* change of position. He in fact wrote that the only movement he recognized is that which is 'plus aisé à concevoir que les lignes des Geometres: qui fait que les corps passent d'vn lieu en vn autre, & occupent successivement tous les espaces qui sont entre-deux' (*Le Monde*, D, XI, p. 40). See also passage quoted above, p. 50, n. 16.

[27] *Le Monde*, D, XI, p. 41: '. . . quand vn corps en pousse vn autre, il ne sçauroit luy donner aucun mouvement, qu'il n'en perde en mesme temps autant du sien; ny luy en oster, que le sien ne s'augmente d'autant.'

[28] See above, p. 26, n. 26.

are more firmly grounded in the fact that God is immutable and that He always acts in the same way. The undoubted immutability of God requires that the quantity of motion, deposited by Him in the world at the moment of creation, should remain unaltered—a condition which, in Descartes' view, can be fulfilled only if these two laws are observed.

The third law states that when a body is set in motion it always *tends* to move in a straight line, even though it may be forced to move otherwise. Thus, the 'action' of a moving body, or its 'tendency to move' is different from its actual motion.[29] For example, the action or tendency of a stone in a sling is indicated by the tension of the cord and the fact that the stone begins to move in a straight line as soon as it leaves the sling; the stone is forced to move in a circle, but it is always tending to move in a straight line. Again, this law is supposed to be based on the same metaphysical foundation as the other two: it rests on the fact 'that God conserves every thing by a continuous action, and consequently, that He conserves it, not as it may have been some time before, but precisely as it is at the very moment that He conserves it'.[30] God is therefore the author of all movements in the world in so far as they exist and in so far as they are straight, 'but it is the diverse arrangements of matter which make [these movements] irregular and curved'.[31]

These three laws together with the indubitable truths on which mathematicians found their demonstrations are, in Descartes' view, enough to account for all the visible phenomena of our world.

The formation of the stars and the planets, and the spatial distribution of the elements constituting, respectively, the planets, the heavens and the stars are described as follows.[32] Having supposed that the original matter was actually divided into parts of various magnitudes which have been moved in various ways, and considering that rectilinear movement is impossible in this solid plenum, the parts of the original matter assume a somewhat circular movement, around different centres. In the process of pushing one another some larger parts are separated so as to constitute the bodies of the planets. They are characterized by opacity and are known as the third

---

[29] *Le Monde*, D, XI, p. 44.   [30] Ibid.   [31] Ibid., p. 46.   [32] *Le Monde*, VIII, D, XI, pp. 48–56.

element. Others are crushed and formed into small round particles which make up the second element, a fluid subtle matter which fills the heavens as well as the gaps between the parts of the grosser bodies. Fluidity of the subtle matter simply consists in the fact that its parts are moving in various ways among themselves with great speed. It is characterized by transparency or ability to transmit the action of light. Still smaller parts, forming the first element, are scraped off the subtle matter to fill the interstices between its little globules, and the surplus is pushed to the centres of motion thus forming there the bodies of the sun and the stars. Each of these has a rotational motion in the same direction as its proper heaven which forms a vortex carrying the planets round the centre.

In accordance with the third law, the parts of the circulating matter, and in particular those of the matter of the stars, will tend to move in straight lines. They are, however, prevented from actually so moving by the surrounding matter, and thus they are bound to press against those parts which immediately lie next to them. It is this pressure which constitutes their light; a pressure which spreads throughout the medium formed by the second element along straight lines directed from the centre of the circular movement. In one place Descartes speaks of the particles of the subtle matter as having a 'trembling movement'—a property which he finds 'very suitable for light'.[33] This would make the action of light consist in a succession of shocks received and transmitted by the particles of the second element.

The picture is, as we see, purely mechanical, but why should the action of light be transmitted in an instant? The following is the answer given by Descartes in the *Traité de la Lumière*:

And knowing that the parts of the second Element . . . touch and press one another as much as possible, we cannot doubt that the action by which the first [parts] are pushed, must pass in an instant to the last parts: in the same way as the action by which one end of a stick is pushed, passes to the other end at the same instant.[34]

---

[33] *Le Monde*, D, XI, p. 95.
[34] The analogy with the stick is taken up in the first chapter of the *Dioptric*. Observing how the stick of the blind man serves him as a means of distinguishing the objects around him,

The doctrine of instantaneous propagation is thus a necessary consequence of Descartes' conception of the medium serving as the vehicle of light. The nature of that medium is itself determined by the Cartesian definition of matter. According to Descartes, the nature of corporeal substance consists solely in its being extended, and this extension is the same as that ordinarily attributed to empty space. From this it follows that two equally extended bodies must have the same quantity or substance of matter: 'there cannot be more matter or corporeal substance in a vessel when it is filled with gold or lead, or any other body that is heavy and hard, than when it only contains air and appears to be empty'.[35] It also follows that the same part of matter cannot have variable extension; it may change

---

Descartes continues: 'And to draw an analogy from this, I want you to think that light is, in the bodies which we call luminous, nothing but a certain movement or action that is very quick and very violent which passes to our eyes through the mediation of the air and the other transparent bodies, in the same way as the movement or resistance of bodies, which this blind man encounters, passes to his hand through the mediation of his stick. This will from the very first prevent you from finding it strange that this light could extend its rays in an instant from the sun to us: for you know that the action with which one moves one end of a stick must likewise pass in an instant to the other, and that it should so pass even if the distance between the earth and the heavens were greater than it is' (D, VI, p. 84). The analogy is old. Simplicius in his commentary on Aristotle's *De anima* 'compared the role of the transparent medium to that of a stick transferring the effect of the knock from the hand to the stone' (Sambursky, op. cit., p. 116). We also find it in the *Book of the Ten Treatises on the Eye* attributed to the great ninth-century scholar and translator of Greek medical writings into Arabic, Ḥunayn ibn Ishāq. In the third treatise (On Vision) Ḥunayn first observes that a man walking in the dark can with the help of a stick in his hand detect the objects he meets with. He then expresses the view that 'sight . . . senses the sense-object which moves it through the mediation of the air just as the blind man feels the object by means of the stick' (*The Book of the Ten Treatises on the Eye Ascribed to Hunain ibn Ishāq*, the Arabic text edited . . . with an English translation . . . by Max Meyerhof, Cairo, 1928, p. 109 (Arabic), pp. 36–37 (English); my translation). According to Ḥunayn vision takes place when the 'luminous spirit', rushing from the brain into the eyes, 'meets the surrounding air and strikes it as in a collision' (ibid., p. 110 of the Arabic text). Yet, it must be noted, the result of this 'collision' is in Ḥunayn's view a *qualitative* change that transforms the air into the nature of the spirit; that is to say, the air itself becomes an organ of vision that is continuous and homogeneous with sight (ibid.). J. Hirschberg has shown (cf. *Sitzungsberichte der Königlich preussischen Akademie der Wissenschaften*, 1903, no, 49, pp. 1080–94) that two medieval Latin versions were made of Ḥunayn's book. The translator of one of these, Constantinus Africanus, ascribed the work to himself, as was his habit, and it was accordingly known as '*Liber de oculis Constantini Africani*'. The other translation went under the title, '*Galeni Liber de oculis translatus a Demetrio*', and was printed in different Latin editions of Galen's works, including the nine Juntine editions which appeared in Venice between 1541 and 1625 (cf. M. Meyerhof, op. cit., Introduction, especially p. vii). For the passage referred to above see Constantini *Liber de oculis*' in *Omnia opera Ysaac*, Lugduni, 1515, fol. 173ᵛb; and 'Galeni De oculis' in Galeni *Opera omnia*, VII, Venetiis (apud Iuntas), 1609, fol. 188ᵛ.

[35] *Principles*, II, 19, HR, I, p. 264.

its figure, but its volume must remain constant. As Pierre Duhem put it, the Cartesian matter is 'rigorously incompressible'.[36] In such an incompressible and inelastic medium, Descartes considers, pressure must be transmitted instantaneously.

Newton later expressed the objection that the idea of instantaneous propagation of motion would involve a doctrine of infinite force. At first sight this might appear as a criticism of the Cartesian doctrine. A careful reading of Newton's text, however, shows that his objection was directed against an opinion that was not actually held by Descartes. We read in Query 28 of Newton's *Opticks*:

If Light consisted only in Pression propagated without actual Motion, it would not be able to agitate and heat the Bodies which refract and reflect it. If it consisted in Motion propagated to all distances in an instant, it would require an infinite force every moment, in every shining Particle, to generate that Motion. And if it consisted in Pression or Motion, propagated either in an instant or in time, it would bend into the Shadow.[37]

There are only two objections here against the (Cartesian) doctrine that light consisted in pression propagated in an instant without actual motion: namely that such a pression would fail to account for the observed rectilinear propagation of light and the fact that it heats the bodies on which it falls. On the other hand, Newton remarks that an infinite force would be required in every shining particle, *if light consisted in motion propagated to all distances in an instant*. This last hypothesis was not held by Descartes. It may be noted here that Newton himself adopted the doctrine of instantaneous propagation of pressure through an incompressible medium. Considering, in Bk. II, Prop. XLIII of *Principia*, how pulses generated in a medium by a tremulous body would be propagated, he writes: 'If the medium be not elastic, then, because its parts cannot be condensed by the pressure arising from the vibrating parts of the tremulous body, the motion will be propagated in an instant . . .'[38]

---

[36] P. Duhem, 'Les théories de l'optique', *Revue de Deux Mondes*, CXXIII (1894), pp. 95–96.

[37] Newton, *Opticks*, p. 362. Earlier objections by Newton against the Cartesian view of light as pressure are contained in a passage quoted from Newton's MS. *Quaestiones quaedam dhilosophicae* (c. 1664) by Richard S. Westfall in 'The foundations of Newton's philosophy of nature', *The British Journal for the History of Science*, I (1962), p. 174.

We come now to Descartes' experimental arguments in support of his view. These, as has been remarked before, are contained in his letter to X of 1634. The fact that they do not occur in any of Descartes' formal writings would seem to imply that they had, in Descartes' view, only a secondary importance: they merely confirmed him in a belief which was ultimately based on his primary conception of matter. X, who believed that light had a finite velocity, had suggested the following experiment to decide the issue between him and Descartes. An observer would move a lantern in front of a mirror placed at a certain distance: the interval between moving the lantern and perceiving the reflection of this movement in the mirror would measure the time required by the light from the lantern to cover the distance from the mirror twice.

But, Descartes replied, there was a better experiment, one which had already been made by thousands of 'very exact and very attentive'[39] observers, and it showed that there was no lapse of time between the moment light left the luminous object and the moment it entered the eye. This experiment was provided by the eclipses of the moon.

$$A \qquad\qquad\qquad B \qquad C$$

In the figure, let A, B and C represent the positions of the sun, the earth and the moon respectively; and suppose that from the earth at B the moon is seen eclipsed at C. The eclipse must appear at the moment when the light which has been sent off by the sun at A, and reflected by the moon at C, arrives at B, if it has not been interrupted in its journey from A to C by the interposition of the earth. Assuming that the light takes half an hour to cover the distance BC once, the eclipse will be seen one hour after the light from the sun reaches the earth at B. Moreover, the sun appears at A precisely at the moment when the light coming from A reaches B. Therefore, the moon will not appear eclipsed at C until one hour after the sun is seen at A.

---

[38] Newton, *Principia*, p. 372.
[39] D, I, p. 308; *Lettres*, ed. Clerselier, II, p. 140.

But the unvarying and exact observation of all Astronomers, which is confirmed by many experiments, shows that when the moon is seen from the earth at B to be eclipsed at C, the sun must not be seen at A one hour previously, but at the same instant as the eclipse is seen.[40]

As the lapse of one hour in these observations would be certainly more appreciable than the much shorter interval which would have to be perceived in X's terrestrial experiment, Descartes concluded: 'your experiment is useless, and mine, which is that of all Astronomers, very clearly shows that the light is not perceived after any sensible interval of time.'[41]

In the preceding considerations Descartes was so generous as to assume, for the sake of argument, a velocity of light which was in fact 24 times that envisaged by his correspondent. But he was not

---

[40] D, I, p. 310; *Lettres*, ed. Clerselier, II, p. 142.

[41] D, I, p. 310: 'Ergo & tuum experimentum est inutile, & meum, quod est omnium Astronomorum, longe clarius ostendit, in nullo tempore sensibili lumen videri.' In the French version this was rendered as follows: 'Par consequent, & vostre experience est inutile, & la mienne, qui est celle de tous les Astronomes, monstre clairement, que la lumiere se voit sans aucun interualle de temps sensible, *c'est à dire, comme i'auois soustenu, en vn instant.*' (*Lettres*, ed. Clerselier, II, pp. 142–3.)

The occurrence of the italicized words (added freely by the translator) following the assertion that light is seen 'sans aucun interualle de temps sensible' has led J. F. Scott to suggest (in op. cit., pp. 40–41) that Descartes did not mean the doctrine of instantaneous propagation literally; that by the expression *in an instant* Descartes simply meant, as Scott put it, 'an extremely short interval of time, the twinkling of an eye, the lightning flash!' Scott does not seem to have consulted Descartes' letter in the original Latin where the words on which he bases his interpretation do not occur. He also unjustifiedly identifies Descartes' correspondent as Mersenne (see above, p. 48, n. 14). This interpretation cannot be maintained in the face of the many explicit statements which we find in Descartes' treatises and letters. As an example I quote the following passage from Descartes' letter to Mersenne (27 May 1638), in which Descartes makes quite clear what he means by the expression *in an instant*: 'Et pour la difficulté que vous trouuez en ce qu'elle [i.e. light] se communique en vn instant, il y a de l'équiuoque au mot d'instant; car il semble que vous le considerez comme s'il nioit toute sorte de priorité, en sorte que la lumiere du Soleil pust icy estre produite, sans passer premierement par tout l'espace qui est entre luy & nous; *au lieu que le mot d'instant n'exclud que la priorité du temps,* & n'empesche pas que chacune des parties inferieures du rayon [that is, the parts of the ray farther from the luminous object] ne soit dependante de toutes les superieures, en mesme façon que la fin d'vn mouvement successif depend de toutes ses parties precedentes' (D, II, p. 143; my italics). See also Descartes to Plempius, 3 October 1637 (D, I, pp. 416–17); Descartes to Morin, 13 July 1638 (D, II, p. 215, quoted above, p. 50, n. 16); Descartes to Mersenne, 11 October 1638, with reference to an experiment described by Galileo in the *Dialogues* to measure the speed of light: 'Son experience, pour sçauoir si la lumiere se transmet en vn instant, est inutile: car les Ecclipses de la lune, se rapportant assez exactement au calcul qu'on en fait, le prouuent incomparablement mieux que tout ce qu'on sçauroit esprouuer sur terre' (D, II, p. 384); and *Principles*, III, 64.

generous enough: the first estimation of the velocity of light (by Roemer) gave eleven minutes as the time required for a ray from the sun to arrive at the earth. Using this result Huygens was able to show why the eclipses of the moon were not suitable for discovering the successive movement of light.[42]

It will not be out of place here to consider Descartes' explanation of what Huygens later described as 'one of the most marvellous'[43] properties of light, namely the fact that light rays are not impeded by crossing one another. In the *Dioptric*[44] Descartes compares the light-bearing medium to a vat which is full of half-crushed grapes and which has two apertures at the base. The grapes are compared to the parts of the gross transparent bodies, while the liquid corresponds to the subtle matter which fills the pores of those bodies. The moment the apertures are opened, the parts of the liquid at a given point on the surface simultaneously tend to descend towards *both* apertures in straight lines. Similarly, the parts of the liquid at any other points on the surface simultaneously tend to move towards one aperture and the other 'without any of these actions being impeded by the others, or even by the resistance of the grapes which are in this vat'.[45] Nor are these actions hindered by any movements which the grapes may have. In the same way, the parts of the subtle matter touching the side of the sun facing us, tend to move in straight lines towards our eyes at the same instant as we open them, without the actions along these lines being hindered by one another, or by the stationary or moving parts of the gross transparent bodies, such as the intervening air.

This, in Descartes' view, is possible because light, as it exists in the medium, is not an actual movement, but a tendency to move. For although a body cannot be conceived to move simultaneously in various directions, or even in a straight line when obstacles are placed in its way, it may nevertheless tend to move at the same instant towards different sides, and without this tendency being hindered by obstacles. Considering, then, that it is not so much the movement as the 'action' or tendency to movement of luminous

---

[42] Below, pp. 203ff.
[44] D, VI, pp. 86–88.
[43] Huygens, *Treatise on Light*, pp. 21–22.
[45] Ibid., p. 87.

bodies that should be taken to constitute their light, the rays of light are 'nothing else but the lines along which this action tends'.[46] And, thus, there is an infinite number of such rays proceeding in all directions from each point on the luminous body, without their actions being stifled or hindered when they happen to cross one another in space.

This was perhaps the most unfortunate of Descartes' analogies in the *Dioptric*, but it pointed out a problem which any subsequent wave-theory of light had to reckon with. If, as Huygens later objected, light was a tendency to movement, how could one and the same particle of the medium tend to move in opposite directions, as would be the case when two eyes view each other? To explain this, Huygens was led to introduce elasticity as a property of the medium and, in consequence, renounce the doctrine of instantaneous propagation.[47]

2. Descartes had the idea of writing the *Meteors* long before he conceived the plan of the volume in which it appeared in 1637 together with the *Discourse*, the *Dioptric* and the *Geometry*.[48] The immediate occasion for writing a treatise on the subject of the *Meteors* was a parhelic phenomenon which was observed in Italy in March 1629 and which attracted a great deal of attention. One of Descartes' friends forwarded to him a description of the phenomenon shortly afterwards and asked him how he would account for it.[49]

---

[46] D, VI, p. 88.

[47] Below, pp. 207ff.

[48] In November 1630 Descartes talked about the *Dioptric* as a 'summary' of *Le Monde* which he had started towards the end of 1629 (cf. Letter to Mersenne, 18 December 1629, D, I, pp. 85–86), but he said that the summary would not be ready for a long time; for six months he had been working on a chapter on the nature of colours and light, but even that was 'not yet half finished' (Letter to Mersenne, 25 November 1630, D, I, p. 179; see also D, I, p. 235 and p. 255). It was only in 1635, when the *Dioptric* was ready for printing (D, I, p. 342), that Descartes thought of joining it with the *Meteors* in one volume and of adding a 'preface' to the two treatises (Letter to Huygens, November 1635, D, I, pp. 329–30). This 'preface' later appeared in the 1637 publication as the *Discourse*. Last in the order of composition was the *Geometry* which Descartes wrote while the *Meteors* was being printed in 1636 (Letter to ★ ★ ★, October 1637, D, I, p. 458). Cf. Gilson, *Commentaire*, pp. x–xi; L. Roth, op. cit., pp. 13–27.

[49] Cf. D, I, p. 29, editors' note.

His reaction was typical of him:[50] he felt that in order to explain this particular phenomenon he had 'to examine methodically [*par ordre*] all the Meteors'.[51] By October of the same year he believed himself to be ready to give a satisfactory explanation of atmospheric phenomena and had therefore decided to write 'a little Treatise which will contain the explanation of the colours of the rainbow, which have given me more trouble than all the rest, and generally of all the sublunar Phenomena'.[52] He asked his confidant Mersenne not to mention this planned treatise to anyone 'for I have decided to present it to the public as a sample of my Philosophy, and to hide behind the work to hear what will be said about it.'[53]

We have already quoted Descartes' statement of 1638 in which he made it quite clear that when he published his 1637 volume he was not aiming either to teach 'the whole of my method' in the *Discourse* or to show how he actually employed it to arrive at the results contained in the three treatises that followed.[54] At the same time, however, he described his explanation of the rainbow which he presented in the eighth chapter of the *Meteors*, as 'a sample' of the method outlined in the *Discourse*. Indeed, the opening sentence in that chapter contains the only reference to 'the method' in all three treatises: 'The rainbow is a marvel of nature that is so remarkable, and its cause has always been so eagerly sought by men of good sense and is so little understood, that I could not choose a subject better suited to show how, by the method which I employ, we can arrive at knowledge which has not been attained by those whose writings we have.'[55] It will be noted that when, in 1629, Descartes had not yet thought of writing a methodological preface to the

---

[50] Expressing to Mersenne his opinion of Galileo's *Dialogues* Descartes first remarked, rather condescendingly, that Galileo 'philosophe beaucoup mieux que le vulgaire, en ce qu'il quitte le plus qu'il peut les erreurs de l'Eschole, & tasche a examiner les matieres physiques par des raisons mathematiques'. Then he added the following reservation, indicating how in his view one should go about explaining particular phenomena: 'Mais il me semble qu'il manque beaucoup en ce qu'il fait continuellement des digressions & ne s'areste point a expliquer tout a fait vne matiere; *ce qui monstre qu'il ne les a point examinées par ordre, & que, sans auoir consideré les premieres causes de la nature, il a seulement cherché les raisons de quelques effects particuliers, & ainsy qu'il a basti sans fondement*' (Letter to Mersenne, 11 October 1638, D, II, p. 380, my italics; see also similar remarks on Galileo's *Chief Systems*, D, I, pp. 304–5).
[51] Descartes to Mersenne, 8 October 1629, D, I, p. 23.
[52] Ibid.   [53] Ibid.   [54] See above, p. 17, n. 2.   [55] D, VI, p. 325.

Meteors, he described the same explanation as 'a sample of my philosophy'.

The experimental character of Descartes' investigations of colours shows a different aspect of his activities as a physicist from that which the doctrine of instantaneous propagation of light has revealed. For although he argued in support of that doctrine from astronomical observations, his real reasons for holding it were ultimately drawn from another source, namely his *a priori* conception of matter and the cosmological picture in which light occupied a prominent place. When we watch Descartes dealing with the problem of the rainbow and of colours, however, we find him guessing, experimenting, calculating, testing. It would be a mistake to think that by indulging in these things he was sinning against his theory of method. Experimentation and the use of hypotheses are, as I hope the preceding chapter has made clear, integral parts of what the physicist is, in Descartes' view, obliged to do. This is not to say, however, that in his study of the rainbow, Descartes was following certain *rules* which he had already formulated. Scientific results are just not arrived at in such a linear fashion. Nor does it mean that he approached that phenomenon with a mind free from preconceived ideas; he was expressly seeking an explanation in terms of his initial suppositions concerning the nature of light. His attempt is therefore interesting in that it reveals an important aspect of his work that is often passed over in descriptions of his attitude as a practising scientist and as a methodologist of science.

Having observed, like others before him, that rainbows were produced in fountain sprays, and noting that sprays are composed of small drops of water, Descartes obtained a large spherical glass vessel filled with water to facilitate the experimental study of the phenomenon.[56] He stood with his back to the sun and held up the

---

[56] *Meteors*, VIII, D, VI, pp. 325–44. The same experiment, using the spherical glass vessel filled with water as a model of an individual rain drop, was independently performed in the first decade of the fourteenth century by the Persian Kamāl al-Dīn al-Fārisī (d. c. 1320) and the German Dominican Theodoric of Freiberg (d. c. 1311). Kamāl al-Dīn's work was particularly inspired by Ibn al-Haytham's investigations on the burning sphere and also by remarks in Avicenna's *Meteorologica* on the production of rainbows in sprays (Kamāl al-Dīn al-Fārisī, *Tanqīh al-manāzir*, Hyderabad, 1348 H., II, p. 283. Avicenna's observations were reported and discussed by Averroës in his middle commentary on Aristotle's *Meteorologica*, III, 4, which

glass vessel in the sun's light. By moving it up and down he found that a bright red colour appeared at an inferior part of the vessel by rays emerging from it at an angle of approximately 42° with the direction of the incident light from the sun. When he raised the vessel from this position the red disappeared; but as he *lowered* it a little, all the rainbow colours successively appeared. Further, looking at a higher part of the vessel, a fainter red than in the previous case was observed when the rays emerging from that part made an angle of about 52° with the incident light. The other colours successively appeared when he gradually *raised* the vessel through a certain small angle, and then they vanished when that angle was exceeded. No colours appeared at the same part when he lowered the ball from the initial position.

From this Descartes gathered, as he tells us, that in the place where the rainbow appears, there must be seen a bright red point in each of all those drops from which the lines drawn to the eye make an angle of about 42° with the incident light. All these points being viewed together, must appear as continuous circles of a bright red colour. Similarly, continuous circles of other colours must be formed by the points in all those drops from which the lines drawn to the eye make slightly smaller angles with the incident rays. This explains the primary (interior) bow, with the order of the colours from red above to blue below.

The secondary (exterior) bow he similarly accounted for by considering the points from which the lines drawn to the eye make an angle that is either equal to or slightly larger than 52°. Returning to the glass-vessel, Descartes found that the rays emerged in the

was translated into Latin in the thirteenth century by Michael Scot). Both Kamāl al-Dīn and Theodoric succeeded in explaining the primary bow by two refractions and one reflection, and the secondary bow by two refractions and two reflections—all of these refractions and reflections being understood as taking place in individual drops. It is not impossible that some of Theodoric's ideas may have reached Descartes, but there is no real evidence that they did. Of all the writers on the rainbow that preceded him, Descartes in the *Meteors* referred only to the sixteenth-century Italian Maurolyco whose work on this subject (contained in his *Diaphaneon* of 1553 and *Problemata ad perspectivam et iridem pertinentia* of 1567) was definitely inferior to his (D, VI, p. 340). For a comprehensive account of Theodoric's theory together with a study of the transmission of ideas on the rainbow to the seventeenth century, see A. C. Crombie, *Grosseteste*; also Carl Boyer, *The Rainbow*. An extensive bibliography will be found in these two books.

first case, corresponding to the primary bow, after two refractions (one on entering the vessel near the top and another on leaving it near the bottom) and one reflection (at the farther side of the vessel). In the second case, corresponding to the secondary bow, the rays emerged after two refractions (one on entering near the bottom and the other on leaving near the top) and two reflections. This explained why the order of the colours of the secondary bow was the reverse of that of the primary bow.

But some difficulties still remained to be tackled. One was to explain the production of colours as such. Descartes' procedure to solve this difficulty recalls to mind Bacon's method of exclusion. '. . . I inquired whether there was another subject [suiet] in which [colours] appeared in the same fashion, so that, by comparing one with the other, I might better be able to determine their cause.'[57] Remembering that the rainbow colours could be produced by a triangular glass prism, he turned to the examination of that.

He allowed the sun's rays to fall perpendicularly on one face of the prism so that no appreciable refraction took place at that surface; and he covered the other face through which the light was refracted with an opaque body in which there was a small aperture. The rays emerging through that opening exhibited the rainbow colours on a white sheet of paper.

From this Descartes learnt, first, that the curvature of the surface of the water drop was not necessary for the production of these colours, since both surfaces of the prism were plane. Second, that the angle at which the rays emerged from the prism was irrelevant, for this angle could be altered without changing the colours. Third, that neither reflection nor a multiplicity of refractions were necessary, as here there was only one refraction and no reflection.

But I have judged that at least one [refraction] was necessary, and even one whose effect was not destroyed by a contrary [refraction]; for experience shows that when the surfaces [of the transparent body through which the light passed] . . . were made parallel, the rays, having straightened up at one [of the surfaces] after they have been broken by the other, would not produce these colours.[58]

---

[57] *Meteors*, VIII, D, VI, p. 329.　　　　　　[58] Ibid., pp. 330–1.

Finally, Descartes observed, the light should be limited by a shadow; for when the opaque body was removed the colours disappeared. Now the problem was to know why red always appeared at one end of the image and blue at the other, even though the rays giving rise to these colours were both refracted once and both terminated by a shadow, and to account for the appearance of the intermediate colours in the observed order.

Here Descartes recalls what he has asserted regarding the nature of light in the *Dioptric*, namely that it is a certain action or movement transmitted through the subtle matter whose parts he imagines to be small globules which can roll through the pores of terrestrial bodies. Before the light is refracted, these globules have only one movement, which is their movement in the direction of propagation. When they obliquely strike a refracting surface they acquire a rotatory motion which, disregarding external influences, would be of the same speed as their movement of translation. The rotatory motion of each globule is, however, affected by the velocity of the surrounding globules. Consider, for example, a globule travelling along the extreme red ray in the prism experiment described above. Such a globule, by pressing against the slower globules on the side of the shadow, and being pressed by those on the other side (whose velocity is greater) will suffer an effect similar to that of rolling a sphere between both hands; thus the rotary velocity of the globule will be increased or diminished depending on the sense of the rotation originally acquired at the refracting surface. Clearly, the effect of the shadowed medium bordering on the globules travelling along the extreme blue (or violet) ray will be contrary to that produced on the other side, since all globules are supposed to rotate in the same sense. Descartes considers that the sense of rotation is such that, whereas the rotary motion of globules along the red ray is accelerated, that of the globules along the blue ray is retarded. This accounts for the appearance of red and blue at the sides of the shadow. The other colours are explained by supposing that the globules along their rays are rotating with intermediate velocities ranging between the greatest rotational velocity (for red) and the smallest (for blue).

In Descartes' view, the preceding explanation was in perfect agreement with our experience of colours. For, he argued, the sensation of light being due to the movement or tendency to movement of the subtle matter touching our eyes, it is certain that the various movements of this matter must give rise to different sensations. Considering that the surfaces of the prism could not affect the movements of the globules in any way other than that described, he concluded that sensations of colours must be attributed to the various rotational velocities of the globules.

He admits that refraction and the termination of the light by a shadow are not always necessary for the production of colours, as for example in the case of opaque bodies. The colours of these, however, must be produced by the effect on the rotational motions of the globules that is due to the 'magnitude, figure, situation and movement of the parts' of these bodies when the light is reflected against their surfaces.[59]

It is these highly speculative considerations which, if we take the autobiographical account in the *Meteors* at its face value, finally led Descartes to his most important discovery in connection with the rainbow. For these speculations suggested a problem which Descartes expresses as follows:

. . . I doubted at first whether the colours were produced in [the rainbow] in exactly the same way as in the prism . . . for I did not observe there any shadow terminating the light, and I did not know why they appeared only at certain angles, until I have taken my pen and calculated in detail all the rays falling on the various points of a drop of water. . . .[60]

His purpose was to know at what angles the rays would come to the eye after two refractions and one or two reflections. By applying the law of refraction which he already possessed he found that, after one reflection and two refractions, many more rays would be seen at an angle of $41°$ for blue to $42°$ for red (the angles being made with the incident light) than at any smaller angle; and that no rays would be seen at a greater angle. Also, after two refractions and two

---

[59] *Meteor* VIII, D, VI, p. 335.　　　　[60] Ibid., pp. 335–6.

reflections, the eye would receive many more rays at an angle of 51° (red) to 52° (blue) than at any greater angle; and no rays would proceed to the eye at any smaller angle.

So that there is a shadow bordering on both sides of the light which, after having passed through an infinite number of rain drops illuminated by the sun, comes to the eye at an angle of 42 degrees, or a little less, and gives rise to the first principal rainbow. And there is also a shadow terminating [the light] which comes at an angle of 51 degrees, or a little more, and causes the exterior rainbow; for, not to receive any rays of light in one's eyes, or to receive considerably less of them from one object than from another near it, is to perceive a shadow. This clearly shows that the colours of these arcs are produced by the same cause as those which appear with the aid of the prism. . . .[61]

The calculations involved here constitute Descartes' contribution to the theory of the rainbow.[62] By their means he could determine, for the first time, why the coloured circles appeared at the angles observed, as well as explain the appearance of the secondary bow in a quantitative manner.

His explanation of the formation of colours by a prism, however, was not successful. This task was left to Newton. Nevertheless, Descartes' speculations made a decided advance on previous accounts. Before him writers on the subject were content to explain colours as a result of the mixture of light and darkness, or of a finite number of 'primary' colours, in various proportions. In Descartes' theory the individuality of all colours is strongly suggested by the fact that to each colour on the spectrum there corresponds a definite physical (and, as it happens, periodic) property, the rotational velocity of the corresponding globules.[63] But the weakness lay in the fact that he could not provide a means of measuring or calculating this

---

[61] Ibid., p. 336.

[62] Cf. Carl Boyer, *The Rainbow*, pp. 211–18; J. F. Scott, op. cit., pp. 78–81.

[63] Descartes argued that since the reality of colours consists in their appearing, all colours are equally real: 'Et ie ne sçaurois gouster la distinction des Philosophes, quand ils disent qu'il y a [des couleurs] qui sont vrayes, & d'autres qui ne sont que fausses ou apparentes. Car toute leur vraye nature n'estant que de paroistre, c'est ce me semble, vne contradiction de dire qu'elles sont fausses & qu'elles paroissent' (D, VI, p. 335).

velocity and, to that extent, his explanation of colours remained qualitative, though mechanical. Newton substituted the size of the light corpuscle for the rotational velocity of the globule, but his theory is distinguished by taking into account the experimental fact of unequal refractions of different colours which had escaped the notice of everyone before him.

Chapter Three

DESCARTES' EXPLANATION OF
REFLECTION. FERMAT'S OBJECTIONS

1. The first explicit formulation of the law of reflection, stating the equality of the angles of incidence and reflection of sight or visual rays, is to be found in Euclid's *Optics* (Proposition XIX).[1] Euclid here states the law in connection with a problem concerning the determination of heights. A reference to this law also exists in the Peripatetic *Problemata*,[2] which is supposed to contain ideas mainly deriving from Aristotle. The passage in which this reference occurs is interesting from the point of view of subsequent attempts to explain optical reflection in mechanical terms. The author of this passage seems to be concerned to bring out a *contrast*, rather than an analogy, between the appearance of images in mirrors and the rebound of objects. An object thrown against a resisting surface rebounds in virtue of the movement that has been imparted to it by the thrower—its natural movement, if any, having ceased when the body reached its natural place. Thus in preventing the movement in the straight line of incidence, the surface causes the object to change its direction. We are told that the object returns at an angle equal to the angle of incidence, but we are not told *why* this must be so. As opposed to this mechanical situation, we see the image of an object in a mirror at the end of the line along which our sight travels—as if this line, or our sight, were *not* impeded by the mirror. Although it is difficult to ascertain exactly what the author wanted to convey,[3] he did mention the formation of images in the course of a discussion about mechanical reflection.

---

[1] Cf. Euclidis *Optica*, in *Opera omnia*, VII, ed. I. L. Heiberg, Leipzig, 1895, pp. 28–31; Euclide, *L'Optique et la Catoptrique*, tr. Paul Ver Eecke, Paris, 1959, pp. 13–14.

[2] [Aristotle], *Problemata*, XVI, 13.

[3] See the interpretation given by Carl B. Boyer in 'Aristotelian references to the law of reflection', *Isis*, XXXVI (1945–46), pp. 92–95.

In what has come down to us of the *Catoptrics* of Heron of Alexandria (about A.D. 75) we find a brief account of the mechanical conditions for the reflection of visual rays.[4] Heron observes that hard bodies repel the movement of projectiles whereas soft bodies,

FIG. III. 1.

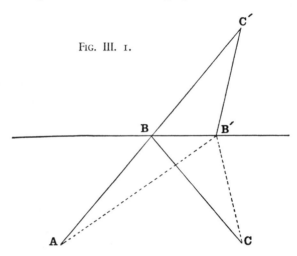

such as wool, arrest it; similarly, visual rays are reflected from polished rather than rough surfaces. For a smooth surface is one whose parts are so compact as not to permit the rays to fall into gaps between them. Glass and water surfaces reflect rays only partially because of the irregularity of these substances: they consist of solid parts which repel the rays falling on them, and of porous parts which allow other rays to pass through. Heron knew the law of the parallelogram of velocities[5] which had been already formulated in the Peripatetic *Mechanica*,[6] but so far as we know he made no attempt to apply it in the derivation of the law of reflection. Instead, he appealed to the widespread dictum that nature does nothing in

---

[4] Cf. Heronis *Catoptrica*, III, in *Herons von Alexandria Mechanik und Katoptrik*, herausgegeben und übersetzt von L. Nix und W. Schmidt (Heronis Alexandrini *Opera quae supersunt omnia*, vol. I, fasc. I, Leipzig, 1900), pp. 322–25 (Latin text with German translation). See English translation of the relevant passage in *A Source Book in Greek Science*, ed. M. R. Cohen and I. E. Drabkin, Cambridge, Mass., 1948, pp. 263–4.

[5] Cf. Heronis *Mechanica*, I, 8, in *Herons von Alexandria Mechanik und Katoptrik* . . ., pp. 18–21 (Arabic text with German translation).

[6] [Aristotle], *Mechanica*, I, 848b.

vain.[7] Interpreting this dictum here as a principle of *minimum path*, he proves that it is presupposed by the equality of the angles of incidence and reflection.[8] For (Fig. III. 1) let *AB*, *BC* be the incident and reflected visual rays making equal angles with the mirror BB', and let B' be any point on the surface of the mirror other than B in the plane of incidence. The actual path *ABC* must be shorter than the hypothetical $AB'C$ which makes unequal angles with the mirror. This is shown by producing *AB* to C', such that $CB = BC'$, and joining $C'B'$. The triangles CBB' and $C'BB'$ are equal and, therefore, $C'B' = B'C$. Since $AB' + B'C'$ is greater than AC', it follows that $AB' + B'C$ is greater than $AB + BC$. The shortest path, which the ray follows from the eye A to arrive by reflection at the object C, is that for which the angles of incidence and reflection are equal. Heron extends this proof to cover the case of spherical convex mirrors.[9]

In addition to proving the law of optical reflection experimentally, Ptolemy tried in his *Optics* to give a theoretical explanation of this law by comparing the reflection of visual rays to the rebound of projectiles. He observes[10] that when a ball (*spera*) is thrown on to a wall at right angles, its movement in that direction is totally prevented. On the other hand the grip of a bow does not impede the movement of the arrow. Thus while an object will prevent a movement to which it is directly opposed, a tangential movement will not be affected by it. This being true of everything that moves, it applies to visual rays. A ray falling perpendicularly on the surface of a mirror is thrown back in the same line, while a ray travelling tangentially to the surface continues to move forward without being deflected. Between these two extreme cases, a ray that falls obliquely to the surface is reflected in a direction symmetrical with that of incidence, that is, in such a way as to make the angle of reflection equal to the angle of incidence. Ptolemy seems to have had in mind here the parallelogram of displacements, but he makes no explicit use

---

[7] We are told this by Damianus and Olympiodorus who had more information about Heron's *Catoptrica* which has not survived in a complete form. Cf. A. Lejeune, *Catoptrique grecque* (1957), pp. 48–49.

[8] Heronis *Catoptrica*, IV, pp. 324–9 (Latin and German).

[9] Ibid., IV, pp. 328–30; *A Source Book in Greek Science*, pp. 264–5.

[10] Cf. Ptolemaei *Optica*, III, 19–20, ed. A. Lejeune, Louvain, 1956, pp. 98–99.

of it, and his text, which has survived only in a Latin translation made from an Arabic version which has itself been lost, is not quite clear.[11]

The ideas of Heron and Ptolemy were raised to a high level of elaboration by Ibn al-Haytham,[12] whose theory deserves to be presented here in some detail for two reasons: first, because it is the most complete mechanical treatment of reflection that we have up until the publication of Descartes' *Dioptric*; and second, because it is very likely that Descartes was acquainted with it.[13] But without necessarily maintaining that Ibn al-Haytham did in fact influence Descartes, it will be interesting to observe the striking similarities, and also the significant differences, between their theories.

Ibn al-Haytham explains in the fourth book of his *Optics*[14] that

---

[11] I have followed Lejeune's interpretation of this passage in ibid., p. 99, n. 33. See also his *Catoptrique grecque*, pp. 34–35.

[12] Avicenna was familiar with the analogy between optical and mechanical reflections, but he explicitly rejected it. He argued that if light were reflected by repulsion, as in the case of a ball, it would be reflected from all impenetrable bodies, even if they are not smooth. He wrote in *De anima*, III, 6, in *Opera*, Venetiis, 1508, fol. 14ʳb: 'Si autem causa reverberatione [lucis] est repercussio retro aut tumor, sicut accidit pille, deberet reverberari ab omni corpore, quid non penetrat, quamvis non esset lene.'

[13] Descartes nowhere refers to Ibn al-Haytham (Alhazen) or to his *Optics*. It would, however, be most implausible to suppose that he ignored a work that was well known to Snell (*Janus*, XXXIX (1935), p. 64), Beeckman (*Journal*, ed. C. de Waard, II, p. 405), Mersenne (e.g. *Quaestiones celeberrimae* (1623), pp. 761, 763, 774, 778) and Fermat (D, I, p. 356), to mention only a very few of his contemporaries; a work, moreover, to which he must have seen many allusions in Kepler's *Paralipomena* (1604, *passim*) as well as in works of other writers. The *Optics* of Ibn al-Haytham was available to Descartes in the volume entitled *Opticae thesaurus* which F. Risner edited in 1572 and which also contained Witelo's *Optics*, which Descartes certainly knew (cf. D, I, p. 239; D, II, p. 142; D, III, p. 483; D, **XI**, p. 646). The fact that Descartes wrote 'Vitellion', the form which we find in the Nuremberg editions of 1535 and 1541 of Witelo's *Optics* (D, I, p. 241) and in Kepler's *Ad Vitellionem Paralipomena*, as distinguished from Risner's 'Vitello', cannot be taken as proof that his knowledge was confined to the Nuremberg editions. Fermat also used the form 'Vitellio' although he was certainly acquainted with Risner's 1572 volume. (D, I, p. 356.)

[14] I have followed in this account the Arabic text (Istanbul MS. Fātih 3215, fols. 66ᵛ9–76ʳ13) which has been somewhat compressed in the Latin translation edited by Risner. Since it is the Latin text that is important for the study of the transmission of Ibn al-Haytham's ideas, I give it here in full. It will be seen that the Latin preserves all the main ideas of the original. Alhazeni *Optica*, IV, 17 and 18, pp. 112–13:

17. *Reflexio lucis & coloris a superficie aspera facta, plerumque fugit visum.*
Amplius, quare ex politis corporibus, non ex asperis fiat reflexio, est: Quoniam lux, ut diximus, non accedit ad corpus, nisi per motum citissimum, et cum pervenerit ad politum, eiicit eam politum a se; corpus vero asperum non potest eam eiicere; quoniam in corpore aspero sunt pori, quos lux subintrat; in politis autem non invenit poros, nec accidit eiectio haec, propter corporis fortitudinem vel duriciem, quia videmus in aqua reflexionem; sed est haec repulsio propria politurae; sicut de natura accidit, quod aliquid ponderosum cadens ab alto super lapidem durum, revertitur in altum; et quanto minor fuerit duricies lapidis, in quem

light is reflected because of a property residing in the reflecting body, namely its 'force of repulsion' or 'opposition' (*mudāfaᶜa, mumānaᶜa: eiectio, repulsio, prohibitio*). This force is stronger in polished than in

ceciderit, tanto regressio cadentis debilior erit; et semper regreditur cadens versus partem, a qua processit. Verum in arena propter eius mollitiem non fit regressio, quae accidit in corpore duro. Si autem in poris asperi corporis sit politio, tamen lux intrans poros non reflectitur; et si eam reflecti acciderit, dispergitur, et propter dispersionem a visu non percipitur. Pari modo si in aspero corpore partes elatiores fuerint politae, fiet reflexa dispersio; et ob hoc occultabitur visui. Si vero eminentia partium adeo sit modica, ut eius quasi sit idem situs cum depressis, comprehendetur eius reflexio tanquam in polito, non aspero, licet minus perfecte.

18. *Radii incidentiae & reflexionis, situs similitudine conveniunt. Itaque anguli incidentiae & reflexionis aequantur.*

Quare autem fiat reflexio lucis secundum lineam eiusdem situs cum linea, per quam accedit ad speculum ipsa lux, est: Quoniam lux motu citissimo movetur, et quando cadit in speculum, non recipitur; sed ei fixio in corpore illo negatur; et cum in ea perseveret adhuc prioris motus vis et natura, reflectitur ad partem, a qua processit, et secundum lineas eundem situm cum prioribus habentes. Huius autem rei simile in naturalibus motibus videre possumus, et etiam in accidentalibus. Si corpus sphaericum ponderosum ab aliqua altitudine descendere permittamus perpendiculariter super politum corpus, videbimus ipsum super perpendicularem reflecti, per quam descenderat. In accidentali motu, si elevetur aliquod speculum secundum aliquam altitudinem hominis, et firmiter in pariete figatur, et in acumine sagittae consolidetur corpus sphaericum, et proiiciatur sagitta per arcum in speculum, hoc modo, ut elevatio sagittae sit aequalis elevationi speculi, et sit sagitta aequidistans horizonti; planum, quod super perpendicularem accedit sagitta ad speculum, et videbitur super eandem perpendicularem eius regressus. Si vero motus sagittae fuerit super lineam declinatam in ipsum, videbitur reflecti non per lineam, per quam venerat, sed per lineam aequidistantem horizonti, sicut et alia erat, et eiusdem situs, respectu speculi cum ea, et respectu perpendicularis in speculo. Quod autem ex prohibitione corporis politi accidat luci motus reflexionis, palam : quia cum fortior fuerit repulsio vel prohibitio, fortior erit lucis reflexio. Quare autem accidat idem motus reflexionis et eius accessus, haec est ratio. Cum descendit corpus ponderosum super perpendicularem, repulsio corporis politi, et motus descendentis ponderosi directe sibi sunt oppositi, nec est ibi motus, nisi perpendicularis; et prohibitio fit per perpendicularem, quare repellitur corpus secundum perpendicularem. Unde perpendiculariter regreditur. Cum vero descenderit corpus super lineam declinatam, cadit quidem linea descensus inter perpendicularem superficiei politi, per ipsum politum transeuntem, et lineam superficiei eius orthogonalem super hanc perpendicularem; et si penetraret motus ultra punctum, in quod cadit, ut liberum inveniret transitum, caderet quidem haec linea inter perpendicularem, transeuntem per politum, et lineam superficiei orthogonalem super perpendicularem, et observaret mensuram situs, respectu perpendicularis transeuntis, et respectu lineae alterius, quae orthogonalis est super perpendicularem illam. Compacta est enim mensura huius motus ex situ ad perpendicularem et situ ad orthogonalem. Repulsio vero per perpendicularem incedens, cum non possit repellere motum secundum mensuram, quam habet ad perpendicularem transeuntem per politum, quia nec modicum intrat; repellit ergo secundum mensuram situs ad perpendicularem, quam habet ad orthogonalem. Et quando motus regressio eadem fuerit mensura situs ad orthogonalem, quae fuit prius ad eandem ex alia parte, erit similiter ei eadem mensura situs ad perpendicularem transeuntem, quae fuit prius. Sed ponderosum corpus in regressu, cum finitur repulsionis motus, ex natura sua descendit, et ad centrum tendit. Lux autem eandem habens reflectendi naturam, cum ei naturale non sit ascendere aut descendere, movetur in reflexione secundum lineam incoeptam usque ad obstaculum, quod sistere faciat motum. Et haec est causa reflexionis.

rough bodies because the parts of the former are so closely packed as not to allow the light to get dissipated among them. Thus it is the compactness of the parts of polished bodies, not their hardness, that is responsible for optical reflection; water, for example, reflects light, but it is not hard. Light may, however, be reflected from some rough bodies because they consist of small polished parts; but these parts are variously situated and consequently they scatter the light falling on them. Also, the gaps between the parts of rough bodies allow some of the light to penetrate the body, and the absorbed light may not emerge again even after it strikes other polished parts within. Strong optical reflection takes place, therefore, when the reflecting body has more polished than rough parts, when the polished parts are similarly (or almost similarly) situated, and when the gaps between them are few and narrow.

Like Heron of Alexandria, Ibn al-Haytham observes that a heavy body falling on an extremely hard body, such as rock or iron, is immediately and strongly repelled; on the other hand, soft bodies, such as wool or sand, arrest the movement of the falling body. He further observes that bodies with intermediate degrees of hardness repel the impinging body with intermediate forces: thus the force with which the impinging body returns is dependent on the degree of hardness with which it meets at impact. Hardness and softness correspond to smoothness and roughness in optical reflection.

He then gives an explanation of the fact that light is reflected at equal angles and in the plane determined by the line of incidence and the normal to the reflecting surface at the point of reflection. He is first concerned to show that the force with which the light is reflected depends both on the force of the incident movement and on the repelling force of the reflecting body. He tries to establish this first with reference to natural movements, that is, movements of freely falling bodies. A small sphere, made of iron or copper, is allowed to fall from variable distances on a plane iron mirror placed horizontally on the ground. The sphere is supposed to be so polished as not to touch the plane surface at more than one point. The greater the distance covered by the sphere in the downward movement, the higher will be the point to which it will rebound. This means that 'the movement of return depends on the movement acquired by the heavy body during its descent, not on the natural movement due to its heaviness' (fol. 70^v).

Ibn al-Haytham then turns to similar experimental considerations of 'accidental' or forced movements. This time the plane iron mirror is set up against a vertical wall and the iron sphere is thrown on to the mirror horizontally by means of a bow. When thrown perpendicularly to the mirror, the sphere returns in the same line before it begins to have a downward movement due to its heaviness. Here Ibn al-Haytham states that a projectile rebounds when opposed because it *acquires* from this opposition a movement in the direction of return; consequently, the greater the opposition, the greater the force with which the projectile rebounds. He states further that the force of opposition increases with the force of impact. In perpendicular projection both the incident movement and the opposition are acting in the same line; therefore, the projectile rebounds in the same perpendicular direction.

A different situation exists when the sphere is thrown at another angle in the horizontal plane. Here the incident movement consists of two parts, one perpendicular to the mirror and the other parallel to it. (He describes the direction of the second part as perpendicular to the first perpendicular part; his aim is obviously to include curved as well as plane surfaces, the perpendicular to the first perpendicular being parallel to the tangent to the surface at the point of reflection.) Thus in inclined projection, the incident movement and the resistance of the mirror are not directly opposed. When the sphere impinges on the mirror, the perpendicular part of the incident movement is 'annulled' because the resisting surface is opposed to the movement of the sphere in this direction. From the incident perpendicular movement *and* the perpendicular opposition, there results an equal movement in the opposite sense. The second, parallel part of the incident movement remains unaltered, being unopposed by the surface. The reflected movement, compounded of the reversed perpendicular part and the unchanged parallel part, will thus be in the plane of these two lines; further, its position will be symmetrical with that of the incident movement, that is, it will make with the normal an angle equal to the angle of incidence. The same is true of the reflection of light, but with one difference: the reflected movement of a body, being composed of a movement due to reflection and another due to the heaviness of the body, will not continue in a straight line. Light, not being itself determined to move in one direction rather than another, has no reason to change the direction given to it by reflection.

Ibn al-Haytham explains that the force of repulsion or opposition in hard bodies consists in the fact that these bodies do not allow of being affected by, or yielding to other impinging bodies. From this it would seem that by 'hardness' he meant rigidity. His account of what happens in mechanical reflection, however, forces us to adopt a different interpretation. He makes three assertions: (1) that the incident perpendicular movement is destroyed at impact; (2) that the reversed movement depends on the incident movement that has been destroyed; and (3) that the reversed movement is acquired from the force of repulsion. It would seem that in order to reconcile (1) with (2) one would have to assume that the initial movement is first communicated to the reflecting body which then imparts it again to the projectile, in accordance with (3). That is to say, the reflecting body acts as a spring, and its force of repulsion is the same as its elastic force.

Before turning to Descartes' treatment of reflection it may be well to observe that before the seventeenth century, one part or another of Ibn al-Haytham's theory had appealed to various Latin authors in varying degrees. The cases of Roger Bacon and Kepler are particularly interesting. Bacon invokes the ball analogy to show that only a perpendicular ray returns in the same line because it is totally resisted by the reflecting body.[15] He also remarks that bodies reflect light because of their density, not hardness or solidity;[16] and

---

[15] *De multiplicatione specierum*, pars II, cap. III, *Opus majus*, ed. Bridges, II, p. 468: 'Quod si corpus recipiens rem cadentem perpendiculariter resistit omnino, tunc res cadens perpendiculariter rediret propter incessus fortitudinem in eandem viam per quam incessit, sicut de pila jacta ad parietem vel ad aliud resistens omnino. Et pila cadens ex obliquo labetur ex altera parte incessus perpendicularis secundum casum obliquum, ut patet ad sensum, et non rediret in viam qua venit propter casus debilitatem.'

[16] *Mult. spec.*, pars II, cap. V, ed., cit. II, p. 479: 'Considerandum etiam ... quod durum et solidum nec faciunt reflexionem nec fractionem, sed solum densum et rarum. Nam crystallus

those which are not perfectly dense refract as well as reflect the light
falling on them.[17] Referring to Ibn al-Haytham, he notes that
light is best reflected from polished, not rough, bodies, because the
polished body repels (*ejicit*) the *species* of light, whereas the rough
body cannot, since there are pores in it which allow some of the
light to enter.[18] In spite of the verb '*ejicit*' in this sentence Bacon
did not in fact conceive of optical reflection as a mechanical process.
He clearly states that the *species* of light is reflected simply because
the opposing surface offers it an *opportunity* to multiply itself in a
direction other than that of incidence.[19] The *species* thus turns away
from the surface by a power all its own; its reflection is not to be
understood as a case of repulsion but of self generation. Since light
is not a body, its reflection cannot be due to the same cause as the
rebound of a ball.[20] This simply amounts to a rejection of Ibn

est dura res et solida, et vitrum et hujusmodi multa, et tamen quia rara sunt, pertransit species
visus et rerum visibilium, et fit bona fractio in eis; quod non contingeret si esset ita densa
sicut sunt dura et solida. Densum enim est quod habet partes propinque jacentes, et rarum
est quod habet partes distanter jacentes, et ideo vitrum est rarum, et crystallum et hujusmodi
non densum perfecte, licet aliquantulum, sed sufficienter sunt rara ut permittant species
lucis transire.'
[17] Ibid., p. 478: 'Sunt etiam alia corpora mediocris densitatis, a quibus fit reflexio simul et
fractio, ut est aqua. Nam videmus pisces et lapides in ea per fractionem, et videmus solem et
lunam per reflexionem, sicut experientia docet. Unde propter densitatem, quae sufficit
reflexioni aliquali, radii omnes reflectuntur, sed propter mediocritatem densitatis, quae non
impedit transitum lucis, franguntur radii in superficie aquae.'
[18] Ibid., p. 479: 'Et Alhazen dicit in quarto libro, quod ex politis corporibus, non ex asperis,
fit bona reflexio, quoniam corpus politum ejicit speciem lucis, sed asperum non potest, quum
in corpore aspero sunt pori, quos subintrat species lucis vel alterius.'
[19] *Perspectiva*, pars III, dist. I, cap. I, in *Opus majus*, ed. Bridges, II, p. 130: 'Nam omne
densum in quantum densum reflectit speciem, sed non quia fiat violentia speciei, immo quia
species sumit occasionem a denso impediente transitum ejus ut per aliam viam se multiplicet
ei possibilem.'
Mult. spec., pars II, cap. II, ed. cit., II, p. 463: 'Si vero corpus secundum non differt in dia-
phaneitate a primo, sed omnino est densum, ita quod potest impedire transitum speciei,
tunc species sumens occasionem a denso in partem alteram redit ex propria virtute; ut cum
non possit se multiplicare in densum corpus multiplicat se in primo corpore faciens angulum,
et dicitur species reflexa proprie, et in communi usu apud omnes; nec tamen repellitur per
violentiam, sed solum sumit occasionem a denso impediente transitum, et vadit per aliam
viam, ut possibile est ei.'
[20] *Mult. spec.*, pars III, cap. I, ed. cit.,'II, p. 505:'Si vero arguitur de reverberatione a corpore,
patet ex dictis quod non fit ex violentia, sed generat se in partem sibi possibilem cum pro-
hibetur transire propter densitatem resistentis. Si enim violenter repercuteretur ut pila a
pariete, oporteret necessario quod esset corpus. Quod si dicatur, cum facit se iterum per
reflexionem in alium locum per suam naturam propriam, medio nec pellente nec movente nec
moto, nec aliquo alio juvante nec faciente quod removeat locum, quod non accidit in umbra,

al-Haytham's mechanical model for reflection. Bacon does not make use of Ibn al-Haytham's application of the parallelogram of velocities for demonstrating the law of equal angles. He attempts instead a geometrical 'proof' which is a *petitio principii*.[21]

A clear and concise use of the parallelogram method, as applied to the problem of reflection, is to be found in Kepler's *Paralipomena* (Ch. I, Proposition XIX). Adopting the analogy between the behaviour of light and of physical objects,[22] Kepler observes that a resisting surface is not directly opposed to oblique movements, and consequently such movements must be reflected in a similarly oblique direction. To prove the equality of the angles, he considers the incident movement to be composed of two parts, one perpendicular and the other parallel to the surface. The equality of the angles follows from supposing the surface to be opposed to the first but not to the second component. There can be no doubt that the source of this proof was Ibn al-Haytham's *Optics* which Kepler frequently cited;[23] and as far as reflection is concerned, Kepler's *Paralipomena*, published in 1604, may have been a link between Ibn al-Haytham and Descartes.

2. Descartes' attempt in the *Dioptric* to explain the reflection of light by comparison with the mechanical reflection of bodies is in a long tradition that had been carried down from antiquity to the seventeenth century. Yet that which was for his predecessors merely an analogy, acquires in his system a new significance. For although he

---

quia ad motum corporis renovatur, videtur quod per se locum occupare debeat; dicendum est, quod locum non quaerit ut corpus sed subjectum, nec tamen unum in numero illud quaerit solum sed diversum semper, propter hoc quod species in una parte medii generata potest facere sibi similem in alia. Et ideo non est ibi acquisitio loci ut corpus acquirit, sed est ibi renovatio speciei per generationem in partibus medii pluribus.'

[21] Cf. *Perspectiva*, pars III, dist. I, cap. I, ed. cit., p. 131; also *Mult. spec.*, pars II, cap. VI, ed. cit., II, p. 484. In preferring to give a geometrical 'proof' Bacon was following the example of al-Kindi, *De aspectibus*, 17, ed. Björnbo and Vogl, pp. 29–30.

[22] Cf. *Ad Vitellionem Paralipomena*, Francofurti, 1604, pp. 14–15.

[23] Kepler refers here to Witelo's *Optica*, V, 20, where in fact the parallelogram method is not used. The reference is corrected in Kepler's *Gesammelte Werke*, II, Munich, 1939, p. 25 to Witelo's *Optica*, V, 10. This is an *experimental* demonstration which is derived from Alhazeni *Optica*, IV, 10.

does not conceive of the propagation of light as an actual movement, he still asserts that light is no more than a mechanical property of the medium transmitting it; it is, as we have seen, a static pressure existing simultaneously in all parts of the subtle matter that pervades all space. Furthermore, Descartes put forward a conception of matter which entailed the exclusion of all forces other than movement. From this point of view, to set a body in motion, another body that is already moving has to come into contact with it and convey to it part or all of its motion. Thus the idea of collision acquires in his physics a new status as the basic and only form of action in the material world to which light belongs.

By looking at optical reflection as a collision phenomenon Descartes was faced with a serious difficulty: if, as he asserted, light was a tendency to movement that was propagated instantaneously, how could it be compared to the successive motion of a projectile? He thought he could overcome this by assuming that the tendency to movement, being a potential movement, follows the same laws as motion itself. Thus he implied that in investigating optical reflection and refraction, one may forget about the actual mechanism of light for a while and study the reflection and refraction of actually moving bodies. What can be proved concerning these will be true of light by virtue of that bridging assumption.

In the Cartesian analogy for reflection a tennis ball takes the place of the small iron sphere in Ibn al-Haytham's model. This was misleading since, as we shall see,[24] it was Descartes' intention to eliminate elasticity from his abstract consideration of mechanical collision. He begins[25] by supposing that the surface of the ground, towards which the ball is pushed with a racquet, is perfectly smooth and perfectly hard (i.e. rigid) and, further, that the ball returns after impact with a speed equal to the speed of incidence. Abstraction is also made from such things as weight, size and figure; and the question, 'What power keeps the ball in motion after it is no longer in contact with the racquet?' is here set aside as irrelevant to the action of light. (The answer to this question, as we know, would

---

[24] Below, p. 81f.
[25] *Dioptric*, II, D, VI, pp. 93–95.

be based on Descartes' first law of motion, the law of inertia.) 'Only it must be noted that the power which continues the motion of this ball, whatever it may be, is different from that which determines it to move in one direction rather than another.'[26] The former, he explains depends on how hard we hit the ball, whereas the latter depends on the position of the racquet at the moment of striking. This distinction

already shows that it is not impossible for the ball to be reflected by the collision with the ground, and thus its former determination to tend towards B [the point of reflection] is changed without this resulting in a change of the force of its motion, as these are two different things. Consequently, one must not think it necessary, as many of our Philosophers do, that the ball should stop for a moment at B before turning back . . . For once its motion were interrupted by this pause, there would be no cause to start it again.[27]

Descartes denies here the medieval doctrine of *quies media*,[28] namely the doctrine that there is a moment of rest between two motions of the same body in opposite or different directions. If the body were to have a moment of rest before it resumed its motion in a new direction, where, he asks, would it derive this motion from? As we have seen, Ibn al-Haytham had tried to answer this question by postulating something like an elastic force which allows the reflecting body to *push* the projectile back after it has deprived it of its motion. Such an explanation would not, however, be acceptable to Descartes. He had his own reasons for believing, and indeed insisting, that the projectile should return by the same 'force' that has always been in it and that had never left it. In other words, he regarded the function of the surface as merely consisting in preventing the ball from continuing its motion in the same direction, while it is the ball itself that decides on the basis of its ever inherent resources where it should go. In *this* respect the situation as viewed here by Descartes is not unlike Roger Bacon's

---

[26] *Dioptric*, II, D, VI, p. 94.
[27] Ibid.
[28] See also Descartes to Mersenne, 18 March 1641, D, III, pp. 338–9.

account of how light is reflected: both the light *species* and the ball return by their own power, and for both the surface is no more than an occasion for a change of direction.

That the preceding remarks express Descartes' own meaning may be clearly seen from statements in his correspondence. Thus he wrote to Mersenne in 1640: 'Your saying that the speed of a hammer stroke takes Nature by surprise, so that she has no time to gather her forces in order to resist, is entirely against my opinion; for she has no forces to gather, nor does she need time for that; rather, she acts in everything Mathematically.'[29]   The following sentence indicates what this means in the case of the rebound of bodies in collision: 'When two balls collide and, as often happens, one of them rebounds, it does so by the same force that was formerly moving it: for the force of the movement and the direction in which it takes place are entirely different things—as I have said in the Dioptric.'[30]

He later conceded in a letter to Mersenne for Hobbes that when a ball impinges upon the ground, both the surface of the ground and of the ball curve in a little at the point of impact and then they regain their shape; but he affirmed his belief that the subsequent rebound is impeded rather than helped by the compression and elasticity of these bodies.[31] Far from conceding that reflection would be impossible if the colliding bodies were supposed incompressible (a view which he found incredible), he maintained that compressibility only results in reflection not taking place at exactly equal angles.

---

[29] Descartes to Mersenne, 11 March 1640, D, III, p. 37.

[30] Ibid.

[31] Descartes to Mersenne for Hobbes, 21 January 1641, D, III, pp. 289–90: 'Concedo tamen libenter partem terrae, in quam pila impingit, aliquantulum vi cedere, vt etiam partem pilae in terram impingentem non nihil introrsum recuruari, ac deinde, quia terra & pila restituunt se post ictum, ex hoc iuuari resultum pilae; sed affirmo hunc resultum magis semper impediri ab istà incuruatione pilae & terrae, quam ab eius restitutione iuuetur; atque ex eo posse demonstrari reflexionem pilae, aliorumque eiusmodi corporum non extremè durorum, nunquam fieri ad angulos accuratè aequales. Sed, absque demonstratione, facilè est experiri pilas molliores non tam altè resilire, nec ad tam magnos angulos, quam duriores. Inde patet quam perperam adducat istam terrae mollitiem ad aequalitatem angulorum demonstrandam; praesertim cum ex eâ sequatur, si terra & pila tam durae essent, vt nullo modo cederent, nullam fore reflexionem; quod est incredibile. Patet etiam quam meritò ego & terram & pilam perfecte duras assumpserim, vt res sub examen mathematicum cadere possit.'

He had therefore assumed incompressibility so that the matter of reflection could be treated mathematically.

Coming back to the same subject in another letter to Mersenne for Hobbes, he flatly asserted that experiment refuted Hobbes's opinion that the rebound of bodies was due to repulsion or elasticity; if this were true, he argued, a ball would be made to rebound by merely pressing it against a hard stone.[32]

What all this boils down to is Descartes' firm belief that the behaviour of bodies in collision can be fully explained solely in terms of conservation of motion. The ball rebounds simply because its motion, understood as an absolute quantity, persists; this is nature's mathematical way. To allow elasticity a role in collision would be to abandon the truly mathematical interpretation of nature! Descartes' treatment of reflection is thus one more example of what Alexandre Koyré has called 'géométrisation à outrance'.

To find the direction of reflection Descartes appeals to the parallelogram method which had been applied to the same problem by Ibn al-Haytham and Kepler: 'it must be remarked that the determination to move in a given direction, as well as motion, and generally every other kind of quantity, can be divided into all the parts of which it can be imagined to be composed.'[33] He imagines the 'determination' along AB (Fig. III. 2) to be composed of two others, one perpendicular to the surface and the other parallel to it. At the moment of impact the ground must hinder (empescher) the first component 'because it occupies all the space that is below CE', but not the second, 'since the ground is not at all opposed to [the ball] in that direction'.[34]

---

[32] Descartes to Mersenne, 18 March 1641, D, III, p. 338: 'Vous . . . parlez de l'opinion de l'Anglois [Hobbes] qui veut que la reflexion des cors ne se face qu'a cause qu'ils sont repoussez, comme par vn ressort, par les autres cors qu'ils rencontrent. Mais cela se peut refuter bien aysement par l'experience. Car s'il estoit vray, il faudroit qu'en pressant vne bale contre vne pierre dure, aussy fort qu'elle frape cete mesme pierre, quand elle est ietée decontre, cete seule pression la pust faire bondir aussy haut que lors qu'elle est ietée. Et cete experience est aysée a faire, en tenant la bale du bout des doigts, & la tirant en bas contre vne pierre qui soit si petite qu'elle puisse estre entre la main & la bale, ainsy que la chorde d'vn arc de bois est entre la main & la fleche, quand on la tire du bout des doigts pour la decocher; mais on verra que cete bale ne reiallira aucunement, si ce n'est peut estre fort peu, en cas que la pierre se plie fort sensiblement comme vn arc.'

[33] Dioptric, II, D, VI, pp. 94–95.

[34] Ibid., p. 95.

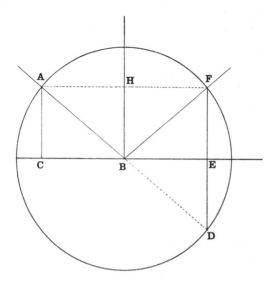

FIG. III. 2.

To find, therefore, exactly in what direction the ball must be reflected, let us describe a circle, passing through A, with B as centre. And we say that, in an equal time to that which it will have taken to move from A to B, it must inevitably return from B to a point on the circumference of the circle; since all the points that are as distant from B as A, are to be found on this circumference; and we have assumed the motion of the ball to be always of equal speed.[35]

Now, to know precisely to which of all the points on the circumference it must return: let us draw three straight lines AC, HB and FE perpendicular to CE, such that the distance between AC and HB is neither greater nor smaller than that between HB and FE. We can then say that in the same time that the ball has taken to advance in the right-hand direction from A, one of the points of the line AC, to B, one of the points of the line HB, it must also advance from the line HB to some point on the line FE. For all the points on the line FE are each as distant, in this direction, from HB as those on the line AC. And the ball is as much determined to advance in this direction as it was before. But it cannot arrive at the same time at a point on FE, and also at a point on the circumference of the circle AFD, unless this is either D or F, these being

---

[35] Ibid., pp. 95–96.

the only two points where the circumference and *FE* intersect. But since the ground prevents the ball from passing through to *D*, it must be concluded that it necessarily goes towards *F*. And thus you easily see how reflection happens, namely at an angle always equal to that which is called the angle of incidence. As when a ray coming from *A* falls at *B* on the surface of the plane mirror *CBE*, it is reflected towards *F*, such that the angle of reflection *FBE* is neither greater nor smaller than the angle of incidence *ABC*.[36]

Descartes distinguishes between what he calls the 'force' of the ball's movement and its 'determination'. His subsequent application of these terms shows that by 'force' he means either the *absolute* quantity of motion or the quantity on which this depends, viz. the *absolute* speed. The term 'determination' proved to be more troublesome. Fermat understood it to mean simply the *direction* in which the ball moved, and this gave rise to difficulties which were not clarified until twenty years after his controversy with Descartes had started, when it was resumed with Clerselier. We need not, however, decide the nature of what Descartes understood by 'determination' in order to follow his proof of the law of reflection. Nor was Fermat's interpretation of it essential to his objections against Descartes' derivation of this law. We shall therefore leave this problem until we come to the discussion of refraction where the mentioned difficulties in fact belong.

In order to deduce the equality of the angles of incidence and of reflection Descartes assumes, first, that the reflecting surface will not affect the 'force' or speed of the ball in any way. That is, if v_i, v_r are the velocities of incidence and reflection, his first assumption is:

(1) $$v_i = v_r.$$

Further, since, as he argues, the surface is not opposed to the parallel motion, the speed in that direction will also be conserved after reflection. His second assumption is:

(2) $$v_i \sin i = v_r \sin r,$$

where *i* and *r* are the angles of incidence and reflection made with the normal.

---

[36] *Dioptric*, II, D, VI, p. 96.

From (1) and (2) it follows that
$$\sin i = \sin r,$$
or,
$$i = r.$$

It should be noted that Descartes does not *postulate* in his proof anything about the perpendicular component of the incident velocity. In particular, he does not assume that it will be *reversed* by the reflecting surface. His conclusion is based on two assumptions, of which one is about the actual speeds, and the other about the parallel speeds, before and after reflection. Not to observe this would not only be to miss the whole point in the proof as a preparation for the derivation of the law of refraction, but would also imply a way of thinking that was not Descartes'.[37] When one says 'reversed' one is thinking of something like Newton's law of action and reaction. We have seen, however, that according to Descartes the ball does not act upon the surface, nor does the surface react upon it. It is not, therefore, an accidental feature of his proof that the conclusion is meant to follow from merely supposing the motion to have been *conserved* in two directions. This point is also essential for appreciating Fermat's objections.

3. The importance of the controversy that took place between Descartes and Fermat after the publication of the *Dioptric* can hardly be exaggerated. Leading as it eventually did to Fermat's formulation of the principle of least time, it has been rightly described by Paul Mouy as 'one of the most interesting and most fruitful in the history of science'.[38] Nevertheless, it has not been studied in any sufficient detail. Gaston Milhaud, who devotes a long chapter[39] to the quarrel between Fermat and Descartes over the former's method of deriving tangents, only refers to the dispute about refractions summarily;[40] and his remarks, which were

---

[37] Compare J. F. Scott, op. cit., p. 34.
[38] Paul Mouy, *Le développement de la physique cartésienne, 1646–1712*, Paris, 1934, pp. 60–61.
[39] Gaston Milhaud, *Descartes savant*, pp. 149–75.
[40] Ibid., pp. 110–13.

concerned with Descartes' proof of the law of refraction, were not always correct. The account, by far the longest, given in Mouy's excellent book, is too general although it is quite instructive.[41]

The circumstances of this controversy were as follows. Descartes' 1637 volume did not arrive in France until the end of that year. Before the printing was completed, however, Descartes had sent a copy to Father Mersenne in Paris, who undertook the task of distributing it in parts to various persons. The *Dioptric* was allotted to Fermat and was accordingly sent to him in Toulouse with the request that he give his opinion of it. As remarked by the editors of Fermat's works, it appears from the end of Fermat's letter to Mersenne of (probably) September 1637[42] that neither the *Discourse* nor the two other treatises, the *Meteors* and the *Geometry*, had been sent to him then; nor did he know that they were to appear together in one volume, or that his reply would be communicated to Descartes.[43] But from that time on, Mersenne played the part of intermediary between the two mathematicians who exchanged their letters through him.

Fermat's arguments, in that first letter of his, were directly concerned with Descartes' proof of the law of reflection, for he believed that similar arguments would equally apply to refraction. Descartes answered by a letter to Mersenne dated (probably) October 1637.[44] Fermat's reply to this, in his letter to Mersenne of (probably) December 1637,[45] took up the question of refraction and in it he composed what he intended as a refutation of Descartes' proof of the sine law. Descartes received Fermat's letter when he was about to defend his own criticism of Fermat's tangent method against Roberval. He consigned his reply to a letter written (probably) on 1 March 1638[46] which was addressed to Mydorge who had been

---

[41] Mouy's account (*Physique cartésienne*, pp. 55–61) is given in the course of a summary of all physical questions discussed in Descartes' correspondence as published by Clerselier in the three volumes which appeared successively in 1657, 1659 and 1667.

[42] F, II, pp. 106–12.

[43] Cf. F, II, pp. 106–7, editors' note.

[44] F, II, pp. 112–15.

[45] F, II, pp. 116–25.

[46] F, IV, pp. 93–99.

taking Descartes' side at the discussions in Father Mersenne's circle. This letter did not reach Fermat until about June of the same year when he was preparing to enter the dispute about tangents. He was not satisfied with Descartes' replies and the discussion concerning the *Dioptric* was suspended.[47]

Twenty years later, in the beginning of 1658, the discussion was resumed between Fermat and Descartes' follower, Clerselier. On May 15 Fermat's last letter to Mersenne for Descartes was made the subject of a special refutation by Rohault,[48] author of the famous *Traité de Physique*, which was the generally accepted text-book of physics until the publication of Newton's *Principia*.[49] In what immediately follows we shall be concerned with the part of the controversy dealing with reflection.

Fermat could not accept Descartes' argument for establishing the law of reflection as a 'legitimate proof and demonstration'.[50] To his mind there was something in this argument that was *logically* wrong; and his attack was directed against the method of resolving motions that was adopted by Descartes in his treatment of both reflection and refraction. For Fermat believed that, in resolving the motion of the incident ball into those particular components, the one perpendicular and the other parallel to the reflecting surface, Descartes had chosen but one out of an infinite number of reference systems which were all valid. But since he had chosen the one which could help him to arrive at the desired, and already known result, his procedure was in Fermat's view completely arbitrary. Accordingly Fermat set out to show that, by adopting a different system of axes, a different result from Descartes' would be obtained—and this, he believed, on the same lines as in Descartes' argument:

---

[47] F, II, p. 125, n. 1. The letter addressed to Mydorge is said by the editors of Fermat's works to have been sent to Mersenne on or about 22 February 1638 (cf. F, II, p. 125, n. 1, p. 126, n. 1). This date cannot be correct if the letter was written on 1 March 1638, as indicated on p. 93 of vol. IV of Fermat's works.

[48] Cf. Rohault's 'Réflexions ou projet de réponse à la lettre de M. de Fermat' [i.e. Fermat's letter of December 1637], 15 May 1658, F, II, pp. 391–6.

[49] For letters exchanged during this second phase of the discussion, see next chapter.

[50] F, II, p. 109.

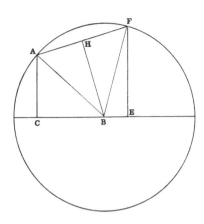

FIG. III. 3

. . . in the attached figure [Fig. III. 3], where *AF* is no longer parallel to *CB*, and the angle *CAF* is obtuse, why cannot we imagine that the determination of the ball moving from *A* to *B* is composed of two others, of which one makes it descend from the line *AF* towards [*vers*] the line *CE* [i.e. in a direction perpendicular to *AF*] and the other makes it advance towards *AF* [i.e. parallel to *AF*]? For it is true to say that, in as much as the ball descends along the line *AB*, it progresses towards [*vers*] *AF* [i.e. parallel to *AF*], and that this progress is to be measured by the perpendiculars [strictly, the distance between the perpendiculars] drawn to the line *AF* from the various points that can be taken between *A* and *B*. And this is to be understood, however, when *AF* makes an acute angle with *AB*; otherwise, if [the angle] were right or obtuse, the ball would not advance towards *AF*, as it is easy to understand.

Having supposed this, we will conclude by the same reasoning of the author [Descartes], that the smoothed body *CE* only hinders the first movement [viz. the movement perpendicular to *AF*], being opposed to it only in that direction; so that, not being in the way of the second [movement], the perpendicular *BH* having been drawn, and *HF* made equal to *HA*, it follows that the ball must be reflected to the point *F*, and thus the angle *FBE* will be greater than *ABC*.

It is therefore evident that, of all the divisions of the determination to motion, which are infinite, the author has taken only that which can serve to give him his conclusion, of which we know as little as before. And it certainly appears that an imaginary division that may be varied in an infinite number of ways never can be the cause of a real effect.

By the same reasoning, we can refute the proof of his foundations for Dioptrics, since they are based on a similar discourse.[51]

Fermat's construction needs clarification. *AB* may be taken to represent the velocity of incidence. From *A* he draws a line *AF* to meet the circumference in *F*, such that the perpendicular *FE* is greater than the perpendicular *AC*. He then draws *BH* perpendicular to *AF* and therefore it bisects *AF* in *H*. *BF* represents the velocity of reflection. The components of *AB* are *AH* and *HB*; and the components of *BF* are *BH* and *HF*.

It is clear in what sense Fermat believed that his figure satisfied Descartes' assumptions for deducing the law of reflection as the latter had expressed them. For in order to justify his assumption about the conservation of the horizontal component, Descartes had relied on the fact that the surface was *not opposed* to the motion of the ball in that direction. But Fermat could make the same claim for his oblique component *AH* which is actually going away from the surface. As for Descartes' other condition, it is fulfilled in Fermat's figure by the equality of *AB* and *BF*.

On the other hand, Fermat's figure does *not* satisfy Descartes' two conditions as we have formulated them. For we have assumed in equation (2) above [52] that *i* and *r* are angles made with the normal to the surface. This, however, is precisely what Fermat would denounce as arbitrary: why should an imaginary line, the normal to the surface, be privileged in the definition of *i* and *r*? If *i* and *r* were angles made instead with any oblique line, such as *HB* in Fermat's figure, their equality would indeed follow, but not the equality of the angles *ABC* and *FBE* which Descartes and Fermat called the angles of incidence and reflection.

Descartes replied[53] by saying that, when he wrote that the motion can be divided into all the parts of which it can be imagined to be composed, he meant that it can be 'really' so divided, and not only by imagination. Any motion can indeed be divided in imagination

---

[51] Fermat to Mersenne, September 1637, F, II, pp. 109–10.
[52] Above, p. 84.
[53] Descartes to Mersenne, October 1637, F, II, p. 114.

into an infinite number of different components. But the surface *CBE* being a real surface, and also being opposed to the motion of the ball in the perpendicular direction alone, it actually divides the motion of the ball into a perpendicular and a parallel component. Fermat's mistake therefore consisted in the choice of the direction of the oblique *HB* (in Fermat's figure). It was thus Descartes' turn to impute arbitrariness to Fermat's own demonstration.

What did Descartes mean by this? In particular, what did he mean by saying that the reflecting surface was opposed to the motion only in the perpendicular direction? For it must be remembered that the reason he gave in the *Dioptric* for regarding the surface as hindering the downward motion was that the ground 'occupies all the space that is below'. If, therefore, by *opposition to motion in a given direction* he meant that the surface does not allow the ball to pass through *in that direction*, then the surface should be considered equally opposed to the motion in the direction of the oblique *HB*. Descartes was obliged to answer the following question: How did he know that the opposition existed only in the perpendicular direction?

His answer would seem to be implied in his use of the word *real*: the *real* surface *really* divides the ball's motion precisely into the perpendicular and parallel components *as a matter of fact*; our knowledge of the manner in which this division takes place is ultimately based on experience. That is to say, he would seem to agree with Fermat that, before the ball strikes the surface, its motion can be imagined to be divided in an infinite number of ways, and consequently there is so far no reason why one system of axes should be preferred to another. Experiment shows, however, that the ball rebounds in such a way that the horizontal speed remains constant, while the perpendicular speed is reversed. Experiment is therefore the foundation of the supposition that the surface is opposed to the ball's motion in the normal direction; and it is also the basis for the choice of a system in which one axis is perpendicular and the other parallel to the reflecting surface. This kind of argument would thus imply that the mathematical device adopted by Descartes, far from dictating what should take place in nature, was in fact itself informed by the actual behaviour of bodies.

Now Descartes did not make it absolutely clear that this is what he had in mind. The preceding interpretation of his argument, however, seems to be supported by the way he expressed the difference between his own argument and Fermat's.[54] He drew a triangle which he divided into four equal and four unequal triangles. The difference as he saw it between himself and Fermat consisted in the following: both he and Fermat agreed that the triangle could be divided into all those triangles of which it could be imagined to be composed; both also agreed that the same triangle could be imagined to be composed of the four equal triangles; but whereas he, Descartes, *having drawn* the lines joining the middle points on the sides of the original triangle, said that it was, 'really and truly',[55] divided into those equal triangles and no others, Fermat was saying that the triangle *could* be divided into another set of unequal triangles. In other words, the decomposition effected by the reflecting surface corresponded to the actual division of the triangle into four equal ones. It is not a mathematical, or imaginary decomposition, but a physical operation; and, as such, it can only be ascertained by the observation of what actually takes place in nature.

These remarks had no effect on Fermat and he accordingly wrote to Mersenne his final word on the disputed question:

In four words I shall cut short our dispute on reflection which, however, I could prolong further and prove that the author has accommodated his *medium* to his conclusion, about the truth of which he had already been certain; for if I were to deny him that his division of the determinations to motion is the one that must be taken, since we have an infinite number of them, I would compel him to prove a proposition which he would find very embarrassing.[56]

This is but a restatement of his original objections, from which it appears that Fermat remained convinced to the end of the discussion about reflection that Descartes' proof did not constitute a 'legitimate demonstration'.[57]

---

[54] F, II, pp. 114–15.   [55] F, II, p. 115.
[56] F, II, p. 117.   [57] F, II, p. 109.

In agreement with a certain *a priori* conception of matter, and in accordance with the decision to deal with natural phenomena in a purely mathematical fashion, Descartes tried to deduce the reflection law from merely assuming the impenetrability of matter and the conservation of the absolute quantity of motion. Fermat's objections showed that there existed an infinite number of laws, each of which would satisfy both of these assumptions. Subsequent writers, as we shall see, had to introduce new elements. Huygens' kinematical (and, to that extent, Cartesian) treatment of reflection made use of a new idea embodied in his principle of secondary waves and he admitted elasticity as a property of the medium. Newton retained the parallelogram method, but he operated explicitly with forces.

Chapter Four

DESCARTES' EXPLANATION OF REFRACTION. FERMAT'S 'REFUTATIONS'

1. Ptolemy, who made the first serious experimental study of refraction, was aware of a connection between the change of direction when light passes from one medium into another and the difference in density between the two media. He observed that in rare-to-dense refraction (as when a beam passes from air into water) the light turns towards the normal, and in the reverse case, away from the normal. The amount of deviation from the original path depended on the degree of density-difference: the greater the difference, the greater the deviation. He also compared optical refraction with the passage of a projectile from one medium into another more or less resisting medium.[1]

These remarks became the basis of an elaborate theory of refraction which Ibn al-Haytham developed nine hundred years after Ptolemy, and in which he attempted to explore mathematically the suggested mechanical analogy for optical refraction.

In Bk. VII of his *Optics*[2] Ibn al-Haytham first explains why light

---

[1] Cf. Ptolemaei *Optica*, V, 19, 31 and 33; A. Lejeune, 'Les tables des réfractions de Ptolémée', *Annales de la Société Scientifiques de Bruxelles*, série I (*Sciences Mathématiques et Physiques*), LX (1940), pp. 93–101; *Euclide et Ptolémée*, pp. 73–74; *Catoptrique grecque*, p. 158. Definition VI of the pseudo-Euclidean *Catoptrica* describes the experiment in which a coin being placed at the bottom of an empty vessel becomes visible from a certain point when water is poured into the vessel. But there is no application of this observation in the book, and it may have been derived from a lost work attributed to Archimedes; see Lejeune, *Catoptrique grecque*, p. 153. The experiment is recorded in Ptolemaei *Optica*, V, 5; see translation of relevant passage in Marshall Clagett, *Archimedes in the Middle Ages*, I, Madison, 1964, p. 634.

[2] Alhazeni *Optica*, VII, 8, in *Opticae thesaurus*, ed. Risner, pp. 240–2: 'Quare autem refringatur lux, quando occurrit corpori diaphano diversae diaphanitatis, causa haec est: quia transitus lucis per corpora diaphana fit per motum velocissimum, ut declaravimus in tractatu secundo. Luces ergo, quae extenduntur per corpora diaphana, extenduntur motu veloci, qui non patet sensui propter suam velocitatem. Praeterea motus earum in subtilibus corporibus, scilicet in illis, quae valde sunt diaphana, velocior est motu earum in iis, quae sunt grossiora illis, scilicet quae minus sunt diaphana. Omne enim corpus diaphanum, cum lux transit in ipsum, resistit luci aliquantulum, secundum quod habet de grossitie. Nam in omni corpore

is refracted when it strikes the surface of a medium denser than the medium of incidence. Light, he says, moves with very great speed (*per motum velocissimum*) in transparent bodies, and its speed is

---

naturali necesse est, ut sit aliqua grossities; nam corpus parvae diaphanitatis non habet finem in imaginatione, quae est imaginatio lucidae diaphanitatis; et omnia corpora naturalia perveniunt ad finem, quem non possunt transire. Corpora ergo naturalia diaphana non possunt evadere aliquam grossitiem. Luces ergo cum transeunt per corpora diaphana, transeunt secundum diaphanitatem, quae est in eis, et sic impediunt lucem secundum grossitiem, quae est in eis. Cum ergo lux transiverit per corpus diaphanum, et occurrit alii corpori grossiori primo, tunc corpus grossius resistit luci vehementius, quam primum resistebat; et omne motum cum movetur ad aliquam partem essentialiter aut accidentaliter, si occurrerit resistenti, necesse est, ut motus eius transmutetur; et si resistentia fuerit fortis, tunc motus ille refringetur ad contrariam partem; si vero debilis, non refringetur ad contrariam partem, nec poterit per illam procedere, per quam incoeperat, sed motus eius mutabitur.

Omnium autem motorum naturaliter, quae recte moventur per aliquod corpus passibile, transitus super perpendicularem, quae est in superficie corporis, in quo est transitus, erit facilior. Et hoc videtur in corporibus naturalibus. Si enim aliquis acceperit tabulam subtilem, et paxillaverit illam super aliquod foramen amplum, et steterit in oppositione tabulae, et acceperit pilam ferream, et eiecerit eam super tabulam fortiter, et observaverit, ut motus pilae sit super perpendicularem super superficiem tabulae; tunc tabula cedet pilae aut frangetur, si tabula subtilis fuerit, et vis, qua sphaera movetur, fuerit fortis. Et si steterit in parte obliqua ab oppositione tabulae, et in illa eadem distantia, in qua prius erat, et eiecerit pilam super tabulam illam eandem, in quam prius eiecerat; tunc sphaera labetur de tabula, si tabula non fuerit valde subtilis, nec movebitur ad illam partem, ad quam primo movebatur, sed declinabit ad aliquam partem aliam. Et similiter, si acceperit ensem, et posuerit coram se lignum, et percusserit cum ense, ita ut ensis sit perpendicularis super superficiem ligni, tunc lignum secabitur magis; et si fuerit obliquus, et percusserit oblique lignum, tunc lignum non secabitur omnino, sed forte secabitur in parte, aut forte ensis errabit deviando; et quanto magis fuerit ensis obliquus, tanto minus aget in lignum; et alia multa sunt similia; ex quibus patet, quod motus super perpendicularem est fortior et facilior, et quod de obliquis motibus, ille, qui vicinior est perpendiculari, est facilior remotiore.

Lux ergo, si occurrit corpori diaphano grossiori illo corpore, in quo existit, tunc impedietur ab eo, ita quod non transibit in partem, in quam movebatur, sed quia non fortiter resistit, non redibit in partem, ad quam movebatur. Si ergo motus lucis transiverit super perpendicularem, transibit recte propter fortitudinem motus super perpendicularem; et si motus eius fuerit super lineam obliquam, tunc non poterit transire propter debilitatem motus; accidit ergo, ut declinetur ad partem motus, in quam facilius movebitur, quam in partem, in quam movebatur; sed facilior motuum est super perpendicularem; et quod vicinius est perpendiculari, est facilius remotiore. Et motus in corpore, in quod transit, si fuerit obliquus super superficiem illius corporis, componitur ex motu in parte perpendicularis transeuntis in corpus, in quo est motus, et ex motu in parte lineae, quae est perpendicularis super perpendicularem, quae transit in ipsum.

Cum ergo lux fuerit mota in corpore diaphano grosso super lineam obliquam, tunc transitus eius in illo corpore diaphano erit per motum compositum ex duobus praedictis motibus. Et quia grossities corporis resistit ei ad verticationem, quam intendebat, et resistentia eius non est valde fortis; ex quo sequeretur, quod declinaret ad partem, ad quam facilius transiret; et motus super perpendicularem est facilimus motuum; necesse est ergo, ut lux, quae extenditur super lineam obliquam, moveatur super perpendicularem, exeuntem a puncto, in quo lux occurrit superficiei corporis diaphani grossi. Et quia motus eius est compositus ex duobus motibus, quorum alter est super lineam perpendicularem super superficiem corporis grossi, et reliquus super lineam perpendicularem super perpendicularem hanc; et motus compositus,

greater in a rare body (such as air) than in a dense body (such as water or glass). He accounts for the variable speed of light as follows: all natural bodies, including transparent bodies, share in the property of grossness, density or opacity; any transparent body can thus be imagined to be more transparent than it is; owing to their density, all transparent bodies resist the movement of light; the denser they are, the greater the resistance they offer.

When, therefore, light strikes the surface of a denser body, its movement must be altered (*transmutatur*). But since the resistance is not strong enough to repel the movement altogether (as is the case in reflection), the movement is only weakened.

Now, he observes, a natural body finds it easier to break through 'passive' bodies along the perpendicular to their surfaces than in any other direction. For example a small iron sphere thrown perpendicularly on to a thin plate (*tabula*) covering a wide opening will in most cases break the plate and pass through, whereas if thrown obliquely with the same force and from the same distance

---

qui est in ipso, non omnino dimittitur, sed solummodo impeditur; necesse est, ut lux declinet ad partem faciliorem parte, ad quam prius movebatur, remanente in ipso motu composito; sed pars facilior parte, ad quam movebatur remanente motu in ipso, est illa pars, quae est vicinor perpendiculari. Unde lux, quae extenditur in corpore diaphano, si occurrit corpori diaphano grossiori corpore, in quo existit, refringetur per lineam propinquiorem perpendiculari, exeunti a puncto, in quo occurrit corpori grossiori, quae extenditur in corpore grossiore per aliam lineam quam sit linea, per quam movebatur. Haec ergo causa est refractionis splendoris in corporibus diaphanis, quae sunt grossiora corporibus diaphanis, in quibus existunt; et ideo refractio proprie est inventa in lucibus obliquis. Cum ergo lux extenditur in corpore diaphano, et occurrerit corpori diaphano diversae diaphanitatis a corpore, in quo existit, et grossiori, et fuerit obliqua super superficiem corporis diaphani cui occurrit, refringetur ad partem perpendicularis super superficiem corporis diaphani extensae in corpore grossiore.

Causa autem, quae facit refractionem lucis a corpore grossiore ad corpus subtilius ad partem contrariam parti perpendicularis, est: quia cum lux mota fuerit in corpore diaphano, repellet eam aliqua repulsione, et corpus grossius repellet eam maiore repulsione, sicut lapis, cum movetur in aere, movetur facilius et velocius, quam si moveretur in aqua; eo quod aqua repellit ipsum maiore repulsione, quam aer. Cum ergo lux exierit a corpore grossiore in subtilius, tunc motus eius erit velocior. Et cum lux fuerit obliqua super duas superficies corporis diaphani, quod est differentia communis ambobus corporibus, tunc motus eius erit super lineam existentem inter perpendicularem, exeuntem a principio motus eius, et inter perpendicularem super lineam perpendicularem, exeuntem etiam a principio motus. Resistentia ergo corporis grossioris erit a parte, ad quam exit secunda perpendicularis. Cum ergo lux exiverit a corpore grossiore, et pervenerit ad corpus subtilius, tunc resistentia corporis subtilioris facta luci, quae est in parte, ad quam secunda exit perpendicularis, erit minor prima resistentia; et fit motus lucis ad partem, a qua resistebatur, maior. Et sic est de luce in corpore subtiliore ad partem contrariam parti perpendicularis.'

The Latin closely follows the Arabic text, Istanbul MS. Fātih 3216, fols. 28ʳ9–31ʳ3.

it will only slide on the surface of the plate without breaking it. Similarly a sword breaks a piece of wood more easily when applied perpendicularly; the more obliquely it strikes the wood the weaker its effect. This helps to establish the general principle 'that movement on the perpendicular is stronger and easier, and of the oblique movements that which is nearer the perpendicular is easier than that which is farther from it'.

Analogically a perpendicular ray enters into a dense body in the same line because of the strength of the movement along this line. The movement of an oblique ray is, however, not strong enough to maintain the same direction. Consequently, the light turns into another direction along which its passage will be easier; that is, it turns towards the normal.

As in his treatment of reflection Ibn al-Haytham considers the incident movement to consist of two perpendicular components; and he makes use of this here to explain why an oblique ray is not always refracted into the perpendicular itself which, according to him, would be the easiest path. He argues that since the compounded (resultant) movement is not completely destroyed at the interface, the light continues to have in the refracting medium both a perpendicular and a parallel component; that is to say it takes a path situated between the original direction and the normal to the surface at the point of incidence.

His application of the parallelogram method to the refraction from a dense into a rare medium is more interesting, though perhaps more problematic. In this case the light meets with less resistance at the common surface, 'just as a stone moves more easily and quickly in air than in water, since the water resists it [*repellit ipsum*] more than the air'. Here Ibn al-Haytham *assumes* that the resistance acts particularly in the direction of the component parallel to the surface. Since the resistance *in that direction* was greater in the denser medium than in the rarer medium of refraction, the light is allowed to turn away from the normal. What is the basis of this assumption? Why should the decrease in resistance, and hence the increase in speed, take place particularly in the parallel direction? As a possible explanation we may note that the observed fact that in the dense-to-rare refraction the light turns away from the normal, and the supposition that the velocity of refraction is in this case greater than the velocity of incidence, together entail the conclusion that the parallel velocity must have increased. It is also interesting to note that he says nothing about what happens in this case to the perpendicular velocity.[3] For (again on the supposition that the actual speed has become greater) the increase in the parallel velocity (when the deflection is farther from the normal) is mathematically compatible with any one of three alternatives: the perpendicular velocity may increase, decrease, or remain constant.

Now let us suppose that Ibn al-Haytham moved one step further and assumed the increase in the parallel velocity to be in a constant ratio. His assumption would have been expressible in this form:

(a) $$v_r \sin r = m \, v_i \sin i,$$

where i = angle of incidence, r = angle of refraction, v_i = velocity of incidence, v_r = velocity of refraction, and m = a constant. Combining this with another assumption that he made, namely that the velocity of light is a property of the medium, or

(b) $$v_r = n \, v_i,$$

where n is a constant, the conclusion follows that

(c) $$\frac{\sin i}{\sin r} = \frac{m}{n} = k, \text{ a constant.}$$

In other words, the sines of the angles of incidence and refraction are in a constant ratio, which is the geometrical statement of the law of refraction. (In the form (c) the refraction index k would not of course depend solely on the constant ratio of the velocities n.)

He did not, however, take that step; but something, I think, may be learnt from our supposition: it was *not impossible* to arrive at the sine law along this path. We shall see that Descartes derived the law from two assumptions, of which one is identical with (b) and the other can be obtained from (a) by putting $m = 1$. This yields the

---

[3] H. J. J. Winter, in 'The optical researches of Ibn al-Haytham', *Centaurus*, III (1954), p. 201, attributes to Ibn al-Haytham the assumption that the velocity along the normal to the refracting surface remains constant. But there is nothing to support this either in the Arabic text or in the Latin translation.

law in the form $\dfrac{\sin\ i}{\sin\ r}=\dfrac{v_r}{v_i}=n$, a constant. While this does not explain

how in fact he discovered the law, his derivation appears less *ad hoc* in the light of Ibn al-Haytham's treatment than it might without it.[4]

Practically all subsequent explanations of refraction, up until the publication of Descartes' *Dioptric*, were almost entirely dependent upon Ibn al-Haytham; and Descartes' explanation itself was largely based on considerations which he may have derived either directly from Risner's edition of Ibn al-Haytham's *Optics*, or indirectly through Witelo and Kepler.

Roger Bacon, Witelo and Kepler all made the assertion that movement along the normal was stronger than oblique movement, and they all attributed the refraction of oblique rays at an interface to the weakness of oblique incidence.[5]

Again, all three of them explained the rare-to-dense refraction towards the perpendicular by reference to the principle that the nearer the movement is to the perpendicular, the stronger, easier and quicker it is. Witelo cites in this connection the principle '*natura . . . frustra nihil agit*',[6] thus linking ideas that were later explored by Fermat.

Bacon and Witelo considered that all transparent bodies being somewhat gross or dense, they all must resist the movement of light with varying degrees; both adopted the view that transparency (or rarity) has no upper limit.

Bacon ignored the resolution of motion into perpendicular components, but it was adopted by Witelo and Kepler.[7] They both asserted that the compounded movement of an oblique incident ray is not totally abolished, but only impeded; hence the movement of

---

[4] There is no foundation for V. Ronchi's statement that 'Descartes demonstrates the law of refraction by the same reasoning as that of Alhazen' (*Histoire de la lumière*, Paris, 1956, p. 115). It is not the case that for Ibn al-Haytham the parallel velocity remains unchanged in refraction (ibid., p. 44).

[5] Cf. R. Bacon, *De multiplicatione specierum*, II, 3, *Opus majus*, ed. Bridges, II, especially pp. 468–9; Vitellonis *Optica*, II, 47, in *Opticae thesaurus*, ed. Risner, pp. 81–83; Kepler, *Paralipomena* (1604), I, Proposition XX, pp. 15–17, and p. 84.

[6] Vitellonis *Optica*, p. 82, line 21 from bottom.

[7] Vitellonis *Optica*, end of p. 82; Kepler, *Paralipomena*, p. 84.

the ray is merely altered, that is, refracted into a new direction that still consists of a perpendicular and a parallel component.

Light striking the surface of a dense medium was compared by Kepler to the impinging of a small sphere (*sphaerula*) on a water surface: the movement is weakened as a result of being somewhat repelled by the surface.[8] This brings to the fore the idea that refraction is due to a surface action.

Witelo explained refraction into a rare medium in a line going away from the perpendicular by the assumption that the resistance becomes less in the direction parallel to the interface.[9]

It will have been noticed that in all this hardly anything was added to what these authors had already found in the *Optics* of Ibn al-Haytham.

2. The law of refraction was published for the first time by Descartes in the *Dioptric* in 1637—fifteen hundred years after Ptolemy had made an attempt to discover the law experimentally. Ptolemy's initiative was not abandoned during that period; his pioneering

---

[8] *Paralipomena*, p. 16.

[9] Vitellonis *Optica*, II, 47, p. 83: 'Cum vero radius *ac* exiverit a corpore grossiore ad subtilius, tunc quia minus habet resistentiae, erit motus eius velociter et magis sui diffusivus. Et quoniam resistentia medii densioris impellit semper lucem obliquam, ut coadunetur ad perpendicularem lineam a puncto incidentiae super superficiem illius corporis productam, quae est *cg*,

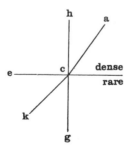

patet, quod in medio rarioris diaphani illa resistentia erit minor quam prima; fit ergo motus lucis ad partem, a qua per resistentiam repellebatur motus maior. Movetur ergo lux in corpore diaphano rariore plus ad partem contrariam parti perpendicularis, ita, quod angulus *gck* sit maior angulo *ach*; fit tamen semper motus lucis *ac* in refractione a corpore secundo rarioris diaphani quam primum, inter lineas *cg* et *ce*; quoniam cum angulus *gce* sit rectus, angulus *gck* nunquam potest fieri rectus.'

99

investigations were continued by such assiduous students of optics as Ibn al-Haytham, Witelo and Kepler, who tried to bring theoretical considerations to bear upon the problem of refraction. There was, however, no success until, apparently almost simultaneously, Snell and Descartes hit upon the correct law.[10] It is not known for certain when exactly Snell made his discovery, and we have no exact idea about how he made it. His work on refraction, seen in manuscript by Jacobus Golius, Isaac Vossius and Christian Huygens, has since been lost;[11] the autograph of the table of contents, containing Snell's own formulation of the law, has been found and published by C. de Waard, but it bears no date.[12] J. A. Vollgraff has suggested

---

[10] Recent research has indicated an earlier experimental determination of the sine law by Thomas Harriott. So far no connection has been established between Harriott's work and that of Snell or Descartes. See J. W. Shirley, 'An early experimental determination of Snell's law', *American Journal of Physics*, XIX (1951), pp. 507–8; Johs. Lohne, 'Thomas Harriott (1560–1621), the Tycho Brahe of optics', *Centaurus*, VI (1959), pp. 113–21.

The important studies on Snell and on the question of Descartes' possible dependence on his work are the following:

P. Kramer, 'Descartes und das Brechungsgesetz des Licthes', *Abhandlungen zur Geschichte der Mathematik*, IV (1882), pp. 233–78; P. van Geer, 'Notice sur la vie et les travaux de Willebrord Snellius', *Archives néerlandaises des sciences exactes et naturelles*, XVIII (1883), pp. 453–68; D.-J. Korteweg, 'Descartes et les manuscrits de Snellius d'après quelques documents nouveaux', *Revue de métaphysique et de morale*, IV (1896), pp. 489–501; Gaston Milhaud, *Descartes savant*, Paris, 1921, pp. 103–23 (Ch. IV: 'Les travaux d'optique de 1620 à 1629', first published as 'Descartes et la loi des sinus', *Revue générale des sciences*, XVIII, 1907;) J. A. Vollgraff, 'Pierre de la Ramée (1515–1572) et Willebrord Snel van Royen (1580–1626)', *Janus*, XVIII (1913), pp. 515–25; 'Snellius' notes on the reflection and refraction of rays', *Osiris*, I (1936), pp. 718–25; C. de Waard, 'Le manuscrit perdu de Snellius sur la réfraction', *Janus*, XXXIX (1935), pp. 51–73; J. F. Scott, *The Scientific Work of René Descartes (1596–1650)*, London, 1952, pp. 36–39 (a summary of Korteweg's article).

[11] Regarding the nature of Snell's research Huygens has briefly stated that Snell arrived at the correct measure of refractions 'after much effort and many experiments [*multo labore multisque experimentis*] . . . without, however, sufficiently understanding what he had discovered' (H, XIII, p. 7). The last words refer to Snell's incorrect explanation of the apparent raising of objects immersed in water by the contraction of vertical rays. C. de Waard points out that the contraction hypothesis was debated in Snell's time, and he refers to discussions taking place in 1604 and 1607 between Kepler and Brengger who held the same opinion as that of Snell; cf. *Janus*, XXXIX (1935), p. 55, n. 2.

[12] Snell expresses the law in the following words: 'Radius verus ad apparentem in uno eodemque medio diverso eandem habent inter se rationem. Ut secans complementi inclinationis in raro ad secantem [complementi] refracti in denso, ita radio apparens ad verum seu incidentiae radium' (C. de Waard, *Janus*, XXXIX, 1935, p. 64). What he calls 'radius verus' is the refracted ray, and the 'radius apparens' is the prolongation of the incident ray from the point of incidence until it meets the normal drawn from the foot of the refracted ray to the interface. This is equivalent to the proportionality: cosec i : cosec $r = k$, a constant (where i and r are the angles of incidence and refraction made with the normal to the surface at the point of incidence).

that Snell already knew the law in 1621, but this was proposed as a mere conjecture unsupported by positive evidence; the dated text of Snell on which it was based deals with reflection, not refraction.[13]

Snell died in 1626. When Golius found his manuscript in 1632, the law of refraction had already been known to several scholars close to Snell, including Golius himself; they all attributed the law to Descartes. No one seems to have been aware of Snell's result until Golius made his discovery.[14] The first publicly to accuse Descartes of plagiarizing the law from Snell's manuscript was Isaac Vossius, who published his accusation in 1662.[15] Christian Huygens had already expressed certain doubts, however. He wrote in his *Dioptrica* (begun in 1653 and first published posthumously in 1703) that he assumed (*accepimus*) that Descartes had seen Snell's manuscript, adding that 'perhaps' (*fortasse*) it was from this

---

[13] In a note appended to Bk. III (*De visione composita reflexa*) of Risner's *Optica*, Snell wrote on 22 December 1621 that 'the place of the image follows in each case a well-defined perpendicular in such a way that always the incident ray observes to the place of the image from the point of incidence its own perpetual proportion'. ('At locus imaginis certas in singulis perpendiculares ita sequiter ut semper radius incidentiae ad locum imaginis a puncto incidentiae in singulis suam et perpetuam analogiam observet.') Taking this to refer to a constant ratio between the distance of the object to the point of incidence and the distance of the image to the same point, and noting the form in which Snell in fact expressed the refraction law (see preceding note), Vollgraff concludes: 'I do not see that the same formula applies—or could be thought by Snellius to apply—to the case of reflection. Notwithstanding the fact that these words occur in a note on reflection, I venture therefore, not to affirm, but to suggest, that at the same time, December 1621, Snellius had already found the law of refraction' (*Osiris*, I, 1936, p. 724). On the next page he remarks: 'The fact that the law of refraction is not mentioned in the marginal notes belonging to Book IV [*De visione refracta*], may easily be explained by admitting that they were written at an earlier date than December 1621.' He further remarks that Snell lectured on optics at Leiden in 1621, and that Descartes is believed to have stayed at the Hague in winter 1621–2. But he adds: 'Unfortunately it is quite impossible to say in what manner Snellius spoke that winter about reflection and refraction, and what of his sayings, if any, may have reached the ears of Descartes' (ibid., p. 724, n. 14). When Descartes began his next long sojourn at Leiden in 1629 (three years after Snell's death), he had already been acquainted with the law; see Van Geer, op. cit., p. 467.

[14] So far as we know, Snell's work on optics was first mentioned by a friend who was present at his death, André Rivet, in a letter to Mersenne written in 1628. It is doubtful that he was acquainted with the contents of Snell's manuscript. Isaac Beeckman was Rivet's and Snell's friend; he learnt of the sine law from Descartes in 1628; when he later alluded to research on the subject of refraction he could only think of Kepler. See letter of Mersenne to André Rivet, 25 December 1628, *Correspondance du P. Marin Mersenne*, II (1936), pp. 157–8, and 161–2; C. de Waard, *Janus*, XXXIX (1935), p. 53. See also, Korteweg, op cit., pp. 495–9.

[15] Cf. Issac Vossius, *De lucis natura et proprietate*, 1662, p. 36. Vossius may have seen Snell's manuscript in the winter of 1661; cf. C. de Waard, *Janus*, XXXIX (1935), p. 55.

manuscript that Descartes had derived the sine law.[16] Later, in 1693, he wrote that the law was 'not the invention of Mr. des Cartes in all likelihood [*selon toutes les apparences*], for it is certain that he has seen the manuscript book of Snellius, which I have also seen'.[17] It is to be noted that in these two passages Huygens was more sure of Descartes' having seen Snell's manuscript than of his having derived the law from it.

There exists in fact other evidence in support of the assumption that Descartes learnt of Snell's discovery at some time during the period 1632–1637. This appears from the following circumstances. It was in arrangement with Descartes that Golius had planned to test Descartes' law experimentally before he discovered Snell's manuscript in 1632.[18] In November of the same year Golius communicated the news of his finding to Constantin Huygens,[19] whose friendship with Descartes continued until the publication of the *Dioptric* in 1637. As Korteweg has remarked, it does not seem likely that Descartes could not have been informed by either Golius or Constantin Huygens of Snell's work. But there are no texts to support this suspicion.

Nevertheless, the assumption that Descartes deliberately suppressed his knowledge of Snell's discovery after he had learnt about it in or shortly after 1632 does not settle the question of plagiarism. For it can be ascertained from Descartes' correspondence that he already had the sine law in 1626 or 1627, when he persuaded his friend Claude Mydorge to have a hyperbolic glass made to test it.[20] That was five or six years before Golius found Snell's manuscript.

Thus, it is still valid to repeat today the conclusion reached by Korteweg in 1896, namely that neither the question of Descartes'

---

[16] Cf. H, XIII, p. 9. Huygens attributed the law to Descartes until at least 8 March 1662. In January 1665 he wrote of the 'principle which has been employed in Dioptrics since Snellius and Monsieur Desc.' He probably saw Snell's manuscript in 1662 or 1663; ibid., p. 9, n. 1; H, V, p. 188.

[17] Cf. H, X, p. 405.

[18] Cf. Descartes to Golius, 2 February 1632, D, I, pp. 236–40.

[19] Golius' letter to Constantin Huygens, dated 1st November 1632, was discovered and published by Korteweg; see *Revue de métaphysique et de morale*, IV (1896), pp. 491–5.

[20] Cf. Descartes to Golius, 2 February 1632, D, I, p. 239; Descartes to Constantin Huygens, December 1635, D, I, p. 335–6. Cf. G. Milhaud, *Descartes savant*, p. 104.

originality nor that of priority can be settled on the evidence available: he may well have discovered the law of refraction independently of Snell (indeed this is more likely than the contrary supposition), and his discovery may have antedated that of Snell.

If, in the absence of any evidence to the contrary, we have to assume that Descartes made an independent discovery of the sine law, the question now to be asked is: how did he make it? As is often the case with similar questions in the history of science it is not easy to give a sure answer. One thing is certain, however: he did not discover the law 'by experiment'. He himself wrote in 1632 that the only experiment he had ever made was with the glass designed by Mydorge in 1626 or 1627 *for the purpose of verifying* the law he had already arrived at. The result of the experiment agreed with what he had predicted. 'Which assured me either that the Craftsman had fortunately failed, or that my reasoning was not false.'[21] What was the nature of the 'reasoning' that had led him to his prediction? The only argument that we have from Descartes for establishing the law of refraction is that presented in the second chapter of the *Dioptric*. It is, as we shall see, an argument based on comparing the refraction of light to the behaviour of a projectile when passing through a surface offering more or less resistance, and in which he applied the method of resolving motions that had been already applied to the same problem by Ibn al-Haytham, Witelo and Kepler. Had he arrived at the law by a different line of thought, what reason would he have had for suppressing it, especially since he was convinced of its correctness, as the passage from the letter to Golius shows?[22]

---

[21] Descartes to Golius, 2 February 1632, D, I, p. 239: 'Si vous n'auez point encore pensé au moyen de faire cette experience, comme ie sçay que vous auez beaucoup de meilleures occupations, peut-estre que celuy-cy vous semblera bien aussi aisé, que l'instrument que dècrit Vitellion. Toutesfois ie puis bien me tromper, car ie ne me suis point seruy ny de l'vn ny de l'autre, & toute l'experience que i'ay iamais faite en cette matiere, est que ie fis tailler vn Verre, il y a enuiron cinq ans, dont M. Mydorge traça luymesme le modelle; & lorsque qu'il fut fait, tous les rayons du Soleil qui passoient au trauers s'assembloient tous en vn point, iustement à la distance que i'auois predite. Ce qui m'assura, on que l'Ouurier auoit heureusement failly, ou que ma ratiocination n'estoit pas fausse.' Korteweg further points out that, given the technological difficulties involved, an experiment performed with a hyperbolic lens could not be decisive; cf. *Revue de métaphysique et de morale*, IV (1896), p. 500.

[22] In 1882 Kramer (op. cit., pp. 233–78) proposed an alternative explanation of how Descartes might have arrived at the sine law. According to his hypothesis, later endorsed by G.

Referring to the deduction of the sine law in the *Dioptric* Paul Tannery has remarked that Descartes was the first to give an example of what theoretical physics should be like, by showing how mathematics could be applied to physical problems. The ancients, Tannery pointed out, had only provided models for statics, the theory of centres of gravity and the principle of Archimedes; and Galileo's *Dialogues* were not published until 1638, that is, after the *Dioptric*.[23] The particular interest and importance of the Cartesian deduction has been recognized even by unsympathetic readers both in his and in our own time. To take only two examples. Fermat, who never accepted Descartes' reasoning for establishing the law of refraction, wrote that he 'much appreciated the spirit and invention of the author'.[24] And Ernst Mach, who described Descartes' explanation of refraction as 'unintelligible and unscientific', admitted that it 'exerted a very stimulating influence' upon subsequent writers.[25] Nevertheless, the general view of Descartes' proof has been that it is open to serious objections. Fermat first denounced the proof from a mathematical point of view, and his opinion seems to have been shared later by Huygens[26] and Leibniz.[27] This agreement among

Milhaud (op. cit., pp. 103–23), Descartes was led to his discovery by the study of conic sections; and this is confirmed by the fact that the experiment suggested to Mydorge for testing the law was with a hyperbolic lens. (See letter of Mydorge to Mersenne of probably February–March 1626, in *Correspondance du P. Marin Mersenne*, I, pp. 404–15.) While there is nothing to refute this conjecture, it would seem quite removed from the way we know Descartes was disposed to think about the problem of refraction, not only in the *Dioptric* and *Le Monde*, but also earlier in the *Regulae* (where the problem of the anaclastic is viewed as a physico-mathematical problem, not to be solved by mathematical means alone, see above, pp. 29ff), and earlier still in the *Cogitationes privatae* (see below, p. 105f).

[23] Cf. P. Tannery, 'Descartes physicien', *Revue de métaphysique et de morale*, 1896, p. 480; *Mémoires*, VI (1926), pp. 307–8.

[24] Fermat to Mersenne, September 1637, F, II, p. 111. Descartes' first public accuser, Isaac Vossius, thought the style of demonstration was new (J. F. Scott, op. cit., p. 37).

[25] E. Mach, *Principles of Physical Optics*, London, 1926, pp. 33 and 34.

[26] 'It has always seemed to me (and to many others besides me) that even Mr. Des Cartes, whose aim has been to treat all the subjects of Physics intelligibly, and who assuredly has succeeded in this better than any one before him, has said nothing that is not full of difficulties, or even inconceivable, in dealing with Light and its properties' (Huygens, *Treatise on Light*, p. 7). The words in parentheses are missing in the English translation; cf. H, XIX, pp. 465 and 467.

[27] Cf. Leibniz, *Discours de métaphysique*, edited by Henri Lestienne, 2nd edition, Paris, 1952, Article XXII, p. 67: 'And the demonstration of the same theorem [i.e., the sine law] which M. des Cartes wished to offer by way of the efficient [causes], is far from being as good [as Fermat's demonstration of the same law by the final causes, i.e. the principle of least

three of the greatest mathematicians of their time has had one effect that is interesting from our point of view: it strengthened the suspicion that Descartes did not himself discover the law. How could he have arrived at the law deductively, when the only argument he has for establishing it is so faulty? Not, of course, that this is impossible, but it would seem rather improbable. Mach, for example, adopts this kind of argument when he remarks that after reading Descartes' 'unintelligible' deductions in the second chapter of the *Dioptric*, 'it will scarcely be assumed . . . that Descartes discovered the law of refraction. It was easy for him as an applied geometrician to bring Snell's law into a *new* form, and as a pupil of the Jesuits to "establish" this form.'[28] The assumption on which this argument rests is not true. For from a mathematical point of view, Descartes' proof is perfectly correct; indeed, no less a mathematician than Newton thought it 'not inelegant'.[29] Moreover, the suggestion that Descartes artificially constructed an argument to establish the law which he had derived in a different form from other sources seems far-fetched. In any case there is no historical evidence to support it.

3. Descartes' interest in optics, and particularly in the problem of refraction, goes back to the early period of 1619–21. The notebook, known as the *Cogitationes privatae*, in which he recorded his thoughts at that time, has the following passage:

Since light can be produced only in matter, where there is more matter it is produced more easily, other things being equal; therefore it penetrates more easily through a denser than through a rarer medium. Whence it comes about that the refraction is made in the latter away from the perpendicular, in the former towards the perpendicular. . . .[30]

---

time]. At least there is reason to suspect that he would not have found it by those means, if he had not learnt in Holland about Snell's discovery.' It is to be noted that already in Leibniz's time the alleged deficiency in Descartes' proof was taken to confirm the view that he did not discover the sine law independently of Snell.

[28] Mach, op. cit., p. 33. See also Van Geer, op. cit., pp. 466–7.

[29] Newton, *Optical Lectures*, London, 1728, p. 47; below, p. 300.

[30] D, X, pp. 242–3: 'Lux quia non nisi in materia potest generari, vbi plus est materiae, ibi facilius generatur, caeteris paribus; ergo facilius penetrat per medium densius quam per

Two assertions are made here, one after the other. The first states a view which is contrary to what had been universally accepted since Aristotle,[31] namely that 'light penetrates more easily through a denser than through a rarer medium'. Descartes bases this assertion on the particular connection he establishes between light and matter, a connection which gained greater plausibility in Newton's theory of refraction.[32]

The second assertion, beginning with 'whence', in fact records an observation that is completely independent of any view regarding the speed of light in different media. Why, then, does Descartes present this second assertion as a consequence of the first? Now such a deduction can be performed if we combine with the first assertion a further assumption expressed in the *Dioptric*, viz. that the horizontal component of the incident velocity is unaltered by refraction. This combination, however, would necessarily yield the sine law. Did Descartes therefore possess the sine law when he wrote the above passage in 1619–21?[33] Was this law already associated at this time with a proof similar to that published in the *Dioptric*? If not, what were his reasons for regarding the deflection from or towards the normal as a result of the decrease or increase in speed respectively? Why 'whence'?

On the other hand, if we do not think it likely that his reasons were connected with the proof published in 1637, then we must at least grant that Descartes' first assertion (the speed of light is greater

---

rarius. Vnde fit vt refractio fiat in hoc a perpendiculari, in alio ad perpendicularem; omnium autem maxima refractio esset in ipsa perpendiculari, si medium esset densissimum; a quo iterum exiens radius egrederetur per eumdem angulum.'

[31] Cf. Aristotle, *Meteorologica*, I, 5, 342^b 5.

[32] In Newton's theory the light rays suffer a greater attraction towards the particles of the denser body than towards those of the rarer body; accordingly the speed must be greater in the former than in the latter. See below, pp. 301ff.

[33] According to G. Milhaud (op. cit., p. 109, n. 1) the proof presented in the *Dioptric* was not formulated until after 1628–29. C. de Waard (*Correspondance du P. Marin Mersenne*, I, 1932, p. 426) finds the earliest evidence of Descartes' use of the method employed in that proof in Isaac Beeckman's notes of 1633. It should be noted, however, that already in 1626 when Mydorge explained his propositions on hyperbolic glasses to Mersenne, he illustrated the sine law (which he had learnt from Descartes) with reference to a diagram from which the constancy of the parallel velocity could be deduced, and which would consequently yield the law in the form $\frac{\sin i}{\sin r} = \frac{v_r}{v_i} =$ a constant, that is the form obtained in the *Dioptric*. Cf. ibid., p. 405.

in denser media) was not simply an embarrassing result of an artificially constructed proof, or a price paid for obtaining a desired
physico-mathematical demonstration of the sine law. The grounds
for this assertion would be totally independent of the form this
demonstration took in the *Dioptric*. We now turn to Descartes'
treatment of refraction in that work.

In the Cartesian model for refraction a tennis ball strikes a frail
canvas which takes the place of the breakable plate in Ibn al-
Haytham's corresponding analogy. Arriving from *A* at *B* (Fig. IV.1)
the ball breaks through the canvas 'thus losing only a part of its
speed, say a half'.[34] That the loss relates to the actual speed along
AB is made clear by further steps in the argument.

Having assumed this, let us observe again, in order to know what course
the ball must follow, that its motion is entirely different from its determination to move in one direction rather than another; from which it
follows that their quantities must be examined separately. And let us
also note that of the two parts of which this determination can be imagined
to be composed, only that which makes the ball tend downwards can in
any way be changed by the encounter with the canvas; and that the other

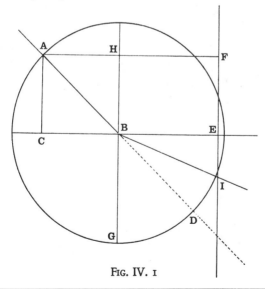

Fig. IV. 1

[34] *Dioptric*, II, D, VI, p. 97.

which makes it tend to the right must always remain the same as it was, because the canvas is not in any way opposed to it in this direction.

Then, having described the circle *AFD* with *B* as centre, and drawn the three straight lines *AC*, *HB*, *FE* at right angles with *CBE*, such that the distance between *FE* and *HB* is twice that between *HB* and *AC*, we shall see that the ball must tend towards the point *I*. For, since the ball loses half of its speed in going through the canvas *CBE*, it must employ twice the time it took above the canvas from *A* to *B*, to travel below from *B* to any point on the circumference of the circle *AFD*. And since it loses nothing whatsoever of its former determination to advance in the right hand direction, in twice the time which it employed to go from the line *AC* to *HB*, it must cover twice the distance in the same direction, and consequently arrive at a point on the straight line *FE* at the same moment as it also reaches a point on the circumference of the circle *ADF*. But this would be impossible if it did not proceed towards *I*, as this is the only point below the canvas *CBE* where the circle *AFD* and the straight line *EF* intersect.[35]

The same considerations apply when, instead of the canvas, the ball strikes a water surface: once the 'force' of the ball is diminished in the same ratio by the opposition of the surface it proceeds in the same straight line:

as to the rest of the water which fills all the space from *B* to *I*, although it resists the ball more or less than did the air which we previously supposed to be there, we should not say for this reason that it deflects it more or less. For the water can open itself up, to make way for the ball, just as easily in one direction as in another, at least if we assume all the time, as we do, that neither the heaviness or lightness of this ball, nor its size, nor figure, nor any other such external cause may change its course.[36]

The more obliquely the ball strikes the surface the more it is deflected towards it, until at a certain inclination it 'must not go through the surface, but rebound . . . just as if it encountered there the surface of the ground. An experiment which was regretfully made sometimes when, firing cannon balls for amusement's sake towards the bottom of the river, one hurt those who were on the other side.'[37]

---

[35] Ibid., pp. 97–98.   [36] Ibid., pp. 98–99.   [37] Ibid., p. 99.

Descartes' figure for this case shows the perpendicular to the surface from *F* outside the circle, thus indicating the impossibility of refraction.[38]

To account for the deflection towards the normal, Descartes supposes that the ball on reaching *B* (Fig. IV. 2)

is pushed once more . . . by the racquet *CBE* which increases the force of its motion, for example by a third, such that it can now travel in two moments a distance which it formerly covered in three. And the same effect would be produced if the ball encountered at *B* a body of such a nature that it would pass through its surface *CBE* a third more easily than through the air.[39]

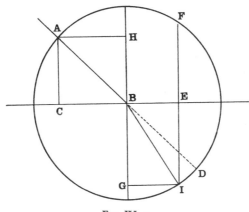

FIG. IV. 2

Accordingly he draws the perpendicular *FE* at a distance $BF=2/3$ *CB*; the direction of refraction is determined by the point *I* at which *FE* cuts the circle.[40]

It is of course assumed throughout that 'the action of light follows in this the same laws as the motion of this ball'.[41] When, therefore,

---

[38] See Fermat's comments, below, pp. 130 ff.

[39] *Dioptric*, II, D, VI, pp. 99–100.

[40] Descartes adds: 'But one may also take the converse of this conclusion and say that, since the ball coming from *A* in straight line to *B* is deflected at *B* and takes its course from there towards *I*, this means that the force or facility with which it enters into the body *CBEI* to that with which it goes out of the body *ACBE*, is as the distance between *AC* and *HB* to that between *HB* and *FI*, i.e. as *CB* to *BE*' (ibid., p. 100).

[41] Ibid., p. 100.

rays of light 'pass obliquely from one transparent body into another which receives them more or less easily than the first, they are deflected in such a way that their inclination to the surface of these bodies is always less on the side of the body which receives them more easily, than on the other side: and this exactly in proportion as it receives them more easily than the other.'[42] The inclinations, Descartes adds, should be measured by lines such as AH and IG, not by angles such as ABH and GBI, 'nor even by the magnitudes of angles similar to DBI which are called the angles of refraction'.[43]

4. In all the cases considered by Descartes he assumes the incident motion to be increased or decreased at refraction in a given ratio. What he immediately infers from this assumption makes it quite clear that what is so increased or decreased is the actual motion or speed. For suppose, as in the first case, that the canvas deprives the ball of exactly half its speed. To deduce from this, as Descartes does, that the ball will reach a point on the circumference in twice the time it formerly employed to cover the equal distance from A to B, implies that he is comparing the speeds along the actual paths of the ball.

This point has sometimes been misunderstood by Descartes' readers and commentators in the seventeenth century and in our own day.[44] Since the parallel velocity is supposed to be unchanged by refraction, it easily comes to the mind that what is increased or decreased by a constant ratio is the *perpendicular* component. But this would invalidate the proof, and we should no longer be in a position to see the problems to which it gave rise. The fact is that, whatever be the force responsible for the change suffered by the perpendicular velocity, Descartes does not provide an independent expression for this force which would allow us to construct the

---

[42] *Dioptric*, II, D, VI, pp. 100–101.

[43] Ibid., p. 101.

[44] This misunderstanding is to be found in the following works: E. Mach, op. cit., pp. 32–33; J. F. Scott, op. cit., pp. 34–35; V. Ronchi, op. cit., p. 115; C. Boyer, *The Rainbow*, p. 203. All these writers mistakenly attribute to Descartes the assumption that the perpendicular velocity is altered in a constant ratio. Their version of Descartes' proof is therefore incorrect.

refracted path. As in his proof of the law of reflection, he arrives at the law of refraction from two assumptions: the one is about the actual speeds, and the other is about the horizontal speeds. In this lies the intended symmetry between the two proofs.

Descartes' first assumption is therefore:

$$(1) \qquad\qquad v_r = n v_i$$

where v_r, v_i are the velocities of refraction and of incidence, and n a constant.

As applied to light, this equation amounts to the assertion that the velocity of light is a property of the medium it is traversing; in other words, the velocity is independent of the angle at which the light enters the refracting medium. The problem of reconciling the physical meaning of this equation with the asserted instantaneous propagation of light will be discussed later.

The second assumption concerning the conservation of speed parallel to the separating surface may be formulated thus:

$$(2) \qquad\qquad v_i \sin i = v_r \sin r.$$

Combining (1) and (2), we get the law:

$$\sin i = n \sin r.$$

Expressed in terms of v_i and v_r, the law reads:

$$\frac{\sin i}{\sin r} = \frac{v_r}{v_i} = n,$$

the sines are in the inverse ratio of the velocities, a result which the corpuscular theory inherited through Newton's adoption of a proof essentially similar to that of Descartes.

Although Descartes is careful to mark the place for the phenomenon of total reflection, his investigation lacks the systematic approach which characterized Kepler's experimental researches.[45] Descartes observes that at normal incidence no refraction takes place, and further that the incident and refracted rays rotate in the same sense. He then mentions total reflection. This, he says, takes place when the line *FE* falls outside the circle.[46] But he does not consider the intermediate critical case when *FE* would be tangential to the circle, and he thus fails to give an expression for the critical angle

---

[45] See E. Mach, op. cit., pp. 30–32.   [46] See above, pp. 108f.

(i.e. the angle of incidence for which the refracted ray makes 90° with the normal) for any two given media.

Such an expression, however, can be obtained in his theory. For, supposing

$$v_i > v_r,$$

there will be an angle of incidence at which the parallel component

$$v_i \sin i = v_r.$$

But this is impossible unless the resultant velocity v_r coincides with the parallel velocity. Call that angle C; we have:

$$v_i \sin C = v_r,$$

$$\sin C = \frac{v_r}{v_i},$$

$$\sin C = n,$$

or,

$$C = \sin^{-1} n.$$

We come now to the physical interpretation of Descartes' two assumptions, and first we consider assumption (1).

It is asserted at the beginning of the *Dioptric* that light, in the luminiferous medium, is a tendency to motion rather than an actual motion. It is further supposed to be transmitted in an instant. The analogy between the action of light and the transmission of pressure through a stick was meant to illustrate and support these assertions. How, then, could the comparison with the moving ball be expected to enlighten the reader as to the mechanism of light in reflection and refraction? Fermat was the first among Descartes' contemporaries to point out this difficulty; he wrote to Mersenne (September, 1637):

it seems that there is a particular disproportionality in that the motion of a ball is more or less violent, according as it is pushed with different forces, whereas light penetrates the diaphanous bodies in an instant and seems to have nothing successive in it.[47]

In view of this apparent lack of analogy Descartes had assumed in the *Dioptric* that the tendency to motion follows the same laws as

---

[47] F, II, p. 109.

motion itself.[48] That is to say, the same equations of motion apply to the action of light, in spite of the absence of the element of time from the latter. He had thus believed to have established a bridge between that branch of physics which studies actual motion and the other branch whose subject is light.

Fermat, however, still felt ill at ease; he found no reason to believe that what is true of motion must also be true of the tendency to motion; for, he objected, there is as much difference between them as between what is actual and what is potential.[49] The appeal to these Aristotelian terms was not a happy one, and it allowed Descartes to give a plausible answer. In perfect accord with the Aristotelian doctrine, he said, he was only asserting that the actual must also exist in the potential, and not vice versa.[50]

The problem becomes clearer when we concentrate on Descartes' first equation: How should we interpret v_i and v_r in the equation? What is it in the action of light that corresponds to the speed of the ball? Descartes would reply: 'its force'. But what is this force with which light *passes* more or less easily from one medium into another? How are we to understand this *passing* as a process without introducing the element of time into the picture? It appears that, in order to have any clear idea of what the first assumption means, one is forced to operate explicitly with speeds, and thus dispose of the oppressive assertion of the instantaneous transmission of light. It must be emphasized, however, that Descartes never actually took that step; he never used the word 'speed' in connection with light, although he sometimes used 'force' to mean practically the same thing, for instance, when he spoke of the force of the moving ball.

This only natural step was taken later by Fermat and Huygens; and Descartes' first assumption, thus understood, forms an essential part of their theories of refraction. But already during the discussion with which we are here concerned, Fermat used 'force' and 'speed' synonymously—a practice which at least did not provoke an objection from Descartes and was readily adopted by his defenders, Clerselier and Rohault.

---

[48] Cf. D, VI, p. 89; see also pp. 78 ff above. [49] Cf. F, II, pp. 108–9.
[50] Descartes to Mersenne, October 1637, F, II, p. 113.

The only physical meaning, therefore, that can be attached to the equation

$$v_r = n v_i$$

is that the velocity of light is characteristic of the medium it is traversing. As we have seen, this idea had been employed by Latin writers since the translation of Ibn al-Haytham's *Optics*, and is already to be found in Ptolemy. When Descartes first formulated it in 1619–21 he departed from the generally accepted view by asserting the speed to be greater in denser media.

In an interesting passage of the *Dioptric*,[51] Descartes attempts to explain why a ray of light passing from air into a denser medium such as water or glass turns towards the normal, whereas a ball falling into water is deflected away from the normal. We know from his proof that this means, in the case of light, that the speed has increased by refraction, while the contrary has taken place in the case of the ball. The problem is to account for the fact that the 'force' or speed of light increases when the medium is denser and consequently offers greater resistance. Descartes first remarks that a ball loses more of its speed when it impinges on a soft body than when it strikes a hard one, and it moves more easily on a polished hard table than on a carpet. The reason, he explains, is that a soft body yields to the shock of the ball and thus deprives it of a part of its speed (or momentum) whereas a hard body offering greater resistance takes away only a little of the momentum. In collision, therefore, the lost force, speed or momentum is less by how much the resistance is greater. Now, Descartes continues, light is 'a certain movement or action in a very subtle matter which fills the pores of other bodies'. As the particles of water or glass are 'harder and firmer' than those of air, they offer greater resistance to the action of the subtle matter, and as a result, allow the light to 'pass more easily' than the ball in water. 'For this light does not have to push from their places any of these particles [and consequently communicate to them at least part of its momentum], as a ball has to push away the particles of water to find a passage among them.'[52]

---

[51] Cf. D, VI, pp. 102–3.   [52] Ibid., p. 103.

These considerations in fact avoid a real problem to which Descartes' mechanical explanation of refraction gives rise. For imagine two rays travelling with the same speed in air. The one strikes a water surface which it penetrates, while the other enters into a denser medium *than water*, say glass. Descartes would inform us that, owing to the greater resistance of the glass, the velocity in glass will be greater than in water. This he would attempt to explain by the fact that the ray which has entered into the glass loses less of its momentum than the one which has passed into water. But he does not explain the fact that the velocity in glass (and in water) will be greater than the velocity of incidence, that is, in air. Such increase, however, as takes place in refraction from a rare into a dense medium is implied by his mathematical assumptions. In other words, whereas his mechanical picture may be taken, as far as it goes, to account for the loss of momentum when the refraction is from a dense into a rare medium, it utterly fails to explain the increase in momentum when the refraction is reversed.

Whatever Descartes' reasons were for asserting the velocity of light to be greater in the denser medium, and however weak was his argument for establishing this assertion in the *Dioptric* we know that he formulated it at a much earlier time, in 1619–21. We are now led to ask: on what grounds did he make the second assumption '$v_i \sin i = v_r \sin r$'? For the view that the velocity increases when the light is deflected towards the normal (as happens when the light passes into a denser medium) does not necessarily imply this assumption although it is compatible with it. We shall see that this problem was in fact brought up during the discussion between Fermat and Clerselier.

But how are we to account for this assumption in Descartes? The question is important, for it is precisely the apparently *ad hoc* character of this assumption which has led several writers to suspect Descartes of having artificially constructed his deductive proof of the sine law after he had obtained this law by different means. If these means were valid, why would he have suppressed them? Ought we therefore to conclude (with Leibniz and Mach) that he must have derived the law in the first place from Snell's manuscript?

It would seem, however, that the question about the origin of Descartes' assumption '$v_i \sin i = v_r \sin r$' is capable of quite a simple and straightforward answer: this equation is a direct consequence of the model employed. Refraction is explained by Descartes with reference to the same collision model which Ibn al-Haytham had used before him for the same purpose. This was no accident in Descartes' treatment of refraction, since in his view all phenomena must be reduced to collision phenomena. Now in every collision between two bodies the velocity component parallel to their common tangent at the point of contact must remain constant. Hence Descartes' assumption, which means nothing more than that refraction is to be treated as a special case of reflection, a case in which the perpendicular speed is altered *at the surface*. It might be objected that if this were as natural as it looks, then why do we not find the same assumption in Ibn al-Haytham? But again the answer is close to hand: to assume the constancy of the parallel velocity would be in conflict with Ibn al-Haytham's belief that the actual velocity decreases in the denser medium where the refracted ray bends towards the normal. Descartes, however, had from the beginning (i.e. in 1619–21) adopted the opposite view which is perfectly compatible with his assumption. It is not therefore unlikely that Descartes' discovery of the sine law may have been the result of bringing two unrelated assumptions together for the first time. That this is in fact how he arrived at the refraction law independently of other scientists or other methods must remain a conjecture, but a conjecture which is not, as has been usually supposed, implausible.

5. In 1664 Fermat wrote a letter to an unknown person[53] describing what appeared to him, in retrospect, as his first reaction to Descartes' *Dioptric*. Having examined Descartes' proportion (that the sines are in constant ratio), he became suspicious of the proof—for the following reasons. First, it was based on a mere analogy, and

---

[53] Cf. Fermat to M. de ***, 1664, F, II, pp. 485–9.

analogies do not found proofs. Second, it presupposed that the passage of light was easier in denser media, a proposition which 'seems shocking to common sense'.[54] And third, it assumed that the horizontal motion was unchanged by refraction—an unjustifiable assumption in Fermat's view.[55]

Apart from the first reason, it would be a mistake to believe that the latter two were really among the reasons for Fermat's rejection of the proof in the first place. The point involved in the second reason was never raised in the correspondence with Descartes; and it is more than probable that Fermat did not formulate this objection until after he had found that a contrary view to that of Descartes (regarding the speed of light in different media) was essential to his own principle of least time. It is significant to observe in this connection that this so-called presupposition was brought into the discussion by Fermat only after he had effected (in 1661–1662) the demonstration of the relation which gives the velocities in direct proportion to the sines. What we learn from Fermat's correspondence about the third reason is that he expressed his objection to Descartes' assumption concerning the horizontal component for the first time in 1658, in a letter to Clerselier,[56] that is, in the second phase of the discussion about the *Dioptric*.

We should not therefore take what Fermat said in the letter to M. de *** of 1664 as a true account of his own views about the *Dioptric* in the years immediately following the publication of that treatise in 1637. The objections he expressed in 1664 were the outcome of a long and winding discussion with Descartes and his defenders. Moreover, they were greatly influenced by a result obtained from the principle of least time about twenty-four years after the discussion started. His first difficulties and, consequently, his first objections were of a completely different kind.

In his letter to Mersenne of December 1637, Fermat undertook to expose what he believed to be the 'paralogisms' of the author of the *Dioptric*. For the sake of argument he accepted Descartes' distinction between the determination and the force of the moving ball, and

---

[54] Cf. Fermat to M. de ***, 1664, F, II, p. 485. [55] Ibid.
[56] Cf. Fermat to Clerselier, 10 March 1658, F, II, p. 371.

set out to show that Descartes was not faithful to his own distinction in his arguments to establish the sine relation.

The way Fermat understood the distinction is clearly revealed in the following words.

if we imagine that the ball is pushed [perpendicularly] from the point *H* to the point *B*, it is evident that, since it falls perpendicularly on the canvas *CBE*, it will traverse it in the line *BG* [*G* being on the perpendicular *HB* extended], and thus its moving force will be weakened, and its motion will be retarded while the determination is unchanged, since... [the ball] continues its movement in the same line *HBG*.[57]

'Determination' is here used to mean something which remains constant when the ball continues to move in the same line even though its speed has changed. That is, determination is identical with direction. Thus understood, it is not surprising that Fermat could find faults with Descartes' arguments. For Descartes maintained that the perpendicular determination of the falling ball was altered when it pierced the canvas and its speed in that direction was thereby changed. But Fermat naturally objected: what does that mean, if the distinction is precisely between the determination and 'the moving force or speed of the motion'?[58] If the distinction is to be maintained, 'one cannot say that, since the movement of the ball has been weakened, the determination which makes it go from above downwards has changed'.[59] Again, if the determination is to be distinguished from the speed, how could Descartes deduce from his assumption about the conservation (after refraction) of the horizontal 'determination' that the speed of the ball will be the same in that direction as before?[60] Understood in this manner, it is obvious that Descartes' arguments would lead to absurdities.[61]

One might be inclined to say that Fermat's difficulty lay in the fact that he took Descartes' words too literally; that the so-called distinction aside, one can still follow Descartes' deduction in terms of speed and direction alone. For what Descartes really meant was that the speed of the incident ball (or ray) is conserved in the

---

[57] Fermat to Mersenne, December 1637, F, II, pp. 117–18.
[58] Ibid., p. 118.　　[59] Ibid.　　[60] Ibid., p. 119.　　[61] Ibid., pp. 119–20.

horizontal direction after refraction, while the speed along the refracted path maintains a constant ratio to the speed of incidence. From these two assertions the sine relation follows immediately. But this suggestion presupposes that we already understand Descartes' proof. Fermat was not in this happy position; and his difficulty lay precisely in understanding the distinction here in question. Also, it would still remain for us to explain the point in Descartes' differentiation between what he called 'determination' and the speed.

Besides, the view has been expressed that Fermat's difficulty (at the time when he made these objections, i.e. in 1637) lay elsewhere; that he had not only accepted Descartes' distinction but understood it better than Descartes. For example, Gaston Milhaud writes about this point:

we see that Descartes radically distinguishes the motion or quantity of motion and consequently, the speed on which it depends, from the determination or tendency to move in one direction or another. For instance, in the case of the reflection of a ball from a hard body, Descartes saw the determination change, but not the motion or the speed. But it is not this distinction which would be sufficient to trouble us. Fermat, too, will draw it more rigorously than Descartes himself and still find this demonstration of the sine law absolutely incomprehensible.[62]

Paul Mouy also follows Fermat in his interpretation of 'determination' as meaning the direction of motion.[63]

Is direction the only meaning that can be attached to 'determination', as Fermat, Milhaud and Mouy have assumed? Did Fermat really draw the distinction more rigorously than Descartes, when he saw it as between direction and speed? If one answers these questions in the affirmative, one is also forced to accept Fermat's conclusion, namely that Descartes' arguments which were based on this distinction are beyond understanding, that they are in fact merely a confusion of that alleged distinction.

This, however, would be an improbable conclusion. It would seem rather strange that Descartes should not only introduce a

---

[62] Milhaud, op. cit., p. 110.
[63] Cf. Paul Mouy, Le développement de la physique cartésienne, p. 55.

quantity, the determination, that was not at all necessarily required for performing his proofs (as these could be formulated in terms of speed and direction alone), but also take pains to establish a fundamental and seemingly elementary distinction which he so shamefully failed to manage. It would also be strange that, in spite of Fermat's disruptive criticism, which has seemed so convincing to recent commentators, the same Cartesian distinction should appear later at the basis of the rules of motion as formulated in the *Principles* in 1644, that is about six years after Descartes had read and considered Fermat's remarks.[64] The question, therefore, should be asked again: what was Descartes' distinction about?

The answer is in fact contained in a passage by Descartes which appears to have escaped attention. In a letter to Mydorge of (probably) 1 March 1638, Descartes pointed out that Fermat was arguing as if

I had supposed such difference between the determination to move in this or that direction [*ça ou là*] and the speed, that they were not to be found together, nor that they could be diminished by the same cause, namely by the canvas *CBE*. . . : which is against my meaning, and against the truth; seeing that this determination cannot be without some speed, although the same speed can have different determinations, and the same determination can be combined with various speeds.[65]

One thing at least comes out clearly after a first reading of this passage—namely that, for Descartes, the determination to move in this or that direction is *not*, as Fermat understood it, identical with direction ('this determination cannot be without some speed'). This means that when the speed of a moving body is altered, its determination is also altered, even though the body still moves in the same direction. If, for example, the canvas diminishes the speed of the perpendicularly falling ball, as in the case envisaged by Fermat, it thereby diminishes the determination of the ball to move in the perpendicular direction, even though the motion is continued in the same line. We may thus understand Descartes' assertion that the

---

[64] Cf. Descartes, *Principles*, II, 44.   [65] F, IV, pp. 94–95.

speed and the determination can be 'diminished by the same cause'. Also, since determination involves direction, when a moving body changes its direction, while its speed remains unaltered, its determination should be considered to have changed. We may thus understand Descartes' statement that 'the same speed can have different determinations'; a more fortunate rendering of this statement should, however, read: different determinations can have the same speed. As to the last sentence in Descartes' passage, 'the same determination can be combined with various speeds', it should not present any difficulty. For example, in his proof of the sine law, the same horizontal determination is combined, both with the actual speed along the actual path of the incident ball (or ray), and with the speed along the actual path of the refracted ball (or ray); these two actual speeds being different. (It is to be noted that Descartes does not say that the same determination can have different speeds.) Again, if the determination is a quantity depending on both direction and speed, when Descartes infers from the conservation of the horizontal determination that the speed in that direction is unchanged, his inference is of course correct. It becomes inconclusive only if, following Fermat's interpretation, we identify determination with direction.

In the light of the preceding passage we are therefore able to give 'determination' a meaning which fits all of Descartes' sentences and arguments which were expressed in terms of that quantity. This meaning can be defined as follows: two bodies A and B have the same determination if, and only if, they move (or tend to move) in the same direction and with the same speed. Thus understood, Descartes' distinction between the determination of a moving body and its speed should appear to us perfectly valid; it corresponds in fact to the distinction between vector and scalar quantities.

But in order to 'destroy completely'[66] Descartes' law of refraction, Fermat composed in the same letter to Mersenne (of December 1637) a refutation of the proof in the *Dioptric*; he actually deduced from what he believed to be Descartes' assumptions a different sine relation from that of Descartes.[67]

---

[66] F, II, p. 120. [67] Cf. F, II, pp. 120–4.

Fermat considers for this purpose the case in which the speed is increased by refraction, and he composes the velocity of the refracted ball from the incident velocity and a given perpendicular component received at the surface of separation. His demonstration is essentially as follows:

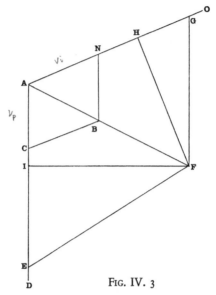

Fig. IV. 3

Suppose that (in Fig. IV. 3)

$$AN = v_i, \text{ the velocity of incidence;}$$

and

$AC = v_p$, the given perpendicular velocity added to the perpendicular component of v_i at the refracting surface.

We thus have the assumption:

(1) $$\frac{AN}{AC} = \frac{v_i}{v_p} = n, \text{ a constant.}$$

If NB and CB are drawn parallel to AC and AN respectively, then

$$AB = v_r, \text{ the velocity of refraction.}$$

The angle DAO is any angle corresponding to the angle of incidence i (the fact that DAO is obtuse in Fermat's figure does not essentially alter the argument). The angle DAB corresponds to the angle of refraction r. It follows that

$$BAO = i - r.$$

F is any point taken on AB extended, from which FG and FE are drawn parallel to BN and BC; and FH, FI are drawn perpendicular to AG, AD, respectively.

From consideration of similar triangles, it is seen that

(2)
$$\frac{FE}{FG} = \frac{BC}{BN},$$

and, the triangles FGH, FIE being similar, we have

(3)
$$\frac{FE}{FG} = \frac{FI}{FH} = \frac{\sin r}{\sin (i-r)}.$$

From (1), (2), and (3), it follows that

$$\frac{\sin r}{\sin (i-r)} = \frac{BC}{BN} = \frac{AN}{AC} = \frac{v_i}{v_p} = n.$$

We see the difference between Fermat's demonstration and that of Descartes: for it is clear from Fermat's figure that AB will vary in length with the angle DAO; in other words, v_r will vary with the angle of incidence i. Fermat's assumption (1) thus contradicts Descartes' requirement that the ratio of v_i and v_r should remain constant.

Are we here faced with a trivial case of misunderstanding, a case of misreading, for example? Hardly; for Fermat saw the contradiction and pointed it out himself in the same letter; though he saw it in a particular way:

the principal reason of the author's [Descartes'] demonstration is founded on his belief that the composite motion on BI [the direction of the refracted ray] is always of equal speed, although the angle GBD [equal to the angle of incidence] that is comprised between the lines of direction of the two moving forces [the one acting in the direction of the *incident* ray and the other in the perpendicular direction] has changed: which is false, as we have already fully demonstrated.[68]

Thus Fermat sees the contradiction as existing, not between two different sets of assumptions, the one belonging to Descartes and the other being his own (which certainly would not continue a refutation of Descartes' proof), but as a contradiction contained in Descartes' theory itself. In other words, he takes Descartes to be

---

[68] F, II, p. 124.

assuming that both the speed along the direction of refraction (v_r) and the perpendicular velocity received at the refracting surface (v_p) are independent of the angle of incidence.

Where did Fermat get the second assumption? How did he understand it from the *Dioptric* where it is not stated? I believe that the answer is to be found in the comparison used by Descartes to illustrate the case which Fermat has chosen for his refutation. Descartes compares the action of the refracting surface in this case with that of a racquet hitting the incident ball perpendicularly, thus increasing its perpendicular velocity. Fermat added the assumption that the racquet always hit the ball with equal force. Consequently, he constructed the velocity of refraction from the incident velocity and the constant perpendicular velocity supplied by the racquet (or refracting surface), and correctly deduced his own sine relation.

In order, therefore, to protect his analogy, Descartes would have to state explicitly that the racquet is supposed to hit the ball perpendicularly with varying degrees of force. But once this implicit supposition is brought to light, the artificial character of the comparison with the racquet becomes manifest. For every time the racquet hits the ball at the point of incidence, the amount of perpendicular speed thus imparted to the ball must be such that v_r (the resultant velocity) bears a constant ratio to v_i. What is this force which acts at the refracting surface and either increases or diminishes the perpendicular component of the incident velocity in such a manner that the value of v_r is always the same? Descartes' theory does not provide an answer to this question. No answer was forthcoming until the publication of Newton's *Principia*. Fermat had thus unintentionally exposed in Descartes' treatment of refraction a serious defect: the failure to provide an independent expression governing the force responsible for refraction.

It is important to realize that Fermat's difficulty in understanding Descartes' proof had nothing to do with the latter's assumption concerning the horizontal velocity. In the letter we have been discussing there is no objection, explicit or implicit, against that assumption. Fermat's arguments have been interpreted differently, however, by Milhaud and the editors of Fermat's works.

Milhaud thought that Fermat's demonstration not only replaces the constancy of v_r by that of v_p, but also fails to satisfy Descartes' condition that the horizontal component of the velocity is conserved after refraction. This remark of Milhaud's is in accord with his view that Fermat's difficulties were, from the beginning, connected with that assumption (and would therefore confirm his interpretation of Fermat's attitude towards Descartes' theory). He writes the following concerning Fermat's demonstration:

Fermat departs . . . from Descartes' hypotheses in that he relates the acceleration or retardation, which are determined by the media, to the normal determination of the motion and not to the direction, which is primarily unknown, of the refracted ray; *and, on the other hand, in that he does not take into account that the determination parallel to the surfaces of separation remains constant.* As he argues with Descartes, and later with Clerselier, he varies his objection; never did he resolve to adopt the postulates of Descartes. It is these postulates *and especially the postulate concerning the parallel determination which he absolutely refuses to accept.*[69]

The first remark in this passage concerning the parallel determination is not correct. For Descartes' condition, expressed in the postulate in question, is satisfied in Fermat's demonstration by the very fact that he composes the resultant velocity of refraction v_r from v_i and a perpendicular component v_p. It follows from this that the component of v_r parallel to the refracting surface is equal to the corresponding component of v_i, that is $v_r \sin r = v_i \sin i$, which is Descartes' postulate.

An opinion similar to Milhaud's had been expressed by the editors of Fermat's works, P. Tannery and C. Henry, in their comment on Fermat's refutation. This comment may have been the origin of Milhaud's remarks. It is as follows:

Fermat concludes that if one should consider, with Descartes, the motion along the refracted ray as the resultant of the motion along the incident ray and an action along the normal, the proportionality must exist not between the sines of the angles of incidence and of refraction, sin i and sin r, but between sin $(i-r)$ and sin r. To this effect, he implicitly supposes the normal action to be independent of the incidence. [That is true:

---

[69] Milhaud, op. cit., pp. 111–12; italics added.

v_p is constant in Fermat's demonstration.] The hypothesis of Descartes is, on the contrary, that the component parallel to the surface of refraction keeps the same value after refraction as before. It is clear that one cannot decide *a priori* between these two suppositions.[70]

Thus it is thought here that the constancy of v_p is not compatible with the conservation of the horizontal speed after refraction; that Fermat's refutation ignores Descartes' postulate about that horizontal speed since it postulates the constancy of the normal action; and consequently that the problem between Descartes and Fermat was to decide between these two postulates, which could not be done *a priori*. In view of what has been remarked above, all this must be based on a misunderstanding.

The conclusion that we reached regarding Fermat's first attitude towards Descartes' proof is therefore justified: his difficulties lay outside the assumption concerning the horizontal speed. He misinterpreted the distinction between the determination and the speed and, consequently, could not follow Descartes' application of it; and he was misled by the comparison of the racquet into believing that the increase (or decrease) in the normal speed was supposed constant.

Descartes' reply in the letter to Mydorge of (probably) 1 March 1638 dealt mainly with two points in Fermat's letter (of December 1637): first, with the latter's interpretation of the distinction between the determination and speed; and second, with Fermat's demonstration itself. Fermat's objections to that distinction and Descartes' comments on them have already been discussed. These comments seem to have left no impression on Fermat as it appears from his letter to Clerselier of 10 March 1658. This letter was written shortly after Fermat had received, from Clerselier, another copy of Descartes' letter to Mydorge, which letter had formerly been forwarded to Fermat by Clerselier.[71] The point in Descartes' distinc-

---

[70] F, II (1894), p. 124, editors' note 1.

[71] Cf. F, II, p. 365, note 3. Examples showing Fermat's persistence in his own interpretation of Descartes' distinction are the following phrases from his letter to Clerselier, 10 March 1658: 'One of these aims (*visées*) or determinations [of the ball]'; or, thinking that he was merely expressing Descartes' intentions, 'this determination from left to right remains the same after refraction, that is to say, it conserves the same aim (*visée*) or direction'; or again, 'the determination or direction of the motion differs from its speed'. Ibid., pp. 371 and 373.

tion was not in fact made clear in Fermat's mind until after he had read the explanation offered by Clerselier and Rohault.[72] As to Fermat's demonstration, this appeared to Descartes as nothing more than a series of paralogisms. His answer, however, was insufficient in that it failed to point out clearly the crucial difference between his own demonstration and Fermat's, namely, that in the one an assumption is asserted which contradicts the corresponding assumption in the other. In any case, it was not to be expected that Fermat would get any satisfaction from Descartes' remarks, since these were expressed in terms of that fundamental distinction of which Fermat could not make sense. It was again Rohault who clearly pointed out the misunderstanding in Fermat's refutation. Rohault wrote in his *Projet* of 15 May 1658:

M. de Fermat considers that, on page 20 of the *Dioptric*, the supposition of M. Descartes is that the increase of one third of the motion which the ball suffers is from above downwards or along the [perpendicular] line *BG*, whereas it is measured in the line which the ball actually describes. And this is easy to understand, because, if that [i.e. what Fermat assumed] were the case, M. Descartes would not have supposed, as he did, that the force of the ball's motion is increased by a third, but that the determination from above downwards is increased by a third. One should not therefore say [with Fermat] that, in his [Descartes'] sense, the motion of the ball along *BI* [the direction of refraction] is composed of the motion which it had along *BD* [the direction of incidence], and another along [the perpendicular] *BG* whose quantity one would want him [Descartes] to have supposed to be greater than the motion along *AB* [i.e. the velocity of incidence] by a third. One should rather say that the actual motion of the ball [after refraction] is faster by a third than before, and leave it to calculation to determine what change must thereby result in the determination from above downwards.[73]

6. When the controversy was later resumed between Fermat and Clerselier,[74] the main point at issue between them was Descartes' assumption concerning the conservation of the horizontal speed and whether it was compatible with his other assumption, stating the

---

[72] See below, p. 129, n. 77.   [73] F, II, p. 395.   [74] Above, pp. 86f.

actual speeds to be determined by the media. Throughout the dispute with Descartes, Fermat's objections had mainly been due to misunderstandings, though they were interesting misunderstandings which sometimes had their roots in Descartes' theory itself. When, twenty years later, Fermat resumed the discussion with Clerselier, his objections began to take a different line. In order to understand correctly Fermat's arguments, it should be borne in mind that his attitude was always that of a mathematician, not of a physicist. If by his patient and sincere analysis he helped to expose certain physical difficulties, that was only by implication. His objections against Descartes' arguments, and those of his defenders, were always of a mathematical or logical character.

In 1658, Fermat still denied, as he always did before and afterwards, that the argument in the *Dioptric* to establish the refraction law constituted a demonstration or even approached a demonstration.[75] Accordingly, his criticisms were directed against the Cartesian proof in so far as it claimed to be a demonstration. With this attitude Fermat imagined, in his letter to Clerselier of 3 March 1658, a 'scrupulous sceptic' who wants to examine Descartes' reasonings from a new angle. This imaginary sceptic is willing (no doubt for the sake of argument) to grant their conclusiveness as applied to reflection, but he questions the validity of their being extended to refraction. For, the sceptic protests, in reflection the speed of the ball is the same after striking the surface as before; the ball is all the time moving in the same medium; and the surface is totally opposed to the perpendicular determination:

But, in refraction, everything is different. Are we to obtain the consent of our sceptic here without proof? Will the determination from left to right remain the same when the reasons which have persuaded him of that in reflection have all vanished? But this is not all: he has reason to be afraid of ambiguity; and, when he will have accorded that this determination from left to right remains the same, he has reason to suspect that the author will confuse him over the explanation of this term. For, although he [the author] has protested that the determination is different from the

---

[75] Cf. Fermat's letter to Clerselier, 3 March 1658, F, II, p. 366.

moving force, and that their quantity should be examined separately, if our sceptic grants him at this point that this determination from left to right remains the same in refraction, that is to say, that it preserves the same aim or direction, it seems that the author will want to force him afterwards to grant him that the ball, whose determination towards the right has not changed, advances towards the right with the same speed as before, although the speed [i.e. the actual speed] and the medium have changed.[76]

We note that Fermat still identifies the determination with the direction of motion. Yet it now seems that his interpretation of Descartes' distinction is irrelevant to the kind of objection he has in mind. For his argument, as expressed in this passage (and as later developed in another letter to Clerselier of 2 June 1658 which he wrote after reading the explanations offered by Clerselier and Rohault concerning the distinction in question)[77] can be put as follows: Granting that the horizontal determination remains unchanged after refraction, that is (according to his first understanding of that term) granting that the ball (or ray) remains to have a motion in that direction, it does not follow that the speed of that motion will be the same; but if Descartes is simply assuming (as he was informed by Rohault and Clerselier) that the horizontal speed is unaltered by refraction, then the question is: on what grounds should that assumption be accepted?

Not that this proposition cannot be true, but it is so only if the conclusion, which M. Descartes draws from it, is true, that is to say, if the ratio or proportion for measuring refractions had been legitimately and truthfully

---

[76] Ibid., p. 371.

[77] Fermat did not come to apprehend that distinction as Descartes had intended it until after it was explained to him by Clerselier and Rohault in May 1658 (cf. F, II, pp. 383–4, and 392; see also F, II, p. 475). The remarks of Clerselier and Rohault were a development of Descartes' brief comment in his letter to Mydorge (see above, p. 120). In agreement with the definition formulated above (see p. 121) they explained that the determination should be regarded as having changed even when the direction remains the same and only the quantity of the determination measured by the speed has increased or decreased. In his letter to Clerselier of 2 June 1658, Fermat confessed that he had not understood Descartes' distinction in this manner, although he still was not satisfied with it (cf. F, II, p. 397). He did not give reasons for his dissatisfaction.

determined by him. He has therefore only proved it [the assumption] by a proposition [the sine relation] that is so doubtful and so little admissible.[78]

Fermat here rejects the Cartesian proof of the sine law, not any longer because it is not conclusive, as he believed twenty years earlier, but simply because it now appears to him to be founded on an assumption that is 'neither an axiom, nor is . . . legitimately deduced from any primary truth'. The Cartesian proof is therefore not a 'demonstration' since 'demonstrations which do not force belief cannot bear this name'.[79]

There are two distinct problems involved in Fermat's remarks, and it is important to know which one of these he had in mind. First, there is a mathematical question which may be expressed as follows: can the equation

$$v_i \sin i = v_r \sin r$$

be true for all values of i when v_i does not equal v_r? It will be shown presently that this is the kind of question he wanted to ask. Second, there is the question of whether a physical interpretation can be given which would satisfy Descartes' equations for deducing the sine law. It is this second question which constituted a really serious problem for the Cartesian theory. And although it is doubtful that Fermat ever conceived it clearly, his objections and suspicions ultimately led Clerselier to construct a mechanical picture which, he hoped, would accommodate Descartes' assumptions. Whether Clerselier succeeded in this will be examined later.

Concerning the mathematical question, Clerselier replied by simply repeating what Descartes had already remarked in his letter to Mydorge: namely that the same (horizontal) determination can be combined with different speeds (in the actual direction of motion).[80] In reply to this, Fermat proceeded to construct, as a challenge to Clerselier and his friends, a counter-example to which the statement

---

[78] Fermat to Clerselier, 10 March 1658, F, II, p. 373. This passage indicates that Fermat had not yet accepted the sine law when he wrote the present letter to Clerselier, that is twenty years after the publication of the *Dioptric*. (See also in the same letter: 'I maintain that the true ratio or proportion of refractions is not yet known.' Ibid., p. 374.) In fact he did not accept the law until after he had derived it himself from his own principle of least time; see next chapter.

[79] Ibid., p. 373.   [80] Cf. F, II, p. 384.

cannot apply. He introduced an imaginary case,[81] and this already indicates that he was concerned with a purely mathematical problem. He pictured a ball falling obliquely on a resisting surface which the ball cannot penetrate. On reaching the surface, the ball loses part of its actual speed, owing to the interference of an imaginary agent; but the horizontal speed remained the same. There is thus an essential parallelism between this case and what Descartes considered to take place in refraction; except, of course, that in Fermat's case the ball does not go through the surface. (This shows again that the intended parallelism is merely mathematical.) If, therefore, v_i and v_l are the velocities before and after the ball strikes the surface, we have by supposition:

(1) $$\frac{v_i}{v_l} = n, \text{ a constant less than unity};$$

(2) $$v_i \sin i = v_l \sin l.$$

Suppose now that the ball falls at an angle i greater than the angle whose sine is equal to n. At such an angle,

(3) $$v_i \sin i > v_l.$$

And it follows from (2) and (3) that, at the same angle of incidence,

$$v_l \sin l > v_l,$$

which is impossible, since $\sin l$ cannot be greater than unity. Therefore, Fermat concluded, the ball will not be reflected. But, by supposition, it will not be refracted either. Where will it go? The problem for Clerselier and Descartes' friends was 'to provide the ball with a passport and to mark for it the way out of this fatal point'.[82]

Fermat's deductions are, of course, correct. But what did he want to prove by deriving that impossibility? That on the assumptions made, there is a case in which the direction of v_l cannot be constructed; and that case is realized when the angle of incidence is greater than the angle whose sine equals $\frac{v_l}{v_i}$. What does this mean when it is applied to the actual case of refraction? The same thing of course: namely, if $v_i > v_r$ and $v_i \sin i = v_r \sin r$, there is a case in

---

[81] Cf. Fermat to Clerselier, 2 June 1658, F, II, pp. 399–402.   [82] Ibid., pp. 401–2.

which the direction of v_r cannot be constructed; and that case is realized when the angle of incidence is greater than the angle whose sine equals $\frac{v_r}{v_i}$. But why should this constitute an objection against Descartes' theory? Descartes himself was aware of the situation described by Fermat, and he explicitly pointed it out in the *Dioptric*.[83] For the case envisaged by Fermat is precisely the case in which the light is *not* refracted but totally reflected. The fact that, when the refraction is from a dense into a rare medium, there is a case in which the refracted ray cannot be constructed from Descartes' assumptions, should be counted for, not against, Descartes' theory. It means that the theory makes room for an actual property of light.

Fermat, however, would insist on considering his own imaginary case, where the ball was abandoned in its strife to leave the impenetrable surface 'for the honour of M. Descartes'.[84] Descartes could interpret the impossibility of constructing the direction of refraction on the given assumptions, by the fact that the ball (or ray) will in this case be reflected; but the ball, in Fermat's case, is in a different situation, since it cannot go anywhere. How, then, is it going to employ the quantity of motion which it still has? Fermat himself believed that ball will be reflected, and at an angle equal to the angle of incidence.[85] This would of course mean that both the vertical and horizontal components of the ball's motion will be reduced in the same proportion as the actual speed. In other words, Descartes' assumption concerning the conservation of the horizontal speed would have to be abandoned, if the ball were to move at all.

Now Fermat was right in drawing that last conclusion regarding the special situation in his own imaginary case. But, as Clerselier objected in his answer to Fermat,[86] that was of no relevance to Descartes' treatment of refraction. The fact that Descartes' assumption concerning the horizontal speed has to be abandoned in Fermat's imaginary situation does not mean that the same assumption cannot

---

[83] See above, pp. 108f.
[84] Cf. Fermat to Clerselier, 16 June 1658, F, II, p. 408.
[85] Ibid., pp. 409–10.
[86] Cf. Clerselier to Fermat, 21 August 1658, F, II, pp. 414ff.

be maintained in the consideration of refractions. We have seen that, when a similar situation to Fermat's arises in the actual refraction of light, one can conclude that the ray will not penetrate the refracting surface, and this conclusion is confirmed by experiment. But, by incessantly challenging the basis of Descartes' assumptions, Fermat's remarks finally led Clerselier to envisage a mechanical situation which, as a rival to Fermat's imaginary case, would be in agreement with Descartes' equations. The examination of Clerselier's attempt brings us to the physical problem mentioned before.

Clerselier's attempt (contained in his letter to Fermat of 13 May 1662) was no doubt inspired by some of Descartes' remarks in the *Dioptric*.[87] It had the merit, however, of going beyond the analogies with which Descartes had been satisfied; instead of Descartes' frail canvas, we have here a ballistic model.[88]

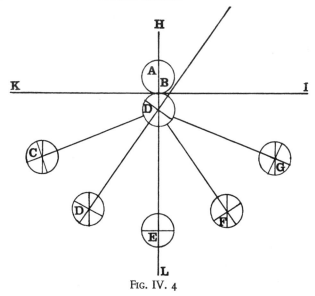

FIG. IV. 4

A is an elastic sphere stationed at *B* (Fig. IV. 4). Relying on experiment, Clerselier observes that if any of the spheres *C*, *D*, *E*, *F* or *G* impinges on *A* at the point *B* (i.e. in such a way that the normal common to *A* and the impinging sphere is perpendicular to

---

[87] See above, pp. 107ff.   [88] Cf. F, II, pp. 477–82.

KI), *A* will move in the (perpendicular) direction of *BH*, whatever be the angle of incidence. This means that the motion (momentum) imparted to *A* has been taken away from the normal component of the motion of the incident sphere, the horizontal component (parallel to *KI*) being unaltered. After impact the incident sphere (if it loses only part of its initial momentum) will thus have at *B* two motions, one parallel to *KI* and equal to its motion in that direction before impact, and the other perpendicular to *KI*; and both of these will determine the speed and direction of the sphere at the moment of collision. If now the medium that is supposed on the upper side of *KI* is such that it will resist the motion of the sphere, that, in Clerselier's view, can only affect the speed of the sphere, and not its direction to which the medium is totally indifferent. According to Clerselier, this explains the refraction away from the normal when the speed is reduced at the refracting surface. Similarly is 'explained' the refraction towards the normal, when the velocity is increased; only in this case *A* is supposed to yield more easily to the impinging sphere and, 'so to speak, drags it towards *H*'.[89]

Clerselier's model, in spite of its elaborations, only succeeds in making more manifest the difficulties which it was proposed to remove. It in fact fails to satisfy the main condition which it was required to fulfil: namely, the conservation of the horizontal speed. For although this condition is preserved at the moment of impact, if the further resistance of the medium subsequently changes the actual speed without altering the direction (as is supposed by Clerselier), both the perpendicular and the horizontal components of the motion will be diminished in the same proportion. Moreover, the model fails to explain, even in a general way, how the velocity can be increased by refraction. To suppose that the ball *A* yields more easily to the impinging ball does not mean that the latter will thereby acquire new momentum. And to say, as Clerselier did, that the incident sphere can be supposed to be 'dragged' into the refracting medium would be to introduce something like an attractive force which would seem completely out of place in this

---

[89] F, II, p. 479.

mechanical picture. (It is interesting to note, however, that Clerselier was led to introduce an idea which Newton later used in his explanation of refraction.)

Clerselier's model thus raises more problems than it solves, and it shows that no solution of the fundamental physical difficulties was forthcoming along the lines suggested by him and earlier by Descartes. Not that Descartes' two fundamental equations contradicted one another; they are both asserted in Newton's theory of refraction. But the dynamical interpretation they received from Newton would be completely repugnant to Descartes. The latter's dream was to build the whole of physics solely on the idea of motion, to the exclusion of all forces. Descartes' favoured idea of collision was not itself to be abandoned altogether in the later development of optics. It is the fundamental idea underlying Huygens' theory of light. But in Huygens' theory there is no place for Descartes' assumption of the conservation of the parallel speed.

There were no comments from Fermat on Clerselier's mechanical considerations. Before he read them he had arrived at Descartes' relation (in the form: sin $i = n$ sin r) from his own principle of least time. In the face of this unexpected result, he was willing to abandon the battlefield, as he said, leaving to Descartes the glory of having first made the discovery of an important truth, and himself being content to have provided the first real demonstration of it. With this conditional declaration of peace in his last letter to Clerselier of 21 May 1662, the discussion came to an end.[90]

---

[90] Cf. Fermat to La Chambre, 1 January 1662, F, II, p. 462; and Fermat to Clerselier, 21 May 1662, F, II, p. 484.

Chapter Five

FERMAT'S PRINCIPLE OF LEAST TIME

1. The discussion in the preceding chapter of Fermat's controversy with Descartes and Clerselier has revealed a negative aspect of the Cartesian theory of refraction, namely its failure to provide a plausible physical interpretation fitting the mathematical assumptions used by Descartes in his deduction of the sine law. Fermat himself went further than that: he doubted the validity of applying the method of the parallelogram of velocities to the problem of refraction, and this led him to doubt the truth of Descartes' law itself. To discover the true law of refraction, Fermat appealed to his own method of maxima and minima which he had invented about eight years before the beginning of the controversy with Descartes in 1637. In order to bring this method to bear on the problem under investigation, however, he relied on the metaphysical principle that nature performed its actions in the simplest and most economical ways. His law of least time was to him a justifiable specification of this principle. Thus, in the hands of Fermat, geometry and metaphysics joined forces in an attempt to defeat Descartes and his followers[1] by going beyond mere opposition to their arguments and actually producing *the* true law of refraction.

Fermat's attempt might appear preposterous and unscientific. But why? Had not the principle of economy been accepted, in one form or another, by many philosophers and scientists since antiquity? Had not an early specification of this principle, the law of shortest distance, been successfully applied by Heron of Alexandria to some cases of optical reflection?[2] Had not considerations of simplicity

---

[1] Thus Fermat wrote to La Chambre in 1657: '. . . si vous souffrez que je joigne un peu de ma géométrie à votre physique, nous ferons un travail à frais communs qui nous mettra d'abord en défense contre M. Descartes et tous ses amis' (F, II, p. 354). The nature of the 'physics' here referred to is revealed by the immediately following sentence: 'Je reconnois premièrement avec vous la vérité de ce principe, que la nature agit toujours par les voies les plus courtes'.

[2] Above, pp. 70f.

and economy guided the steps of a great seventeenth-century scientist, Galileo, towards the discovery of the important law of freely falling bodies?[3] Why should not Fermat make a similar attack on the problem of refraction when the appropriate mathematical tools were in his hands?[4]

It might be protested that in the case of refraction, a law had already been proposed; why, then, not consult the experiments before embarking on an endeavour that might well prove chimerical? Such a question would imply, however, that experiment enjoys a certain authority with which Fermat did not invest it. For, as we shall see, he was in fact informed, before he performed his calculations, that the sine law had been confirmed experimentally, and he believed his informants. Fortunately, however, this discouraged him only temporarily. He then reasoned that the true ratio for refractions might still not be identical with that proposed by Descartes, though so near to it as to deceive the keenest observer; there was, therefore, room for a fresh attack on the problem *in spite of* the apparent verdict of experiment.

Fermat was disappointed in his expectations in so far as he discovered that the principle of least time leads to the very sine ratio which Descartes had laid down. But this principle also requires the velocity of light to be greater in rarer media, contrary to Descartes' view; and this result has since been supported by experiment. Fermat has been rewarded for his bold attempt by the fact that his principle (in a modified form) has remained to this day an accepted theorem in optics; a theorem which, as Huygens later showed, in fact followed from wave considerations.[5]

2. In 1657, about six months before Fermat resumed the discussion concerning the *Dioptric* with Clerselier, he received a copy of a

---

[3] Below, p. 157, n. 57.

[4] The idea of 'easier' and 'quicker' path was the basis of Ibn al-Haytham's explanation of refraction from a rare into a dense medium, and Witelo, who adopted the same idea, quoted in this connection the principle of economy; see above pp. 95f, and p. 98. Fermat referred to the work of these two writers on refraction; cf. F, II, p. 107.

[5] Below, pp. 218f.

treatise on light that had been published in the same year by a physician called Cureau de la Chambre.[6] The author, who sent the copy to Fermat, held the view that light was a quality, not a substance; that it was capable of a local motion which, nevertheless, took place in an instant; and he offered as the general causes of reflection and refraction the 'animosity' and 'natural antipathy' between light and matter.[7] Fermat, in his letter of acknowledgement to La Chambre (August 1657),[8] does not appear to have been convinced by the author's argument in support of these views, although he described it, presumably out of politeness, as 'very solid and very subtle'.[9] He found in the physician's book, however, one principle which he could accept without hesitation: 'I first recognize with you the truth of this principle, that nature always acts by the shortest courses [*que la nature agit toujours par les voies les plus courtes*].'[10] La Chambre had recalled the successful application of this principle to the reflection of light at plane surfaces, and his book contained a geometrical demonstration of the following theorem: if A and B be any two points on the incident and reflected rays, and C a point on the plane surface where the light is reflected at equal angles, then, if D be any other point on the surface in the plane of incidence, the distance ACB is shorter than ADB.[11] But he mentioned two difficulties in the way of consistently implementing his adopted principle.[12] First, there was the objection that in reflection at concave surfaces, the light sometimes followed a longer path than that required by the principle; this was illustrated by the case in which the two points related as object and image are taken on a spherical concave surface. And second, the principle was not applicable to the behaviour of light in refraction: obviously, if the light always followed the shortest path, it should always go in a

---

[6] *La Lumière a Monseignevr L'Eminentissime Cardinal Mazarin, par Le Sieur De La Chambre, Conseiller du Roy en ses Conseils, & son Médecin ordinaire*, Paris, 1657.

[7] Cf. ibid., pp. 276–81, 296–302 and 335–41.

[8] Cf. F, II, pp. 354–9.

[9] Ibid., p. 354. Later in the same letter (p. 357), however, Fermat poured doubt on the opinion that the movement of light was not successive.

[10] Ibid., p. 354.

[11] La Chambre, op. cit., pp. 312–13.

[12] Ibid., pp. 313–15.

straight line from one medium into another. La Chambre's own answer to these difficulties was that the light was not 'free' to act in these exceptional cases in accordance with its 'natural tendency'.[13] For him, therefore, the principle of shortest course should be understood to govern, not the actual behaviour of light, but rather its hidden tendency which manifested itself only in the case of reflection at plane surfaces!

Looking at these difficulties from a different angle, the mathematician Fermat could not fail to see the possibility of a more satisfactory answer, at least as far as refraction was concerned:[14] if instead of shortest path one postulated *easiest course*, the principle might be found to agree with the behaviour of light when passing from one medium into another. For, assuming that different media offer different resistances, it might be the case that the straight line joining two points in the two media is not the path for which the resistance is a minimum.

For example, suppose that DB (Fig. V.1) separates the medium of incidence below from the medium of refraction above; and let the resistance of the lower medium be less than that of the upper medium, say by a half. The sum of the resistances along the incident ray CB and the refracted ray BA may, according to Fermat, be represented by $CB + 2BA$. In the same way, the resistance along the straight line CDA can be represented by $CD + 2DA$. Now although $CB + BA$ is greater than CDA, it may be the case that, for a certain position of B, $CB + 2BA$ is less than $CD + 2DA$.

Assuming, therefore, that light follows the easiest path, the problem of refraction is in this case reduced to the following

---

[13] Ibid. pp. 322–4.

[14] Concerning the difficulty connected with reflection at concave surfaces, Fermat simply suggested that the light should be understood to act, not on the concave surface as such, but on the tangent to that surface at the point of reflection in the plane of incidence. The measurement of the light's path should, therefore, be considered in relation to the tangent and not to the concave surface itself. On this suggestion the problem would be reduced to that of reflection at plane surfaces and Heron's theorem would hold. In Fermat's view this suggestion was justified by the 'physical principle' that 'nature performs its movements by the *simplest courses*' (F, II, pp. 354–5, my italics), since the straight line (the tangent with which the light should be concerned) is simpler than the curved surface. It is interesting to note the transformations that the principle of economy had to undergo in order to suit the various cases.

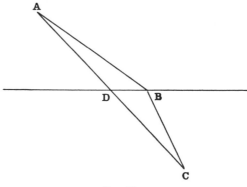

FIG. V. I

mathematical problem: given C and A, to find the point of re-fraction B such that $CB+2BA$ is a minimum. Concerning this problem, Fermat wrote to La Chambre in the same letter of 1657:

I confess to you that this problem is not one of the easiest; but, since nature solves it in all refractions so that she may not deviate from her ordinary way of acting, why cannot we undertake the [solution] ?[15]

He ended his letter by promising to supply the solution whenever it would please his correspondent to receive it; and that he would be able to deduce such consequences as would 'firmly establish the truth'[16] of the accepted principle by their exact agreement with all the relevant observations.

That promise remained unfulfilled for more than four years, and it will be interesting to follow the development of Fermat's thought during that period.

He could first easily prove on a numerical example that the straight line was not necessarily the most economical path for refracted light. To show this to La Chambre in 1662,[17] he readily interpreted the principle of easiest course as a principle of *least time*: Let the diameter AB (Fig. V. 2) represent the interface. It is sup-posed that the light can pass twice as 'easily' in the rare medium above as in the dense medium below:

[15] F, II, p. 358.   [16] Ibid., p. 359.
[17] Fermat to La Chambre, 1 January 1662, F, II, pp. 457–63.

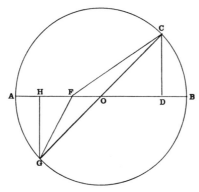

FIG. V. 2

It follows from this supposition that the time employed by the moving thing or the light from C to O is less then the time taken from O to G, and that the time of the movement from C to O in the rarer medium is but half the time of the movement from O to G. And consequently the measure of the total movement along the straight lines CO and OG can be represented by half CO added to OG; similarly, if you take another line such as F, the time of the movement along the straight lines CF and FG can be represented by half CF added to FG.[18]

Taking $CO=OG=10$, $HO=OD=8$, and $OF=1$, he found by calculation that $\frac{1}{2}CF+FG$ was less than $\frac{1}{2}CO+OG$ (the latter sum being equal to 15, and the former less than 59/4, i.e. less than 15). Since these two sums measured the times of the movements through F and O respectively, he concluded that in this case the time taken along the straight line COG was more than that taken along the sides CF and FG:[19]

I arrived at that result without much effort, but the investigation had to be carried farther; for, in order to satisfy my principle, it is not sufficient

---

[18] Ibid., p. 459.

[19] To La Chambre, for whom the movement of light was instantaneous, Fermat wrote: 'If you continue to deny successive movement to light, and to maintain that it is performed in an instant, you need only compare either the facility, or the aversion and resistance which become greater or less according as the media change. For this facility, or this resistance, being more, or less in different media, and this in diverse proportion according as the media differ more, they can be considered to be in a given ratio and, consequently, they fall under the calculation as well as the time of the movement, and my demonstration will apply to them in the same manner' (ibid. p. 463).

to have found a certain point F through which the natural movement is performed more quickly, more easily and in less time than along the straight line COG, but it was necessary to find the point through which the movement would be made in less time than through any other point taken on either side. For this purpose I had to appeal to my method of *maxima* and *minima* which expedites the solution of this sort of problem with much success.[20]

Before he undertook the task of applying his method to the problem, however, a certain obstacle appeared in his way and nearly discouraged him from making the attempt. He had never accepted Descartes' method for establishing the law of refraction and, as a result, he had never believed that Descartes' law was true.[21] He had thus assumed that, by means of his own method of maxima and minima, the principle of easiest course would lead to a different law from the one that had been proposed by Descartes; and, of course, he had no doubts that that 'well established'[22] principle would yield *the* correct law. Before he undertook the calculations, however, he was informed by various persons whom he greatly esteemed that the experiments exactly agreed with the Cartesian ratio of sines. It therefore appeared to him at first a useless task to try to invent a new law that was bound to be born dead, especially as the calculations involved were not easy to make. Fortunately for the history of science, he got over that obstacle by a hopeful resolution:

I disposed of [that] . . . obstacle by my knowledge that there is an infinite number of proportions which, though all different from the true one, approach to it so insensibly that they can deceive the most able and the most exact observers.[23]

---

[20] Fermat to La Chambre, 1 January 1662, F, II, p. 460.

[21] Thus we read in his first letter to La Chambre of August 1657: 'The use of these composed movements is a very delicate matter which must not be handled and employed except with very great precaution. I compare them to some of your medicines which serve as poison if they are not well and duly prepared. It is therefore sufficient for me to say in this place that M. Descartes has proved nothing, and that I share your opinion in that you reject his' (F, II, p. 356). See also his letter to La Chambre 1 January 1662 (F, II, pp. 457–8), and his letter to M. de ★★★, 1664, (F, II, pp. 485–6).

[22] F, II, p. 458.

[23] F, II, p. 461. See also ibid., pp. 486–7.

That is to say, he hoped that the true proportion might be one that was very near to Descartes' though not identical with it. Such a proportion would be as much in agreement with experiment as the one proposed by Descartes, since the difference would be too small to be detected. But, then, how would Fermat know that his own proportion would be 'the true' one? Because 'nothing was so plausible and so obvious'[24] as the principle from which he was going to deduce it, whereas Descartes' proportion was based on a pseudo-demonstration. The question of the *truth* or *falsity* of the law of refraction was thus, in Fermat's view, not to be decided by experiment alone; the law had to be based on a more solid foundation by correctly deducing it from an assumption whose truth was, for him, beyond doubt. This attitude towards establishing an experimental law was not peculiar to Fermat; we shall see[25] that Newton also was not content to accept the truth of the sine law *as far as it appeared from experiments,* but wanted to prove its *accurate truth* by deducing it mathematically from an assumption yielding the exact ratio.

Having got over that obstacle, Fermat then set out to attack the technical difficulties connected with his problem. As he first conceived it, the problem presented four lines by their square roots and thus involved four asymmetries which would have required calculations of great length. After repeated attempts he finally succeeded in reducing the four asymmetries to only two, 'which relieved me considerably'.[26] When he reached the end of his calculations he found himself face to face with Descartes' sine ratio:

the reward for my effort has been the most extraordinary, the most unforeseen and the happiest that ever was. For, after having gone through all the equations, multiplications, antitheses and other operations of my method, and having finally concluded the problem which you will see in a separate paper, I have found that my principle gave exactly and precisely the same proportion for refractions as that which M. Descartes has laid down.

I have been so surprised by this unexpected result that I find it difficult to recover from my astonishment. I have repeated my algebraic calculations several times and the result has always been the same.[27]

---

[24] F, II, p. 460.　　[25] Below, p. 299.　　[26] F, II, p. 461.　　[27] Ibid., pp. 461–2.

3. Before I give an account of Fermat's proof as presented in the work referred to here, it will perhaps be useful to describe briefly his method of maxima and minima.

Suppose[28] that, in a given problem, a is the unknown to be determined by the condition that a certain quantity is maximum or minimum. To find a, Fermat first expresses the maximum or minimum quantity in terms of a—let us call this expression M. He substitutes $a+e$ for a in M (where e is a very small quantity) and thereby obtains another expression M'. The two expressions M and M' may then be compared as if they were equal. To denote this operation Fermat used a term (*adaequare*) meaning approximate equation.[29] The quasi-equation $M=M'$ is then divided by e or by one of its higher powers so as to make e completely disappear from at least one of the terms. Omitting the terms in which e remains after division, the result will be an equation whose solution gives the value of a.[30]

To illustrate this by a simple example given by Fermat, suppose that b is a straight line to be divided into two segments a and $b-a$, such that the product

$$a(b-a) = ba - a^2 = M, \text{ a maximum.}$$

Let the first segment equal $a+e$; the second will be equal to $b-a-e$; and their product will be given by:

$$ab - a^2 + be - 2ae - e^2 = M'.$$

Equating M and M', and cancelling the common terms:

$$be = 2ae + e^2 \text{ approx.}$$

---

[28] See Fermat's account (written before 1638) entitled '*Methodus ad Disquirendam Maximam et Minimam*', in F, I, pp. 133–6; French translation, F, III, pp. 121–3.

[29] On the relation of Fermat's term to Diophantus' παρισότης, see F, I, p. 140 and p. 133, editors' note 2. Cf. T. L. Heath, *Diophantus of Alexandria*, 2nd ed., Cambridge, 1910, pp. 95–98.

[30] It is seen that Fermat's method contains the fundamental idea of the differential calculus. He principally made use of this method in the construction of tangents to curves. An account of it was published for the first time in 1644 (see Pierre Hérigone, *Cursus mathematicus*, VI (Paris, 1644), pp. 59–69 of the *Supplementum* beginning after p. 466 with separate pagination). We gather from Fermat's correspondence that he invented his method in, or shortly before, 1629; see Paul Tannery, 'Sur la date des principales découvertes de Fermat', in *Bulletin des sciences mathématiques et astronomiques*, 2ᵉ série, VII (1883), tome XVIII de la collection, première partie, p. 120. It had not been known for certain whether Newton was acquainted with Fermat's ideas until L. T. More brought attention, in 1934, to a letter of Newton's in which he wrote: 'I had the hint of this method [of fluxions] from Fermat's way of drawing tangents, and by applying it to abstract equations, directly and invertedly, I made it general.' Cf. Louis Trenchard More, *Isaac Newton*, A biography, New York and London, 1934, p. 185, note 35. See also Newton, *Correspondence*, ed. Turnbull, II, p. 167, n. 4; III, pp. 182 and 183.

Dividing by e, we have:
$$b = 2a + e \text{ approx.}$$
Omitting e, we get:
$$b = 2a.$$
That is, on the condition supposed, the line should be divided in the middle.

It took Fermat more than four years and a great deal of effort before he could see his way to applying this method to the problem of refraction. The result was the following demonstration.[31]

In the figure (Fig. V. 3) the diameter AB represents the interface between the rare medium above and the dense medium below. The incident ray CD meets AB at the middle point D; CF is perpendicular to AB; m is a line given outside the circle.

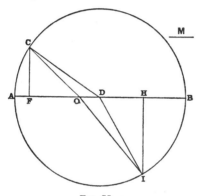

FIG. V. 3

If R_i and R_r be the 'resistances' of the medium of incidence and of refraction respectively, it is assumed that

(1) $$\frac{R_r}{R_i} = \frac{DF}{m} = K > 1.$$

On the supposition that the light follows the *easiest* path, the problem of constructing the refracted ray consists in finding a point H on the diameter AB such that, having drawn the perpendicular HI and joined ID, the sum
$$CD.m + DI.DF$$

---

[31] See Fermat's '*Analysis ad Refractiones*', in F, I, pp, 170–2; French translation, F, III, pp. 149–51.

is a minimum. For this sum may be taken to represent the 'move-ment' along CDI.

Let $\quad\quad\quad\quad n = CD = DI,$

$\quad\quad\quad\quad\quad\quad b = DF,$

and $\quad\quad\quad\quad a = DH.$

a is determined by the condition:

(2) $\quad\quad\quad\quad nm + nb = $ a minimum.

To find a, Fermat takes a point O on the diameter AB such that:

$\quad\quad\quad DO = e = $ a very small quantity.

Joining CO and OI, we have:

(3) $\quad\quad\quad CO^2 = n^2 + e^2 - 2be,$ and,

(4) $\quad\quad\quad OI^2 = n^2 + e^2 + 2ae.$

Taking the square roots of (3) and (4); and multiplying (3) by m, and, (4) by b:

(3') $\quad\quad\quad CO.m = \sqrt{m^2n^2 + m^2e^2 - 2m^2be,}$

(4') $\quad\quad\quad IO.b = \sqrt{b^2n^2 + b^2e^2 + 2b^2ae.}$

Now, according to Fermat's method, we may write:

$\quad\quad\quad CO.m + IO.b = nm + nb.$

Therefore,

$$\sqrt{m^2n^2 + m^2e^2 - 2m^2be} + \sqrt{b^2n^2 + b^2e^2 + 2b^2ae} = nm + nb.$$

From this equation the following is obtained by repeated squaring and cancelling of common terms:[32]

$$8b^2e^2m^2n^2 \quad\;\; - 8b^3m^2n^2e$$
$$+ 2e^4b^2m^2 \quad\;\; - 4b^3e^3m^2$$
$$+ 8b^2m^2n^2ae + 4e^3b^2m^2a$$
$$\quad\quad - 8b^3e^2m^2a = -4m^3e^2n^2b + 8m^3b^2n^2e$$
$$\quad\quad\quad\quad\quad\quad\quad\quad - 4b^3e^2n^2m - 8b^3n^2aem$$
$$\quad\quad\quad\quad\quad\quad\quad\quad - 4m^4e^3b \quad + 4b^4e^3a$$
$$\quad\quad\quad\quad\quad\quad\quad\quad\quad + e^4m^4 + 4m^4b^2e^2$$
$$\quad\quad\quad\quad\quad\quad\quad\quad\quad + b^4e^4 + 4b^4a^2e^2.$$

Dividing by e, and omitting the terms in which e (or one of its higher powers) remains, we get:

$$8b^2m^2n^2a - 8b^3m^2n^2 = 8m^3b^2n^2 - 8b^3n^2am,$$

---

[32] The following deductions are not given by Fermat; he describes, however, the steps leading to conclusion (5).

$$8mb^2n^2(ma-bm)=8mb^2n^2(m^2-ba),$$
$$ma-bm=m^2-ba,$$
$$a(m+b)=m(m+b),$$
$$a=m.$$

Therefore, by (1),

$$\frac{DF}{a}=K, \text{ a constant.}$$

Since

$$\frac{DF}{CD}=\sin i,$$

and

$$\frac{a}{DI}=\frac{DH}{DI}=\sin r$$

(i and r being, respectively, the angles made by CD and DI with the normal to AB at D), it follows that

(5) $$\frac{\sin i}{\sin r}=K,$$

'which absolutely agrees with the theorem discovered by Descartes; the above analysis, derived from our principle, therefore gives a rigorously exact demonstration of this theorem.'[33]

It will have been observed that in the above demonstration the word *time* does not occur. Perhaps Fermat wanted to formulate his proof in terms that would be acceptable to his friend La Chambre.[34] But suppose now we write—with Descartes:

$$\frac{R_r}{R_i}=\frac{v_r}{v_i},$$

that is, the resistances are *directly* proportional to the velocities. It would then follow, by assumption (1) and conclusion (5) in Fermat's demonstration, that

$$\frac{\sin i}{\sin r}=\frac{v_r}{v_i}=K.$$

This result is identical with that obtained in the Cartesian proof of the sine law; it means that the velocity of light is greater in denser (and more resisting) media. Had Fermat adopted Descartes' opinion concerning the velocity of light in different media, his

---

[33] F, I, p. 172; F, III, p. 151. [34] See above, p. 141, n. 19.

principle of *easiest course* would have been identical with a principle that was proposed by Leibniz in 1682 *against* the principle of least time. Leibniz suggested[35] that light follows the path of least 'resistance', that is, the path for which the sum of the distances covered, each multiplied by the 'resistance' of the medium, is a minimum. In the case of two homogeneous media, if s_i and s_r are the paths covered in the medium of incidence and of refraction, and R_i, R_r the 'resistances' of the media, then, according to him:

$$R_i.s_i + R_r.s_r = \text{a minimum.}$$

If we let R_i and R_r be represented, in Fermat's figure (Fig. V. 3), by m and DF respectively, and let CD and DI represent s_i and s_r, it is seen that Leibniz's principle yields the same condition as Fermat's condition (2) above in accordance with which the latter deduced the sine law. But Leibniz further assumed, however, in further agreement with Descartes, that light travels faster in denser and more resisting media. He argued that as the particles of such media are close together, they prevent the light from being diffused and thus cause its flow to be accelerated, just as a stream of water runs faster as the passage becomes narrower.

This, however, was not Fermat's intention in the preceding *Analysis*. To him, 'resistance' simply meant the reciprocal of velocity. His initial assumption (1) was, therefore, that the velocity of the light in the (rare) medium of incidence was greater than the velocity in the (dense) medium of refraction; and the quantity assumed to be a minimum—see condition (2)—simply measured the sum of the *time* along the incident and refracted rays. Consequently, his surprise at obtaining Descartes' ratio was all the greater since, to his understanding, Descartes had 'supposed' in his deductions that the velocity was greater in denser media.[36]

This way of looking at his own and Descartes' demonstrations had the result of confirming him in his belief that the Cartesian proof was logically incorrect. Thus, at the beginning of his *Synthesis for*

---

[35] Cf. Leibniz, '*Unicum opticae, catoptricae, & dioptricae principium*', in *Acta eruditorum, Leipzig,* 1682, pp. 185–90; R. Dugas, *Histoire de la mécanique,* p. 249.
[36] Fermat to La Chambre, 1 January 1662, F, II, p. 462.

Refractions,[37] after having remarked that he and Descartes had started from contrary suppositions regarding the velocity of light in different media, he asked whether it was possible 'to arrive, without paralogism, at one and the same truth by two diametrically opposed ways'.[38] This question he left in that place for the consideration of the 'more subtle and more rigorous Geometers', so that he might not involve himself in 'vain discussions' and 'useless quarrels'.[39] Later, however, in the letter to M. de *** of 1664, he made known his own answer to his query: after he had arrived at Descartes' ratio (from the principle of least time and consistently with the supposition that the velocity was greater in the rarer medium) and having made sure that there was no mistake in his deductions, he became convinced of two things, 'the one, that the opinion of M. Descartes on the proportion of refractions is very true; and the other, that his demonstration is very faulty and full of paralogisms'.[40]

Fermat was certainly mistaken in drawing the latter conclusion. What he overlooked was, strangely enough, that his principle does *not*, strictly speaking, yield the same law which is obtained in Descartes' proof, though both laws assert the constancy of the ratio of the sines. Thus in his *Synthesis for Refractions*, he attributed to Descartes the theorem which can be expressed as follows:[41]

(A)
$$\frac{\sin i}{\sin r} = \frac{v_i}{v_r} = n.$$

As we have seen in the preceding chapter, Descartes' theorem is on the contrary:

(B)
$$\frac{\sin i}{\sin r} = \frac{v_r}{v_i} = n.^{42}$$

---

[37] Cf. Fermat, '*Synthesis ad Refractiones*', in F, I, pp. 173–9; French translation, F, III, pp 151–6. This work was written after the '*Analysis ad Refractiones*', and was sent to La Chambre in February 1662. See Fermat to La Chambre, 1 January 1662, F, II, p. 463 (last sentence in P.S.); and F, I, p. 173, editors' note 1.

[38] F, I, p. 173; F, III, p. 152.

[39] Ibid.

[40] F, II, p. 488.

[41] Cf. F, I, p. 175; F, III, p. 153.

[42] It is curious to note that Fermat thus failed to realize that he had in fact achieved what he had originally hoped for—namely, to deduce a law of refraction different from that of Descartes, though the difference was not of the kind he had expected.

Now (A) is a necessary result of Fermat's principle of least time;[43] and since light in going from a rare into a dense medium is bent towards the normal (the angle of refraction being thereby made smaller than the angle of incidence, and its sine less than the sine of incidence), one has to conclude (rather than assume), in accordance with (A), that the velocity in the rare medium of incidence (v_i) must be greater than the velocity in the dense medium of refraction (v_r). For the same reason the contrary conclusion has to be drawn from (B) with which this conclusion is, therefore, perfectly consistent.

4. The purpose of Fermat's *Synthesis* is to deduce from (A) the proposition that the actual path of light in refraction is that for which the time of the movement is a minimum, while being consistent with the supposition—itself a consequence of (A)—that the velocity is greater in rarer media. The undoubted validity of Fermat's deductions does not, however, imply—as he thought—that Descartes' proof is logically inconclusive. The following is an account of Fermat's *Synthesis*.

The diameter ANB (Fig. V. 4) separates the rare medium of incidence above from the dense medium below; MN is an incident ray that is refracted into NH; MD and HS are perpendiculars.

R is any point (other than N) of the diameter AB—taken, for example, on the radius NB.

---

[43] This is already clear from the preceding demonstration by 'analysis'. Bearing in mind what has been remarked above concerning Fermat's understanding of resistance, his assumption (1), expressed in terms of v_i and v_r, should be written thus:

$$\text{(i)} \quad \frac{v_i}{v_r} = \frac{DF}{m} = n.$$

Joining this with his conclusion (5), we get:

$$\text{(ii)} \quad \frac{\sin i}{\sin r} = \frac{v_i}{v_r} = n.$$

It is to be noted that in deriving this result it is not logically necessary to postulate the condition, stated in Fermat's assumption (1), that n (v_i/v_r) is greater than unity. This condition is in fact obtained as a result of comparing (ii) with the actual behaviour of light in going from a rare into a dense medium.

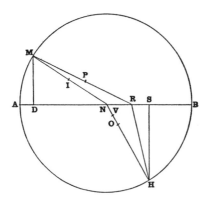

FIG. V. 4

It is assumed that

(1)
$$\frac{DN}{NS}=\frac{\sin i}{\sin r}=\frac{v_i}{v_r}=\frac{MN}{NI}=\frac{MR}{RP}=n>1$$

(where i and r are the angles respectively made by MN and NH with the normal to AB at N; and v_i, v_r, the velocities of incidence and of refraction).

It is to be proved that the time along MNH is less than the time along MRH.

Since

$$\frac{\text{time along } MN}{\text{time along } NH}=\frac{MN}{NH}\times\frac{v_r}{v_i}=\frac{MN}{NH}\times\frac{NI}{MN}=\frac{NI}{NH},$$

and

$$\frac{\text{time along } MR}{\text{time along } RH}=\frac{MR}{RH}\times\frac{v_r}{v_i}=\frac{MR}{RH}\times\frac{RP}{MR}=\frac{RP}{RH},$$

therefore,

$$\frac{\text{time along } MNH}{\text{time along } MRH}=\frac{NI+NH}{RP+RH}.$$

It is now to be proved that
$$RP+RH>NI+NH.$$

Let

(2)
$$\frac{MN}{DN}=\frac{RN}{NO}\text{ and }\frac{DN}{NS}=\frac{NO}{NV}.$$

Since, by construction,
$$DN<MN\text{ and }NS<DN,$$

we have:

(3) $$NO < RN \text{ and } NV < NO.$$

But,

$$MR^2 = MN^2 + RN^2 + 2DN.RN.$$

Therefore, by (2),

$$MR^2 = MN^2 + RN^2 + 2MN.NO.$$

From which it follows, by (3), that

$$MR^2 > MN^2 + NO^2 + 2MN.NO = (MN + NO)^2.$$

Therefore,

(4) $$MR > MN + NO.$$

But, from (1) and (2),

$$\frac{DN}{NS} = \frac{MN}{NI} = \frac{NO}{NV} = \frac{MN + NO}{NI + NV} = \frac{MR}{RP}.$$

Therefore, by (4),

$$RP > NI + NV.$$

It now remains to be proved that

$$RH > HV,$$

for then it will follow that

$$RP + RH > NI + NV + HV,$$
$$RP + RH > NI + NH.$$

In the triangle NHR,

(5) $$RH^2 = NH^2 + NR^2 - 2NS.NR.$$

From (2)

$$MN(= NH).NO = DN.NR,$$
$$NS.NO = DN.NV.$$

Therefore,

(6) $$\frac{NH}{NS} = \frac{NR}{NV}.$$

From (5) and (6),

$$RH^2 = NH^2 + NR^2 - 2NH.NV$$

But since, by (3),

$$NR > NV, \quad NR^2 > NV^2,$$

it follows that

$$RH^2 > HN^2 + NV^2 - 2NH.NV,$$

and we finally have:

$$RH > HN - NV, \quad RH > HV,$$

which was to be proved.

Fermat provides a similar proof for the case in which R is one of the points on the radius AN.

Having satisfied himself as to the truth of Descartes' ratio, Fermat offered to make peace with the Cartesians on the following terms:[44] Descartes had made an important discovery; this, however, he had achieved 'without the aid of any demonstration'; nature, as it were, had not dared to resist 'this great genius'; as in those battles which were won on the sole strength of the conqueror's reputation, nature had surrendered her secret to him without being forced by demonstration; the glory of victory was Descartes' and Fermat was 'content that M. Clerselier should at least let me enter into the society of the proof of this very important truth'.

Peace under these conditions was not, of course, acceptable to Clerselier.[45] He admitted that Fermat's proof was logically valid; but so, he (rightly) maintained, was Descartes'. The difference between them, in his view, was that whereas Descartes founded his proof on physical considerations, the principle which formed the basis of Fermat's demonstration was 'only a moral and not a physical principle, which is not and cannot be the cause of any natural effect'.[46] To hold this principle was to ascribe knowledge and choice to nature, whereas nature in fact acted by the 'force' residing in all things and the dispositions of bodies to receive the action of this force—in short, by a 'necessary determinism'.[47] Once a body was set in motion, its speed and 'determination' depended solely on the 'force' of its movement and the 'disposition' of that force.[48] Only motion in a straight line was naturally determined (in accordance with the law of inertia formulated by Descartes). Thus a ray of light would always travel in the same direction so long as it remained in the same medium. It would change direction at the meeting of a refracting surface, not for the sake of Fermat's principle (for, then, the ray would have to remember its point of departure, choose a certain destination, and perform the necessary calculations!), but

---

[44] Fermat to La Chambre, 1 January 1662, F, II, p. 462.
[45] Cf. Clerselier's letter to Fermat, 6 May 1662, F, II, pp. 464–72. He wrote this letter after having seen Fermat's letter to La Chambre (1 January 1662) and also Fermat's 'Synthesis for Refractions'. [46] F, II, p. 465. [47] Ibid. [48] Ibid., p. 466.

because the different disposition of the refracting body would of necessity alter the ray's force and determination—in the manner that Descartes had shown. As time was not a cause of movement, the shortness of time along the lines of incidence and of refraction could not be the cause of the light's following those lines. Fermat's principle was, therefore, metaphysical, imaginary and had no foundation in physics. Moreover, his view that light passed more easily through rarer media mistakenly ignored Descartes' physical argument in support of the opposite view. Thus, far from Descartes' having conquered nature without the force of demonstration, it would, according to Clerselier, be more true to say that Fermat had forced geometry to adapt itself to false opinions.

It was in answer to this that Fermat wrote the following passage in his last letter to Clerselier of 21 May 1662:[49]

I believe that I have often said both to M. de la Chambre and to you that I do not pretend, nor have I ever pretended to be in the inner confidence of Nature. She has obscure and hidden ways which I have never undertaken to penetrate. I would have only offered her a little geometrical aid on the subject of refraction, should she have been in need of it. But since you assure me, Sir, that she can manage her affairs without it, and that she is content to follow the way that has been prescribed to her by M. Descartes, I willingly hand over to you my alleged conquest of physics; and I am satisfied that you allow me to keep my geometrical problem—pure and *in abstracto*, by means of which one can find the path of a thing moving through two different media and seeking to complete its movement as soon as it can.[50]

5. It would be a mistake to conclude from the preceding passage that Fermat really looked at his efforts as having resulted merely in the solution of an abstract geometrical problem. The ironical tone of his words is in fact more telling about his real beliefs than what his statements might appear to say. If he expressed his willingness to hand over his conquest of physics to Clerselier, that was only in order to end a controversy which had been getting to be increasingly unpleasant. In fact, he himself indicated immediately after the above

---

[49] Cf. F, II, pp. 482-4.   [50] Ibid., p. 483.

passage that he still found it improbable that he could have arrived, so unexpectedly, at Descartes' true ratio from a false principle, while making an assumption that was directly opposed to the corresponding assumption of Descartes. It must also be remembered that Fermat originally set out to discover *the true* law of refraction; he was not concerned to deduce an arbitrary law from an arbitrary assumption. When, to his great surprise, he finally obtained Descartes' law (in the form sin $i = n$ sin r), he interpreted his calculations as a 'demonstration' of an important truth. As we have seen[51], a demonstration for Fermat was not simply a valid inference from supposed premises, but a deduction from *true* principles. It is very doubtful that he would have embarked on the difficult task of applying his method of maxima and minima to the problem of refraction, had he not been convinced that the general principle of economy, which inspired this application, was a true law of nature. Indeed he was not convinced of the truth of the refraction law itself, even after it had been confirmed by experiments, until he had derived it from his principle of least time. It would not, therefore, be correct to suppose, as his preceding passage might lead us to do, that he performed his deductions in the interests of pure geometry alone. Rather, he undertook the solution of the problem suggested to him by the principle of shortest course as a mathematician whom chance seemed to have presented with an opportunity to render an important service to physics, to found the science of dioptrics on a sound basis, and not merely to provide it with a tool for calculation. In accordance with this aim, he had understood that principle as a *true physical principle*, not as a mathematical hypothesis.

Fermat started from a principle of economy which had had a long history before him. In its most general form, it asserted that nature did nothing in vain. As applied to reflection at plane surfaces it had been interpreted as a principle of *shortest distance*. Since this interpretation, however, could not be made to fit the phenomenon of refraction, Fermat substituted a new specification, namely that of *easiest course* which he understood to mean *least time*. Heron's theorem

---

[51] Above, pp. 129f, 142f.

would then be shown to be only a special case of Fermat's principle and, therefore, in agreement with it. But this was not the only meaning which Fermat attached to his principle. In order to answer the objection that reflection at concave surfaces did not satisfy the law of shortest course, he had to appeal to a different interpretation in which considerations of *simplicity* replaced the shortness of distance and of time.[52] The suggestions he made in this connection could only help to evade the problem which was a real one. In fact, Fermat's principle is now accepted as generally true only if understood as a principle of extremum, that is, of maximum or minimum.[53] He was bound to avoid this formulation as its admission would have conflicted with the metaphysical reasoning which had determined his approach.

Supposing, however, that the principle of economy was generally true, there was still no *a priori* reason why time, rather than any other quantity, should be the minimum. As it happened, however, Huygens later showed in his *Traité de la Lumière* (1690) that the principle of least time was deducible from his own law of secondary waves.[54] Fermat's principle thus received a physical foundation by being shown to follow from wave considerations. Before the appearance of Huygens' *Traité*, Newton had already published in *Principia* a dynamical explanation of refraction which required the velocity of light to be greater in denser media[55]—in contradiction to Huygens' theory and Fermat's principle. In the eighteenth century Newton's explanation was accepted as part of a corpuscular theory of light which had completely overshadowed Huygens' theory; and Fermat's principle was regarded as a detached metaphysical principle to be examined on its own merits, or, rather, demerits. Final causes had not then completely disappeared from physical considerations; but the adherents of Newton's theory could point to an arbitrary character of Fermat's principle: if light could not at once travel in

---

[52] See above, p. 139, n. 14.

[53] There are also cases in which the path of reflected light is neither a minimum ncr a maximum but stationary, as when the reflecting surface is an ellipsoid of revolution and the two points related as object and image are the foci.

[54] Below, pp. 218f.

[55] Below, pp. 301ff.

the shortest time by the shortest path, why should it choose one of these courses rather than the other?—that was Maupertuis' argument in 1744. Instead of the shortness of either distance or time, he proposed that the path really followed by light is that for which the *quantity of action* is a minimum.[56] Thus Maupertuis accepted the principle of economy, but he offered a new interpretation. By *action* he understood a quantity proportional to the sum of the distances covered, each multiplied by the velocity with which it is traversed; and he showed that his principle of least action (as he called it) leads to the Newtonian law of refraction giving the sines of incidence and of refraction in the inverse ratio of the velocities.

Maupertuis' procedure was no less metaphysical than Fermat's. In a sense it was even more arbitrary. For whereas Fermat could have at least argued that time was a natural and basic concept whose intimate relation to motion should recommend it as the minimum quantity,[57] Maupertuis' 'action' was an artificially constructed quantity introduceed to fit the already formulated Newtonian law. Similarly to what had happened with Fermat's principle, however, Euler showed (in 1744) that the principle of least action was applicable in several important cases of particle dynamics, as, for example, when a body moves under the action of a central force.[58] And later,

---

[56] Cf. R. Dugas, *Histoire de la mécanique*, p. 252.

[57] Galileo had made use of such an argument in connection with his law of naturally accelerated motion. Having stated that nature employs in her various processes 'only those means which are most common, simple and easy', he concluded: 'When, therefore I observe a stone initially at rest falling from an elevated position and continually acquiring new increments of speed, why should I not believe that such increases take place in a manner which is exceedingly simple and rather obvious to everybody? If now we examine the matter carefully we find no addition or increment more simple than that which repeats itself always in the same manner. *This we readily understand when we consider the intimate relationship between ime and motion*; forj ust as uniformity of motion is defined by and conceived through equal times and equal spaces . . . so also we may, in a similar manner, through equal time-intervals, conceive additions of speed as taking place without complications . . .' *Dialogues Concerning Two New Sciences*, ed. H. Crew and A. de Salvio, New York, 1914, p. 161; my italics. Fermat himself quoted the example of Galileo twice in support of his principle of least time; cf. Fermat to La Chambre, August 1657, F, II, p. 359; 'Synthesis ad Refractiones', F, I, p. 173 French translation, F, III, p. 152.

[58] Cf. R. Dugas, *Histoire de la mécanique*, Ch. V, Sec. 11. Euler in fact started from the assertion that 'all the effects of Nature follow some law of maximum or minimum' (ibid., p. 262), in violation of the principle of economy. But although he presented his principle of extremum as a law to be verified *a posteriori*, he still sought for metaphysical justifications (cf. ibid., p. 263).

Lagrange extended Euler's applications of Maupertuis' principle (as a law of extremum) which he regarded, not as a metaphysical explanation, but as a 'simple and general result of the laws of mechanics'.[59]

Thus, as far as optical refraction is concerned, the incompatibility of Fermat's principle with that of Maupertuis became a conflict, not between two interpretations of a metaphysical view of nature, but between the wave and the dynamical (or corpuscular) theories of light in which these two principles were respectively verified. Until 1850 there had been no experiment to decide between the wave and the corpuscular explanations of refraction. When, in that year, Foucault performed his famous experiment, he found that the velocity of light was greater in air than in water—in agreement with Huygens' explanation and against the Newtonian view.

---

[59] Ibid., p. 327.

Chapter Six

HUYGENS' CARTESIANISM AND HIS THEORY
OF CONJECTURAL EXPLANATION

1. Huygens' commitment to a Cartesian point of view in physics was most clearly expressed in the course of a controversy which took place at the *Académie Royale des Sciences* in 1669. Although this controversy was concerned with the cause of gravity, not with light, a discussion of Huygens' views on the former subject will be of help towards a better understanding of the statements relating to physical explanation and hypotheses which later appeared in his *Traité de la Lumière*.

The chief opponent of Huygens was the great enemy of Descartes, Roberval. His memoir, which served as the basis of discussion, was delivered on 7 August 1669.[1] It explained that the gravity of a body is 'that which inclines this body to descend towards a centre naturally and without artifice.'[2] According to Roberval one may, in this generalized sense, recognize gravities other than terrestrial, for example, lunar, or solar, or jovial. This does not mean, however, that he conceived the idea of universal gravitation: he did not speak of the earth, the sun, and the other heavenly bodies as gravitating towards one another. Rather, the gravities which he had in mind were between bodies belonging to one whole; that is, terrestrial bodies gravitate towards the earth, solar bodies towards the sun, and so forth.[3]

---

[1] Cf. H, XIX, pp. 628–30.

[2] Ibid., p. 628.

[3] Thus we read on the first page of his *Traité de méchanique* (1636): '. . . the line of direction of a freely falling body is the line drawn from the centre of gravity of the same heavy body to the natural centre of ponderable things which, for terrestrial bodies, is the centre of the earth.' The idea that there is a *natural* centre for heavy bodies which varies according as the body is terrestrial, or jovial, etc., cannot be reconciled with the idea of universal gravitation. For it would imply, for example, that a terrestrial body does not gravitate towards Jupiter or towards the sun. Roberval's concept, which is to be distinguished also from Aristotle's conception of heavy bodies tending towards the *centre of the world*, as a point determined independently of the nature of the body occupying it, is immediately derived from Copernicus.

As Newton later found it necessary to point out in the *Principia*, Roberval remarked that one did not have to attribute a certain virtue to the centre 'which is but a point';[4] it was sufficient to understand that all the parts of a body tended to be united together so as to form a single body. 'For there will result from that a centre of gravity towards which all these parts will be directed with greater or lesser force, according to their proper nature: and it is in this force that gravity consists.'[5]

He then distinguished three views regarding the nature of gravity. According to the first view, gravity resides only in the gravitating body owing to a certain quality which drives it downwards (that is, towards a natural centre). On the second view, gravity is a quality common to all parts of a body, and in virtue of which all those parts attract one another. The third view ascribed gravity to the action of an external agent, a subtle matter, which either causes the heaviness of a body by pushing it downwards, or its lightness by pushing it upwards. The first two opinions agreed in that they made gravity or heaviness the first and independent cause of the movement downwards, and lightness the first and independent cause of the movement upwards. As distinguished from these, the third opinion considered *motion* to be the cause of both heaviness and lightness.

Roberval granted that all three opinions had no foundation other than in the imagination of their authors. Yet, the first two appeared to him more easily tenable, and he himself preferred the second. His main reason was that experiments could be made to test it. For example, if it were true that bodies mutually attracted one another, then it would follow that the same body would weigh less near the centre of the earth than at the circumference (where, according to him, the weight would be greater than at any other point on either side of it).[6] This, he pointed out, could be tested by a spring balance.

Cf. on these distinctions Alexandre Koyré, *La gravitation universelle de Képler à Newton*, Paris, 1951.

[4] H, XIX, p. 628. Compare Newton, *Principia*, Definition VIII, pp. 5–6: 'the reader is not to imagine . . . that I attribute forces, in a true and physical sense, to certain centres (which are only mathematical points); when at any time I happen to speak of centres as attracting or as endowed with attractive powers.'

[5] H, XIX, p. 628.

[6] Cf. ibid., p. 629. Roberval's reason why the weight would decrease as the body approached

It would also follow, in his view, that a huge and very high mountain would deflect a small hanging weight from the vertical direction towards it. But, finding these tests rather difficult to realize, he left the judgement on this matter to the 'spéculatifs'.

Nevertheless, if one admits occult qualities; that is, qualities for which we have no proper and specific sense, this second opinion seems to me to be the most plausible of the three. But it is also possible that the three opinions are all false, and that the true one is unknown to us.[7]

Roberval then concluded his discourse with the following declaration which bears a certain similarity to Newton's views on the same subject as they were expressed in the General Scholium to the *Principia*:

. . . I shall always do my best to imitate Archimedes who, on this occasion of gravity, lays down, as a principle or postulate, the invariable fact which has to the present day been verified throughout the past centuries, that there are heavy bodies, which possess the conditions of which he speaks in the beginning of his treatise on this subject; and on this foundation I shall, as he did, establish my reasonings for Mechanics, without bothering to know at bottom the principles and the causes of gravity, waiting to follow the truth, if it suits her to reveal herself one day clearly and distinctly in my mind.[8]

This, Roberval said, was the maxim which he would always follow in uncertain matters, if he was not obliged to take sides in them.

In contrast to Roberval's attitude, the question for Huygens (who

---

the centre from the circumference was, presumably, that the attraction towards the central parts of the earth would be gradually weakened by the opposite attraction towards the parts near the circumference.

[7] Ibid., p. 629.

[8] Ibid., pp. 629–30. Both Roberval and Newton found it sufficient for their limited purposes to consider that gravity was a fact—Newton, asserting in the General Scholium that 'to us it is enough that gravity does really exist' (*Principia*, p. 547) and Roberval, pointing out that 'there are heavy bodies'. Both preferred not to commit themselves to any view regarding the cause of gravity, either for want of a clear and distinct idea as to the nature of that cause (Roberval), or for lack of experiments (Newton). Neither one nor the other denied that it was within the limits of science to search for such a cause. They differed, however, in that Newton extended gravity to the solar system as a whole (and, indeed, to all parts of matter everywhere), and that, unlike Roberval, he explicitly denied that attraction could be a primary quality of matter.

read his discourse on 28 August 1669)[9] was to look for an 'intelligible cause of gravity' without supposing in nature anything other than 'bodies made up of the same matter, in which one does not recognize any quality, or tendency to approach one another'; that is, bodies which differ merely by their 'magnitudes, figures, and movements'.[10] From these components alone, and without resorting to forces or occult qualities, one had to explain how several bodies 'directly tend towards the same centre, and there remain assembled around it, which is the most ordinary and principal phenomenon of what we call gravity'.[11]

Thus Huygens starts from a criterion of intelligibility which leads him to banish tendencies and attractions from physical considerations.[12] He accepts the Cartesian view according to which explanations in physics ought to be in terms of the 'intelligible', geometrical properties: magnitude, figure and motion. In consequence, one has to assume the existence of certain invisible corpuscles which have only those admissible properties, and which in some way cause the gravity of bodies. And the problem to be solved is then the following: On which of these properties does gravity depend, and how? According to Huygens, the simplicity (or, rather, scarcity) of these principles did not leave much room for choice. For, as it did not appear to him that gravity could be due to the magnitude or figures of the corpuscles, he concluded that the phenomenon, 'being a tendency or inclination to movement, must probably [*vraysemblablement*] be produced by a movement'.[13] What kind of movement? There were only two in nature, the straight and the circular.[14] But, as there was nothing in the nature

---

[9] Cf. H, XIX, pp. 631–40.

[10] Ibid., p. 631.

[11] Ibid.

[12] At a later meeting of the *Académie des Sciences* (23 October 1669), Huygens defended his position, against a previous objection from Roberval, in the following words: 'I exclude attractive and repulsive qualities from nature because I am looking for an intelligible cause of gravity'. H, XIX, p. 642.

[13] H, XIX, p. 631.

[14] Huygens' assertion that there existed a circular motion in nature does not mean, as A. E. Bell understood in his *Christian Huygens* (London, 1947, pp. 64 and 164), that he believed (in 1669) this form of motion to be 'fundamental' or 'natural'. The same misunderstanding was implied in an objection of Roberval made during the session of the *Académie* on 4 September,

of the first or in the laws governing its communication or reflection that appeared to determine bodies to move towards a centre, he concluded that the required property must be looked for in the second. And he found it precisely in the inclination of a circulating body to recede from the centre of its motion. For this inclination of the body (its centrifugal force) increases with the speed of the motion. If now it is supposed that a fluid and very subtle matter circulates with great speed around the earth, the gross bodies which do not participate in this motion (or which participate in it to a lesser degree) will be less inclined to recede from the centre of the earth than the surrounding matter, and will thus be continually replaced by the parts of the fluid which lie next to them towards the centre. In this way the gross bodies which happen to be in the region around the earth that is filled with the circulating fluid, will be driven towards the centre; this accounts for their gravity.[15]

Huygens illustrated these ideas by an improved version of an 'experiment' that had been described by Descartes. The improvement is interesting as it shows Huygens trying to be more Cartesian than Descartes himself. A round vessel is filled with water (or some other liquid). A small sphere, of a substance whose density may be greater than the density of the liquid or equal to it, is immersed near the wall of the vessel. The sphere is hindered from moving freely by two stretched strings that allow it to move only along the diameter of the vessel. The vessel is then placed on a round table, which is then made to rotate. The motion will be gradually communicated to the liquid, but the sphere will be prevented by the strings from participating in the circular motion. If now the table is suddenly stopped, the liquid will continue to circulate for a while, but the

---

1669 (cf. H, XIX, p. 641, Art. 3). To this objection Huygens answered on 23 October 1669 (H, XIX, p. 643): '. . . I have not said that circular motion is natural but that such motion exists in the world, which cannot be denied.' The view attributed to Huygens by Bell would not have been reconcilable with his Cartesian belief that matter naturally tended to move only in straight lines. In fact the explanation given by Huygens (in the same discourse of 28 August 1669) of the circular motion of the fluid matter surrounding the earth is in principle identical with that of Descartes: namely, owing to the existence of 'other bodies' beyond it in space, the fluid matter is prevented from pursuing the rectilinear course which it would otherwise take, and, as a consequence, it is *forced* to take a curvilinear course (H, XIX, p. 634). Cf. Descartes, *Principles*, II, 37 and 39.
[15] H, XIX, pp. 632-3.

sphere will move towards the centre and, once there, will come to a stop.

The difference between this experiment and that of Descartes is that the latter had required the vessel to be filled with fine shots of lead (representing the circulating subtle matter), and the immersed body had been supposed to be pieces of a *lighter* material, wood.[16] But, Huygens objected, the motion of the wood towards the centre would in this case be 'the effect of the difference in gravity of the wood and the lead, whereas one should explain gravity without supposing any, and by considering all bodies to be made of one and the same matter'.[17] In Huygens' experiment, on the other hand, the weight (or gravity) of the sphere is immaterial (as it may be of the same density as the liquid) and, therefore, the effect is here produced, according to him, by 'motion alone'.[18] The intended improvement is thus introduced in the name of the Cartesian doctrine of the perfect homogeneity of all matter; and for Huygens himself, the superiority of his own experiment lay in that it provided for an explanation of gravity in terms of motion alone, as should be demanded by a consistent Cartesian.

Huygens expressed these views in 1669, that is about eighteen years before the appearance of Newton's *Principia* in 1687. It is, however, evident from the *Discours de la Cause de la Pesanteur*, which he published in the same volume with his *Traité de la Lumière* in 1690, that Newton's book had not altered his fundamental position. The *Discours* itself was only an amplified version of the memoir of 1669, apart from a final *Addition* which he wrote after he had read Newton's *Principia*.[19] It is true that in this *Addition* Huygens appears to be aware of more difficulties facing the Cartesian vortex theory than he had realized before. For example, having accepted the Newtonian extension of gravity to the whole of the solar system, he now had to take into account such complicated facts as the constant eccentricity of planetary orbits; the constant

---

[16] Cf. Descartes to Mersenne, 16 October 1639, D, II, pp. 593-4.
[17] H, XIX, p. 634.
[18] H, XIX, p. 633.
[19] The *Discours* together with the *Addition* are printed in H, XXI, pp. 445-88.

obliquity of their planes to one another; the fact that the motion of comets is often contrary to that of the planets; and so forth.[20] All this made him renounce the Cartesian vortices in their original form.[21] But it did not change his opinion as to the necessity of explaining gravity by vortex motions. The reason, no doubt, is that he still maintained his original view concerning the nature of physical explanation, which he had inherited from Descartes. Thus we read in a new Preface to the *Discours*:

Mr Des Cartes has recognized better than those who came before him that nothing would be better understood in physics than that which could be related to Principles not exceeding the capacity of our mind, such as those which depend on bodies, considered without qualities, and on their movements. But, as the greatest difficulty consists in showing how such a great variety of things are the result of these Principles alone, he has not greatly succeeded in that [task] when he proposed to treat of several particular subjects: among which, in my opinion, is that of gravity. . . . Nevertheless I confess that his trials, and his views, though false, have served to open the way for me towards what I have found on this same subject.[22]

And again in the same Preface:

I do not offer [this Discourse] as something beyond the reach of doubt, or as something against which one cannot make objections. It is too difficult to go that far in researches of this nature. It is, however, my belief that if the principal hypothesis on which I stand is not the true one, there is little hope that [such a true hypothesis] can be found, while remaining within the limits of the true and sane Philosophy.[23]

---

[20] Cf. H, XXI, pp. 472–3. For Huygens' proposals to meet these difficulties, see Mouy, *Le développement de la physique cartésienne*, pp. 258–63. See also H, XXI, pp. 429–41, editors' *Avertissement*.

[21] Huygens wrote the following note on comets in 1689, after he had made Newton's acquaintance during a visit to England (June–August, 1689): 'I am now almost of the opinion of Mr Newton who holds that the comets turn in very oblong ellipses round the sun which occupies one of the foci. This becomes very probable after he has removed the vortices of des Cartes which, besides, have failed to agree with several phenomena of planetary motions' H, XIX, p. 310). Cf. on Huygens' meeting with Newton: H, IX, p. 333, note 1; H, XXI, p. 435, note 31; David Brewster, *Memoirs of the Life, Writings, and Discoveries of Sir Isaac Newton* Edinburgh, 1885, I, p. 215; J. Edleston, *Correspondence of Sir Isaac Newton and Professor Cotes*, London, 1850, p. xxxi.

[22] H, XXI, p. 446.

[23] Ibid.

The 'principal hypothesis' mentioned here is one in which an invisible subtle matter is supposed to produce the effects of gravity. As to the 'true and sane Philosophy', this is the mechanical philosophy which aims to explain motion by motion alone, without bringing in attractive or other powers.[24] For although Huygens adopted Newtonian gravitation as extended to the solar system,[25] he still could not accept the idea which, as he pointed out, was assumed in some of Newton's calculations, namely that the constituent parts of different bodies attracted one another or tended to approach one another. His reason for rejecting this idea was 'that the cause of such an attraction is not explicable by any principle of Mechanics, or by the laws of motion,'[26] that is, the laws governing the communication of motion when bodies come into contact with one another.

Thus, up to the publication of his *Traité de la Lumière* in 1690, Huygens' position had remained unaltered: gravity and the other natural phenomena had to be explained in purely mechanical terms.[27] This he maintained as a result of his adoption of the Cartesian doctrine according to which physical explanations ought to be formulated with reference to a conceptual system which is determined by a fixed criterion of intelligibility, a system which would be made up of such words as magnitude, figure and motion. For Huygens, as for Descartes, no explanation was intelligible unless it could be completely expressed in these words. This amounts to the Cartesian requirement that all physical explanations should be in terms of push or contact action, a requirement which, in turn,

---

[24] In his *Traité de la Lumière* Huygens speaks of 'the true Philosophy in which the cause of all natural effects is conceived by mechanical reasons' (H, XIX, p. 461).

[25] Cf. *Addition* to the *Discours*, H, XXI, p. 472: 'I have ... nothing against the *Vis Centripeta*, as Mr. Newton calls it, by which he makes the Planets gravitate towards the Sun, and the Moon towards the Earth ... not only because we know from experience that such a manner of attraction or impulsion exists in nature, but also because it is explained by the laws of motion, as has been seen in what I have written above concerning gravity. For nothing prevents the cause of this *Vis Centripeta* towards the Sun from being similar to that which pushes bodies, that are called heavy, towards the Earth.'

[26] H, XXI, p. 471.

[27] It is gathered from Huygens' correspondence that he in fact persisted in this position until his death (in 1695). Cf. Huygens to Leibniz, 24 August 1694, H, X, p. 669; and H, XXI, p. 439, editors' *Avertissement*.

demands the postulation of some hypothetical matter whose motion or mechanical action is responsible for the production of natural phenomena. Hence Huygens' rejection of attraction as existing between the parts of bodies, while, at the same time, he accepted astronomical gravitation. He believed that whereas the latter could be explained by a complicated system of vortex motions, the former was not, in his view, capable of such an explanation.

Huygens' position does not imply, however, that he accepted the *a priori* part of Descartes' physical doctrine. For Descartes, since he starts from the identification of space and matter, the phrase *empty space* is a contradiction in terms; the existence of the void is thus denied in Descartes' physics as a result of an *a priori* definition giving the essence of matter. But this Huygens could not accept. Having expressed in the *Addition* his readiness to accept the astronomical facts as asserted in Newton's *Principia*, he pointed out one 'difficulty'.[28] Newton, he remarked, would only go so far as to allow the celestial space to contain an 'extremely rare matter',[29] so that the planets and the comets might easily move in them. But if that were assumed, Huygens objected, it would be impossible to explain either the action of gravity or that of light, 'at least by the means which I have made use of'.[30] Therefore, according to him, Newton's condition for the possibility of celestial motions had to be reconsidered. And to avoid misunderstandings, he distinguished two meanings of rarity: either the particles of the ethereal matter are far apart from one another 'with much void' between any two of them (as Newton had required), or they 'touch one another, but the tissue of each is rare and inter-mixed with many small empty spaces'.[31] But since he found it 'entirely impossible' to account for the 'prodigious' velocity of light (which Roemer had estimated to be at least six hundred thousand times greater than that of sound) without assuming contact between the ether particles (so that the action of light might be transmitted through them), he concluded by rejecting rarity in Newton's sense:[32]

---

[28] H, XXI, p. 473.   [29] Ibid.   [30] Ibid.   [31] Ibid.

[32] H, XXI, p. 473. In 1694, Huygens still maintained against Newton that 'such a rapid passage of the [light] corpuscles from the Sun or Jupiter to us' was inadmissible. H, X, p. 613.

As to the void, I accept it without difficulty, and I even believe it necessary for the movement of small particles among themselves, not being of the opinion of Mr Descartes, who holds that extension alone makes the essence of body.[33]

This declaration brings Huygens on the side of Democritus and Gassendi, farther from Descartes. It will be seen later that he also endowed his ether particles with a property, namely that of elasticity, which Descartes was prevented from ascribing to his matter on *a priori* grounds. This property allowed Huygens to postulate a finite velocity of light, where Descartes had to assert an instantaneous propagation.

Thus, although Huygens adopted Descartes' criterion of intelligibility as applied to physical explanation, he did not follow what the latter had regarded as a rigorous application of this criterion. This means, however, that he rejected Cartesian rationalism only in so far as it sought to establish certain physical doctrines *a priori*. He nevertheless accepted Descartes' demand, itself a consequence of his rationalism, that only geometrical concepts should be admitted into physics, a demand which implied the doctrine that all physical actions must be of the nature of impact.

Following Descartes, Huygens attached primary importance to causal explanations in physics and, like Descartes, he would only admit explanations in terms of movement. A physical question would not be completely settled until such an explanation could be found. It was not enough to establish the experimental laws governing a certain phenomenon; for while such laws might be useful in practical applications, and although they formed the basis for further research, they were not sufficient by themselves.

This may be illustrated by his attitude towards Newton's inverse square law, and towards Newton's theory of colours. Huygens wrote in the *Addition*:

I had not thought . . . of this regulated diminution of gravity, namely that it was in inverse ratio to the squares of the distances from the centre:

---

[33] H, XXI, p. 473.

which is a new and remarkable property of gravity, *of which the reason is well worth looking for.*[34]

Huygens here accepts Newton's law; but this suggests to him a further problem which he hastens to point out at the same time as he expresses his admiration for the discovery: What is the cause of this diminution? Newton himself certainly recognized the validity of such a question when he attempted to explain gravity by the action of a rare and elastic ethereal medium.[35] But whereas Newton was prepared at least to consider [36] alternative explanations in terms of what he called 'active principles' of a non-mechanical order, there was for Huygens only one possible form of explanation, namely by the mechanical properties of the subtle matter whose existence he never doubted. In this he was in sharp disagreement with Newton's followers who, like Roger Cotes, went so far as to regard attraction as a primary property of matter—a view which implied that a causal explanation of gravity was pointless.[37]

Again Huygens accepted Newton's discovery (published in 1672) that light rays of different colours were not equally refracted by one and the same medium as a fact 'very well proved by experiment'.[38] At the same time, however, he was of the opinion that the question of colours was one of those matters 'in which no one until now [i.e. 1690] can boast of having succeeded'.[39] At one time (1673) Huygens was inclined towards Hooke's view that there were only two principal colours (red and blue) because he thought it would be easier to find 'an Hypothesis by Motion' to explain two colours, than for an infinite variety of them.[40] It would seem from the words just quoted from the Preface to the *Traité* that, by 1690, Huygens had abandoned Hooke's opinion. But they also indicate that he had not completely reconciled himself to Newton's theory.

---

[34] H, XXI, p. 472; italics added.
[35] Cf., for example, Newton, *Opticks*, Query 21, pp. 350-2.
[36] Cf. ibid., Query 31, pp. 401-2.
[37] Cf. Cotes' Preface to the second edition of Newton's *Principia*, pp. xxvi-xxvii.
[38] Huygens, *Treatise*, p. 106.
[39] Ibid., Preface (dated 8 January 1690), p. vii.
[40] Cf. *Phil. Trans.*, No. 96, 21 July 1673, p. 6086. Huygens in fact subscribed to Hooke dualistic theory until at least 1678. Cf. H, XIX, pp. 385-6, *Avertissement*.

No reasons for this are given in the *Traité*, but we know that Huygens' main objection in 1672 against the Newtonian theory was that it had omitted to explain 'by the mechanical physics wherein this diversity of colours consists'.[41] And there is nothing to suggest that he had withdrawn this objection before (or after) he published the *Traité* in 1690. Huygens' attitude greatly contrasts with Newton's view that his theory of colours was not in need of any mechanical or other explanation to support it. And it was this kind of attitude that Newton had in mind when he first raised his objections against the use, or rather (in his view) the abuse, of hypotheses. For it appeared to him that whereas he was propounding a theory of colours wholly derived from experiments, others (including Huygens) were, unjustly, more interested in seeking mechanical hypotheses that had to conform with their own preconceived ideas.[42] Huygens' commitment to the Cartesian view of physical explanation will thus have to be borne in mind when we study his controversy with Newton over the latter's theory of colours. A proper understanding of Newton's views on hypotheses can only be reached by reference to the mechanistic programme to which Huygens, and other adversaries of Newton's theory, had subscribed.

2. Having seen that Huygens always remained faithful to Descartes' views on the aim of physical science and the character of physical explanation, it might now come as a surprise to learn that, both before delivering the 1669 memoir on gravity and after the publication of the *Discours de la Cause de la Pesanteur* in 1690, he praised the excellency of Bacon's views on scientific research.[43] Since we have seen,[44] however, that Descartes himself recommended Bacon's

---

[41] Cf. H, VII, p. 229. See below, pp. 270f.

[42] See below, pp. 287, 293f.

[43] A. E. Bell, for example, finds Huygens' laudatory attitude (in 1668) towards Bacon's teachings 'the more striking when one reflects that Descartes had so long been his model'. Bell then explains: 'The point here is that while both Bacon and Descartes distrusted formal logic, Descartes scorned empiricism [*sic*] while Bacon apprehended its power' (*Christian Huygens*, p. 61).

[44] See above, p. 36.

suggestions regarding the task of experimentation, Huygens' attitude to Bacon's empiricism should not be too surprising. It will be seen from the following discussion that what impressed Huygens in Bacon's writings in no way conflicted with his Cartesian leanings. On the other hand, Huygens' views on the status of scientific explanations and the role of hypotheses will be found to be completely un-Baconian. This will appear from an examination of some important references in Huygens to the author of the *Novum Organum* and, later, from a comparison of Bacon's inductive method with Huygens' statements in the Preface to his *Traité de la Lumière*.

In (probably) 1668,[45] Huygens drew up a plan for the Physical Assembly of the *Académie Royale des Sciences* in which he set forth his views on the proper way of going about research in physics. It begins as follows:

The principal and most useful occupation of this Assembly should be, in my opinion, to work on the Natural History, somewhat according to the plan of Verulam. This History consists in experiments and observations which are the only means for attaining knowledge of the causes of all that is seen in nature.[46]

Later on we read: 'The collecting of all ["particular experiments"] is always a firm foundation for building a natural philosophy in which one must necessarily proceed from the knowledge of the effects to that of the causes.'[47]

There are, in Huygens' view, two aims of the natural history. The first is to obtain knowledge of 'all' physical things and natural effects that have so far been discovered and verified—'as much for the sake of curiosity as to derive from them every possible utility'.[48] To achieve this, one must collect the history of all plants, all animals and all minerals. The second aim is to seek knowledge of the causes. Such knowledge consists in a 'perfect understanding of the conformation

---

[45] Cf. H, XIX, p. 268, editors' note 1.
[46] H, XIX, p. 268.
[47] Ibid.
[48] H, XIX, p. 269.

of all physical bodies and of the causes of their observed effects.[49] This second task will, in Huygens' view, become infinitely useful when 'one day we shall reach its end',[50] and men will be able to employ the things created in nature for the production of their effects with certainty. For this purpose we must first collect 'all observations of the phenomena which seem to be capable of opening the way towards the knowledge of the conformation of physical things and of the causes of the natural effects that appear before our eyes'.[51]

But since natural history should be collected for every branch of physics separately, Huygens finds it necessary to prepare a general and orderly classification of these branches. His plan divides physical science into six branches corresponding to the following subject matters: the 'so-called' four elements, the meteors, the animals, the plants, the fossils, and the natural effects. Each of these is further divided into chapters. The natural effects, for example, include such phenomena as gravity, magnetism, sound, light, colour and so forth.[52]

The function of this plan, when complete, will be to present the the investigator with 'all the material'[53] from which he can select particular items for his consideration, to register the various observations which occasion or chance may bring to notice, and to facilitate the consultation of whatever will have been recorded concerning a given matter.

As a final practical advice Huygens adds that there should be six portfolios corresponding to the six general divisions of physics, and each should contain a loose book bearing the heads of chapters and

---

[49] H, XIX, p. 269. The term *'conformation'* appears to be Huygens' translation of Bacon's *schematismus* by which Bacon means the configuration of the smaller parts of bodies. Cf. *Novum Organum*, especially Bk. I, 50 and 51; Bk. II, 1 and 7. See Fowler's edition (1889), p. 227, note 83; and p. 354, note 39. According to Bacon (*Novum Organum*, I, 50) a knowledge of the invisible structure of bodies is a necessary requirement for the knowledge of causes and, consequently, for the artificial reproduction of natural effects. He mentions in this connection (ibid., I, 51) the school of Democritus as an example of those who recognized the true method of penetrating the merely sensible properties of things to their ultimate components.

[50] H, XIX, p. 269.

[51] Ibid.

[52] H, XIX, pp. 270–1.

[53] H, XIX, p. 269.

sections included in it. If these sections are separately filed, one will be able to interleave as many of them as will be necessary. There is no evidence from Huygens' papers and writings that he ever followed this advice himself.[54]

Almost all of these suggestions are derived from Bacon. Huygens shares Bacon's optimistic view that the task of physical science is a finite one; he accepts Bacon's opinion that a complete enumeration of the topics of physics, and even of the material required, is both possible and desirable; he attaches great practical value to the classification of data; he emphasizes the importance of experiments and maintains that knowledge of the particular natural effects should be the starting point for the inquiry into their causes. All this, however, does not constitute any serious departure from Descartes' views. Descartes was also impressed by Bacon's idea of enumeration, and the Cartesian method equally requires that when inquiry concerns particular phenomena, one should begin with the effects whose explanation we are then to seek.

It is to be noted that there is no reference in Huygens' suggestions to Baconian induction. When he did refer to Bacon's 'method', he understood by that term something that would have been repudiated by Bacon himself. Thus, in a letter to E. W. von Tschirnhaus (10 March 1687) Huygens says that the difficulties involved in physical research cannot be overcome except by 'starting from experiments . . . and then contriving hypotheses against which the experiments are weighed, in which task the method of Verulam seems to me excellent and deserving of further cultivation'.[55] Huygens here understands Bacon's method as one in which the formation of hypotheses, after the experiments have been made, is a permissible procedure. The same understanding is implied in later correspondence with Leibniz. In a letter to Leibniz (16 November 1691), Huygens complained that too much effort was being spent on pure geometry at the expense of physics. To advance empirical science, he suggested, 'one would have to reason methodically on

---

[54] H, XIX, p. 270, n. 3.
[55] H, IX, p. 124.

experiments, and collect new ones, somewhat according to the plan of Verulam'.[56] In reply, Leibniz agreed that Bacon's plan should be followed—on condition, however, that 'a certain art of guessing, should be adjoined to it.[57] Huygens' answer to this reveals the source of his interpretation of Bacon's method:

It seems to me that Verulam has not omitted this art of guessing in Physics on given experiments, considering the example which he has given on the subject of heat in metallic and other bodies, where he has well succeeded, if only because he has thought of the rapid movement of the very subtle matter which must temporarily maintain the vibration [*le bransle*] of the particles of bodies.[58]

It should be first observed that what Huygens seems to have particularly liked in Bacon's study of heat was the fact that it had resulted in the kind of mechanical explanation which, as a true Cartesian, he was bound to favour. Consciously or unconsciously, he readily introduced into Bacon's account the expression 'subtle matter' which he borrowed, not from Bacon, but from Descartes. Bacon had in fact simply proposed the following assertion: '*Calor est motus expansivus, cohibitus, et nitens per partes minores.*'[59]

Now to come to the question of hypotheses. It is true that Bacon gives the above definition of heat as a conjecture, a tentative hypothesis, or, as he called it, a 'First Vintage'. It would not be correct, however, to consider this example to be characteristic of Bacon's method in general, as Huygens seems to have thought. In fact Huygens' own doctrine of the essential role of hypotheses (as he had expressed it before he wrote the preceding words to Leibniz) is the exact opposite of Bacon's views and cannot be reconciled with Bacon's conception of induction. To show this, a brief account of Bacon's method is needed in order to contrast it afterwards with Huygens' opinions.

---

[56] H, X, p. 190.
[57] Leibniz to Huygens, 8 January 1692, H, X, p. 228: '*Je suis de vôtre sentiment, qu'il faudroit suivre les projects de Verulamius sur la physique en y joignant pourtant un certain art de deviner, car autrement on n'avancera gueres.*'
[58] Huygens to Leibniz, 4 February 1692, H, X, 239.
[59] *Novum Organum*, II, 20; Fowler's edition (1889), p. 412.

Bacon called his task 'interpretation of nature' in contrast to the peripatetic method which he termed 'anticipation of nature'. In the latter, Bacon explains, one hastens, in a premature and precipitate fashion, from a vague and incomplete information of the senses, to the most general principles. The syllogism is then applied to these rashly acquired principles to deduce consequences from them with the help of middle terms.[60] As opposed to this discreditable procedure, the interpretation of nature would consist in laying down 'stages of certainty' [certitudinis gradus] through which the mind would slowly and surely advance from the lowest level generalizations (or axioms, as Bacon called them) through higher and higher ones until the highest principle would be reached in due course.[61] After one or more axioms have been formed, one may start to move in the opposite direction to see whether they point to 'new particulars'.[62] Thus, whereas the greater attention in the method of anticipation is concentrated on deducing what is in agreement with the principles, the primary concern of the interpretation is to discover the principles themselves and to establish them on a firm basis.[63]

---

[60] Cf. Plan of the *Great Instauration* (the work of which the *Novum Organum* was conceived as the second part), B, IV, p. 25. Bacon used the expressions '*anticipatio naturae*' and '*anticipatio mentis*' synonymously; cf. B, I, pp. 154 and 161. For an illuminating analysis of Bacon's concept of *interpretatio*, see Karl R. Popper, *Conjectures and Refutations*, London, 1963, pp. 13–15.

[61] Cf. ibid.; Preface to *Novum Organum*, B, IV, p. 40; *Novum Organum*, I, 104, B, IV, p. 97.

[62] Cf. *Novum Organum*, I, 106 B, IV, p. 98; II, 10 B, IV, pp. 126–7.

[63] Cf. Plan of the *Great Instauration*, B, IV, p. 24, and Preface to *Novum Organum*, B, IV, p. 42.

Bacon's ascending-descending ladder of axioms must not be understood as a hypothetico-deductive scheme. The axioms from which we deduce new 'particulars' (facts, phenomena, experiments) are *not* hypotheses (i.e. conjectures that may turn out to be false), but propositions whose certain truth has already been established on the basis of previously performed observations and experiments. Thus, for example, Bacon writes: '. . . the true method of experience . . . first lights the candle, and then by means of the candle shows the way; commencing as it does with experience duly ordered and digested, not bungling or erratic, and from it educing axioms, and from established axioms [*axiomatibus constitutis*] again new experiments . . .' (*Novum Organum*, I, 82, B, IV, p. 81; B, II, p. 190). Again, in *Novum Organum*, I, 103, Bacon speaks of the 'axioms, which having been educed from those particulars by a certain method and rule [*certa via et regula*] shall in their turn point out the way again to new particulars . . .' (B, IV, p. 96; B, I, p. 204). Propositions derived 'by a certain method and rule' cannot be called hypotheses. Accordingly, Bacon does not descend down the ladder of axioms in order to *test* them, i.e. to see whether they are true or false—he already knows they are true. His only aim is to find out whether or not they point to new phenomena. If they do, then they are fruitful; if they do not, then they are trivial but still true. He does not envisage the possibility of counter instances. Cf. *Novum Organum*, I, 106: 'But in establishing axioms

In order to achieve this Bacon proposes a new *form of demonstration* corresponding to the syllogistic form in the old method. This new form he calls *induction*. It is the method which allows one to move from lower to higher axioms on the ladder of generalizations—with certainty.

Although therefore I leave to the syllogism and these famous and boasted modes of demonstration their jurisdiction over popular arts and such as are matter of opinion (in which department I leave all as it is), yet in dealing with the nature of things I use induction throughout, and that in the minor propositions as well as in the major. For I consider induction to be that form of demonstration which upholds the sense, and closes with nature, and comes to the very brink of operation, if it does not actually deal with it.[64]

That the contrast here is between *probable* opinion and *certain* knowledge is made clear by Bacon's distinction of his form of induction from Aristotelian induction:

But the greatest change I introduce is in the form itself of induction and the judgment made thereby. For the induction of which the logicians speak, which proceeds by simple enumeration, is a puerile thing; concludes at hazard [*precario concludit*] ; is always liable to be upset by a contradictory instance; takes into account only what is known and ordinary; and leads to no result.

Now what the sciences stand in need of is a form of induction which shall analyse experience [*experientiam solvat*] and take it to pieces, and by a due process of exclusion and rejection lead to an inevitable conclusion [*necessario concludat*].[65]

---

by this kind of induction, we must also examine and try whether the axiom so established be framed to the measure of those particulars only from which it is derived, or whether it be larger and wider. And if it be larger and wider, we must observe whether by indicating to us new particulars it confirm that wideness and largeness as by a collateral security; that we may not either stick fast in things already known, or loosely grasp at shadows and abstract forms; not at things solid and realised in matter' (B, IV, p. 98). The only possibilities contemplated here by Bacon are the following: (1) the axiom is made to the measure of particulars from which it was derived; (2) the axiom is wider in a real sense, i.e. it points to new particulars actually realized in matter; (3) the axiom is wider in an illusory sense, i.e. it proposes only shadows and abstract forms and therefore has no bearing on the empirical world.

[64] Plan of the *Great Instauration*, B, IV, pp. 24-5.
[65] Ibid., B ,IV, p. 25; B, I, p. 137.

The certainty of inductive knowledge is thus derived from two sources. First, from the solid basis of experience on which it is founded as on a rock bottom;[66] this, in Bacon's view, distinguishes his induction from the syllogism whose ultimate components, the terms, are often only signs of 'notions' that are 'improperly and over-hastily abstracted from facts' that are vague and not sufficiently defined.[67] And second, from the sure method of exclusion which distinguishes Baconian induction from the Aristotelian induction by simple enumeration. In view of the first source it naturally appeared to Bacon that an experimental history of phenomena should constitute the foundation of interpretation.[68] In this primary stage observations and experiments are collected, verified and recorded in a somewhat passive manner without interference from the mind, that is, without allowing the mind to anticipate or indulge in its natural weakness for hasty generalizations.

But this is only a preparatory stage which, strictly, does not form part of the inductive procedure. For, according to Bacon, induction (being a method of demonstration) does not begin with natural history, nor even with the classification of what has been collected into tables of presence, of absence and of comparison. The function of these tables merely consists in presenting the collected material to the mind in an orderly form.[69] Induction proper begins when the mind starts to act upon these tables, that is, when the method of exclusion begins to operate.

What, then, is the method of exclusion? Suppose that heat is the phenomenon or, in Bacon's language, the nature whose form, cause or law (these words being used by him as equivalents) is to be discovered. The Table of Presence should contain all instances

---

[66] *Novum Organum*, I, 70: 'But the best demonstration by far is experience, if it go not beyond the actual experiment' (B, IV, p. 70).

[67] Cf. Plan of the *Great Instauration*, B, IV, p. 24.

[68] *Novum Organum*, II, 10: 'For first of all we must prepare a *Natural and Experimental History*, sufficient and good; and this is the foundation of all; for we are not to imagine or suppose, but to discover, what nature does or may be made to do [neque enim fingendum aut excogitandum sed inveniendum, quid natura faciat aut ferat]' (B, IV, p. 127; B, I, p. 236).

[69] *Novum Organum*, II, 15: 'The work and office of these three tables I call the Presentation of Instances to the Understanding. Which presentation having been made, Induction itself must be set at work' (B, IV, p. 145).

(that is, things, conditions and situations) in which heat occurs. The Table of Absence should include instances in which heat would be expected to occur although this is not in fact the case. The Table of Comparison includes instances in which some other nature suffers an increase or decrease when the same happens to the nature under consideration. Induction then begins by forming the Table of Exclusion in which are recorded all the natures which should be rejected from the form of heat by a comparative examination of every item in each of the first three tables individually. For example, from the information (recorded in the Table of Presence) that heat is produced by the sun's rays, we exclude elementary nature (earth, common fire, etc.). And on account of subterranean fire, we reject celestial nature. And so forth.[70]

There are other aids that come to the help of the mind during the inductive procedure. Such, for example, are the twenty-seven Prerogative Instances which occupy the greater part of the second book of the *Novum Organum*. They are distinguished for the extraordinary help which, according to Bacon, they can offer to the mind on different levels of the investigation. Outstanding among the Prerogative Instances are the *instantiae crucis*. The particularly important role which these have in the inductive process is apparent from the various names which Bacon has given them: Decisive Instances, Judicial Instances, Instances of the Oracle, Instances of Command.[71] These instances may be readily available in nature and, perhaps, already noted in the Tables of Presentation, 'but for the most part they are new, and are expressly and designedly sought for and applied, and discovered only by earnest and active diligence'.[72] Their function is to decide between two or more opinions as to the cause or form of a certain nature on account of its ordinary occurrence with two or more other natures. An *instantia crucis*, exhibiting the nature whose form is to be discovered together with only one of the other concurrent natures, will, according to Bacon, decide which of these natures should be rejected and which should be included in

---

[70] *Novum Organum*, II, 18, B, IV, pp. 147-9.
[71] *Novum Organum*, II, 36, B, IV, p. 180.
[72] Ibid.

the form sought. And so, instances of this kind, unlike any of the other Prerogative Instances, 'afford very great light, and are of high authority, the course of interpretation sometimes ending in them and being completed'.[73]

The method of exclusion in general and the *instantiae crucis* in particular are thus assigned a role similar to that of the indirect proof in mathematics.[74] For Bacon believed that a complete enumeration of the particulars of nature could eventually be achieved.[75] Assuming, therefore, that the Tables of Presentation (relating to the investigation of a particular nature) have, at one stage or another, been perfected, one may, after a sufficient number of rejections, conclude in the affirmative regarding the form of the nature in question. If the process of elimination has been properly conducted, and the various aids of induction duly applied, the conclusion reached must, in Bacon's view, be necessarily true. This, indeed, is the aim of the Baconian science of induction as a '*forma demonstrativa*', that is a form that leads from certain premises derived from experience to certain conclusions necessarily entailed by them.

But, then, what place does the First Vintage have in this general scheme? This brings us to Bacon's study of heat to which Huygens referred. Having performed fourteen rejections in the Table of Exclusion for heat, Bacon writes: 'There are other natures besides these; for these tables are not perfect, but only meant for examples'.[76] Moreover, Bacon continues, some of the natures rejected in that

---

[73] Ibid.

[74] This has been noted by Pierre Duhem; see his *Théorie physique, son objet, sa structure*, Paris, 1914, p. 286.

[75] *Novum Organum*, I, 112:' . . . let no man be alarmed at the multitude of particulars, but let this rather encourage him to hope. For the particular phenomena of art and nature are but a handful to the inventions of the wit, when disjoined and separated from the evidence of things. Moreover this road has an issue in the open ground and not far off; the other has no issue at all, but endless entanglement. For men hitherto have made but short stay with experience, but passing her lightly by, have wasted an infinity of time on meditations and glosses of the wit. But if some one were by that could answer our questions and tell us in each case what the fact in nature is, the discovery of all causes and sciences would be but the work of a few years' (B, IV, pp. 101–2). See also *Novum Organum*, II, 15 and 16, where the exclusion of negative instances is viewed as a finite operation; and ibid., II, 21, where Bacon indicates the possibility of making a 'Synopsis of all Natures in the Universe' (B, IV, p. 155).

[76] *Novum Organum*, II, 18, B, IV, p. 148.

table, as the notion of elementary and celestial nature, are 'vague and ill-defined'.[77] The matter must therefore be pushed further and 'more powerful aids'[78] (such as the Prerogative Instances) must come to the help of the understanding:

And assuredly in the Interpretation of Nature the mind should by all means be so prepared and disposed, that while it rests and finds footing in due stages and degrees of certainty, it may remember withal (especially at the beginning) that what it has before it depends in great measure upon what remains behind.[79]

Then, Bacon adds,

And yet since truth will sooner come out from error than from confusion, I think it expedient that the understanding should have permission after the three Tables of First Presentation (such as I have exhibited) have been made and weighed, to make an essay of the Interpretation of Nature in the affirmative way; on the strength both of the instances given in the tables, and of any others it may meet with elsewhere. Which kind of essay I call the *Indulgence of the Understanding* [*permissionem intellectus*], or the *Commencement of Interpretation*, or the *First Vintage*.[80]

From this it is clear that the First Vintage is attempted by Bacon only *faute de mieux* ('truth', he says, 'will sooner come out from error than from confusion'). He gives his mind permission to venture a tentative definition of the form of heat simply because, at this stage of inquiry, when the table of exclusion is not yet completed and the natures mentioned in it not yet clearly defined, nothing better can be offered to complete his example. But when the natural history of heat is completed and the various aids of induction duly applied to a perfected table of exclusion, and when the real task of interpretation comes to an end, all hypotheses will have disappeared. Only certain and necessary knowledge will remain. As R. L. Ellis has pointed out,

---

[77] *Novum Organum*, II, 19, B, IV, p. 149.
[78] Ibid.
[79] Ibid., II, 20, B, IV, p. 149.
[80] Ibid.

the Vindemiatio prima, though it is the closing member of the example which Bacon makes use of, is not to be taken as the type of the final conclusion of any investigation which he would recognise as just and legitimate. It is only a parenthesis in the general method, whereas the Exclusiva, given the eighteenth aphorism of the second book [of the *Novum Organum*], is a type or paradigm of the process on which every true induction (inductio vera) must in all cases depend.[81]

The influence of Bacon's illustrative study of heat can be seen in the views on hypothesis generally held by seventeenth-century members of the Royal Society. Taking advantage of Bacon's *permissio intellectus*, while at the same time remaining faithful to his general method, they were willing to allow the use of hypotheses in the course of inquiry—subject to the following conditions: (1) that experiments have to be made first; (2) that if one has to invent hypotheses they should be either fitted to explain the already discovered facts or *temporarily* used to predict new facts; (3) that hypotheses formed for these purposes should be proposed as conjectures, queries or problems, not as asserted doctrines or theories; (4) that the ultimate aim should be to deduce from a sufficient number of experiments a theory or theories in which hypothetical elements no longer exist. Boyle, for example, quoted Bacon's sentence (that truth may be more easily extricated from error than confusion) to justify the use of hypotheses for making new experimental discoveries while agreeing with him that all hypotheses, as such, will have been discarded at the end of investigation.[82] And Newton's views on this question were not essentially different from Boyle's, as will be seen below.[83]

How does Bacon's method, and the views adopted by his followers, compare with Huygens' doctrine of hypotheses? The following is what Huygens wrote on 8 January 1690 in the Preface to his *Traité de la Lumière*:

There will be seen in it [the *Treatise*] demonstrations of those kinds which do not produce as great a certitude as those of Geometry, and which even

---

[81] B, I, p. 365. Quoted by Fowler in his edition of *Novum Organum* (1889), p. 404, note 96.
[82] See below, p. 322; see also Hooke's passage on p. 187 below.
[83] See also above, p. 31.

differ much therefrom, since whereas the Geometers prove their Propositions by fixed [*certains*] and incontestable Principles, here the Principles are verified by the conclusions to be drawn from them; the nature of these things not allowing of this being done otherwise. It is always possible to attain thereby to a degree of probability [*un degré de vraisemblance*] which very often is scarcely less than complete proof [*qui bien souvent ne cède guère à une évidence entière*]. To wit, when things which have been demonstrated by the Principles that have been assumed correspond perfectly to the phenomena which experiment has brought under observation; especially when there are a great number of them, and further, principally, when one can imagine and foresee new phenomena which ought to follow from the hypotheses which one employs, and when one finds that therein the fact corresponds to our prevision. But if all these proofs of probability [*preuves de vraisemblance*] are met with in that which I propose to discuss, as it seems to me they are, this ought to be a very strong confirmation of the success of my inquiry; and it must be ill if the facts are not pretty much [*à peu pres*] as I represent them.[84]

The following may be gathered from this passage. First, there is no question here of an inductive method in the proper Baconian sense. Huygens does not claim that his 'principles' or theories are *deduced* from the experimental propositions which they propose to explain (as Newton, for example, claimed for his own optical theories). Huygens' principles, therefore, do not share the certainty which he might wish to attach to the experimental propositions.[85] Since the direction in the kind of 'demonstrations' he wants to offer is from the principles to the experimental laws and not *vice versa*, the truth of the latter is not thereby transmitted to the former. This is to be contrasted with the Baconian 'form of demonstration' which seeks to establish the truth of the axioms or principles by a deductive movement (effected by the method of exclusion) in the opposite direction.

Second, Huygens does not propose his hypothetico-deductive

---

[84] Huygens, *Treatise*, pp. vi–vii; H, XIX, pp. 454–5.

[85] Huygens, *Treatise*, p. 1: 'As happens in all the sciences in which Geometry is applied to matter, the demonstrations concerning Optics are founded on truths drawn from experience. Such are that the rays of light are propagated in straight lines; that the angles of reflexion and of incidence are equal; and that in refraction the ray is bent according to the law of sines, now so well known, and which is no less certain than the preceding laws.'

method for the same reasons which Bacon adduced in justification of his *permissio intellectus* or the First Vintage. Here the mind is 'permitted' to form hypotheses right from the beginning, not because nothing better can be done at a given moment, or at a certain stage of inquiry, but simply because the 'nature' of the subject matter *does not allow of this being done otherwise*. In other words, what is now a hypothesis will always remain so, however strong is the evidence which has been or will be collected in its favour.

Third, it is true that Huygens maintains that his proofs can attain 'un degré de vraisemblance' which very often is scarcely less than 'une évidence entière', namely, when the number of facts in agreement with the theory is very large and, especially, when the predictions made on the basis of the supposed hypotheses are later confirmed by experiments. But it is important to note that we are always concerned here with 'preuves de vraisemblance', that is, proofs in which we may approach closer and closer to certitude (as the confirming experiments accumulate) without ever attaining complete proofs. The 'évidence entière' of which Huygens speaks is a limiting state of knowledge that is never achieved.

It is thus seen that, for all the naïve views which he communicated to the *Académie Royale des Sciences* in 1668, Huygens was in fact (at least in his more mature years) completely free from Bacon's naïve empiricism. In spite of his 'religious respect for the facts' (as Mouy has described his attitude),[86] he did not assign to them the determinative role which they had in Bacon's method.

Does the passage quoted above also imply that he equally freed himself from Descartes' position?

Descartes, it will be remembered, expressed views similar to those contained in Huygens' passage on more than one occasion.[87] As in Huygens' *Traité de la Lumière*, the explanations given in Descartes' *Dioptric* are presented as 'suppositions' to be verified by their experimental consequences. But whereas Descartes simultaneously believed his suppositions to be deducible from certain *a priori* principles, no such belief is to be found in Huygens' writings. In

---

[86] Cf. P. Mouy, *Développement de la physique cartésienne*, p. 186.
[87] See above, pp. 18ff, 23, 38ff.

fact his main criticism of Descartes was that he should have presented his system of physics as a system of plausible conjectures, not of verities.[88] The implicit assumption in this criticism is no doubt Huygens' belief (expressed at the beginning of his passage above) that in physics no propositions can be established by deducing them from *a priori* truths, that no physical theory (nor any part thereof) can be developed *ordine geometrico*.

And yet Huygens' aim in the *Traité* is explicitly stated as an attempt to explain the properties of light 'in accordance with the principles accepted in the Philosophy of the present day'.[89] This Philosophy he also describes as 'the true Philosophy in which one conceives the causes of all natural effects in terms of mechanical motions [*par des raisons de mécanique*]. This, in my opinion, we must necessarily do, or else renounce all hopes of ever comprehending anything in Physics'.[90] Thus, together with Huygens' assertion that physical explanations should always be regarded as conjectures or hypotheses, we find the simultaneous belief that they can be of only one possible type, namely, that they should be mechanical explanations. A renunciation of mechanism would be, for Huygens, a renunciation of the task of physical science itself, since nothing else would be *intelligible*. This firm belief in a fixed framework within which alone all physical hypotheses have to be formulated is a result of Huygens' adoption of the Cartesian *programme* after it has been detached from the part that is asserted by Descartes on *a priori* grounds.[91]

---

[88] In 1693 Huygens wrote the following in his annotations on Baillet's *Vie de Monsieur Descartes* which appeared in 1691: 'Descartes, who seemed to me to be jealous of the fame of Galileo, had the ambition to be regarded as the author of a new philosophy, to be taught in academies in place of Aristotelianism. He put forward his conjectures [and fictions] as verities, almost as if they could be proved by his affirming them on oath. He ought to have presented his system of physics as an attempt to show what might be anticipated as probable in this science, when no principles but those of mechanics were admitted: this would indeed have been praiseworthy; but he went further, and claimed to have revealed the precise truth, thereby greatly impeding the discovery of genuine knowledge.' H, X, p. 404; quoted and translated by Whittaker in *History of the Theories of Aether and Electricity*, the classical theories, Edinburgh, 1951, pp. 7–8. Cf. also Huygens to Leibniz, 4 February 1692, H, X, p. 239.

[89] Huygens, *Treatise*, p. 2.

[90] Huygens, *Treatise*, p. 3; H, XIX, p. 461.

[91] Above, pp. 24ff.

Chapter Seven

TWO PRECURSORS OF HUYGENS'
WAVE THEORY: HOOKE AND PARDIES

1. Huygens first communicated his *Traité de la Lumière* to the *Académie Royale des Sciences* in 1679.[1] Before that date some progress in optical theory had already been achieved in addition to the discovery of the sine law and of Fermat's least-time principle. An elaborate account of the colours of thin transparent bodies was given by Hooke in his *Micrographia* which appeared in 1665.[2] In the same year Grimaldi's *Physico-mathesis*[3] was posthumously published, and this book contained an elaborate description of some diffraction phenomena. Four years later Erasmus Bartholinus published his new observations on double refraction.[4] In 1672 an account of Newton's theory of prismatic colours appeared in the *Philosophical Transactions*,[5] and this publication was followed by a controversy in which Huygens himself took part. And, finally, in 1676, Roemer communicated to the *Académie des Sciences* his astronomical 'demonstration' that light travelled with a finite speed, and gave his estimation of that speed.[6]

None of these achievements, however, can be said to have contributed to the formation of Huygens' wave theory. Nowhere in his *Traité* does Huygens treat of any colour phenomena; and we have seen that even in 1690 he was still of the opinion that no acceptable physical explanation of colours was forthcoming. Nor

---

[1] The date given by Huygens in the Preface to his *Traité* is 1678. But according to the registers of the *Académie des Sciences* the reading did not take place until 1679. The year 1678 is therefore taken to be that in which the *Traité* was, most probably, completed. Cf. H, VIII, p. 214 (editors' note 3); H, XIX, p. 453 (editors' note 1) and p. 439.

[2] Robert Hooke, *Micrographia*, London, 1665.

[3] Francesco Maria Grimaldi, *Physico-mathesis de lumine, coloribus et iride*, Bononiae, 1665.

[4] Erasmus Bartholinus, *Experimenta crystalli Islandici disdiaclastici, quibus mira & insolita refractio detegitur*, Hafniae, 1669.

[5] *Philosophical Transactions*, No. 80, 19 February 1671-2, pp. 3075-87.

[6] *Mémoires de l'Académie Royale des Sciences, depuis 1666 jusqu' à 1699*, X (Paris, 1730), pp 575-7.

does he mention Grimaldi's name, or his book, or the phenomena of diffraction anywhere in his writings.[7] The *Traité* does contain an explanation of double refraction; but this was a later application of Huygens' theory after its fundamental ideas had already been developed.[8] As to Roemer's discovery, it will be seen that Huygens had adopted the finite velocity of light as a *hypothesis* several years before Roemer announced his results. The fact that Huygens gives an account of Roemer's proof in the beginning of his *Traité* does not mean that his theory was inspired by it. Huygens originally devised his theory to account for precisely those phenomena which Descartes' theory had proposed to explain: namely, rectilinear propagation, the fact that rays of light may cross one another without hindering or impeding one another, reflection, and ordinary refraction in accordance with the sine law. His aim was to give a clearer and more plausible explanation than the unsatisfactory comparisons proposed in Descartes' *Dioptric*; and his starting point was exactly those physical problems which the Cartesian theory had left unsolved.[9]

In his attempt to reform the Cartesian theory Huygens certainly benefited from ideas introduced by previous writers. He specifically mentions in the *Traité* Robert Hooke and the Jesuit Father Ignatius Pardies as two of those who had 'begun to consider the waves of light'.[10] His dependence on these two writers was not accidental. Both Hooke and Pardies had followed Descartes in attempting to explain the properties of light by the action of a subtle matter that fills all space and permeates all matter. Hooke, in particular, had arrived at his ideas through a close examination of Descartes' assumptions, and his work had thus been a natural preparation for

---

[7] It is, however, known for certain that Huygens possessed a copy of Grimaldi's *Physico-mathesis* before he died in 1695, cf. H, XIX, p. 389 (*Avertissement*); XIII, *fasc.* I, p. CII, editors' note 4. It may also be noted that a description and an attempted explanation of some diffraction phenomena were given (under the name 'inflexion') by Newton in his *Principia* (1687).

[8] See below, pp. 221f.

[9] Compare the account in A. Wolf, *A History of Science, Technology, and Philosophy in the 16th and 17th centuries* (London, 1950, p. 260) of how Huygens' theory came to be elaborated out of preceeding ideas 'established' or 'suggested' by observation and experiment.

[10] Huygens, *Treatise*, p. 20.

Huygens' attempt. It will therefore be instructive to look at Hooke's
and Pardies' investigations before we approach Huygens' theory.

2. The principal aim of Hooke's optical investigations in the *Micro-graphia* was to give an explanation of his observations on the colours
of thin transparent bodies. His attempt must have been one of those
excesses to which he refers in the following address to the Royal
Society which we read at the beginning of his book:

The Rules You have prescrib'd Your selves in YOUR Philosophical
Progress do seem the best that have ever been practis'd. And particularly
that of avoiding *Dogmatising*, and the *espousal* of any Hypothesis not
sufficiently grounded and confirm'd by *Experiments*. This way seems the
most excellent and may preserve both *Philosophy* and *Natural History*
from its former corruptions. In saying which, I may seem to condemn my
own Course in this Treatise; in which there may perhaps be some
Expressions, which may seem more *positive* then YOUR Prescriptions
will permit; And though I desire to have them understood only as
Conjectures and *Quaeries* (which YOUR Method does not altogether
disallow) yet if even in those I have exceeded, 'tis fit that I should declare,
that it was not done by YOUR Directions.[11]

Hooke in fact broke the 'Rules' of the Society by *committing* himself
to nothing less than the Cartesian hypothesis of an ethereal medium
serving as the vehicle of light. This is not altogether to be regretted.
For he was in fact able to make proper and fruitful use of his observations by confronting them with already-formulated hypotheses
and bringing them to bear upon already existing problems. In this
way he succeeded in exposing certain difficulties in Descartes'
assumptions, and, by proposing more plausible conjectures, he
pushed the Cartesian theory a step forward. Hooke looked at the
colour phenomena exhibited by thin transparent bodies as providing
an *experimentum crucis* that falsified Descartes' explanation of
colours.[12] While this may have been the starting point of his own
investigations in this field, much more common observations seem

---

[11] Hooke, *Micrographia*, 'TO THE ROYAL SOCIETY' (page not numbered).
[12] See above, pp. 65f.

to have been more decisive in the process of building up his fundamental ideas. The following account will be concerned with his ideas only in so far as they may have contributed to Huygens' theory.

Hooke begins the exposition of his theory by asserting that 'it seems very manifest, that there is no luminous Body but has the parts of it in motion more or less'.[13] Whittaker has interpreted this statement as an attack on 'Descartes' proposition, that light is a tendency to motion rather than an actual motion'.[14] Actually Descartes had distinguished in the *Dioptric* between what constitutes light in the luminous body and in the medium: '. . . light, in the bodies which are called luminous, is nothing but a certain movement, or a very prompt and very violent [viue] action . . . .'[15] The distinction was not, however, clearly stated in the *Dioptric*, as the correspondence between Descartes and Morin indicates.[16] But it was later made clear in the *Principles* where Descartes spoke explicitly of a circular motion of the parts of the sun and the stars, while the spreading of light in straight lines was attributed to the pressure exerted by those parts on the neighbouring medium. It appears therefore that the disagreement between Hooke and Descartes was rather on the *kind* of motion that should constitute light in the luminous body:

for though it be a motion, yet 'tis not every motion that produces it, since we find there are many bodies very violently mov'd, which yet afford not such an effect; and there are other bodies, which to our other senses, seem not mov'd so much, which yet shine.[17]

Descartes' theory could not plausibly apply to a wide range of circumstances in which light was produced. He had constructed his theory mainly to account for the light proceeding from the stars. But what about flames and the shining of a diamond in the dark

---

[13] *Micrographia*, p. 54.
[14] Whittaker, op. cit., p. 14.
[15] D, VI, p. 84. Also, *Le Monde*, D, XI, p. 8.
[16] Cf. Descartes to Morin, 13 July 1638 (D, II, p. 204), where Descartes makes the distinction more explicit by calling the actual motion of light in the luminous body *lux*, and the tendency to motion in the medium *lumen*.
[17] *Micrographia*, p. 55.

when it is rubbed? It was natural for Hooke to ask such questions. The example of the diamond appeared to him particularly instructive: it provided him with an argument against Descartes' attribution of light to a circular motion of the parts of the luminous object, and, having eliminated other hypotheses, he was finally led to propose a vibrating movement. Hooke's reasoning to establish this conclusion was as follows:

. . . the newly mention'd *Diamond* affords us a good argument: since if the motion of the parts did not return, the Diamond must after many rubbings decay and be wasted: but we have no reason to suspect the latter, especially if we consider the exceeding difficulty that is found in cutting or wearing away a Diamond. And a Circular motion of the parts is much more improbable, since, if that were granted, and they be suppos'd irregular and Angular parts, I see not how the parts of the Diamond should hold so firmly together, or remain in the same sensible dimensions, which yet they do. Next, if they be *Globular*, and mov'd only with a *turbinated* motion, I know not any cause that can impress that motion upon the *pellucid medium*, which yet is done. Thirdly, any other *irregular* motion of the parts one amongst another, must necessarily make the body of a fluid consistence, from which it is far enough. It must therefore be a *Vibrating* motion.[18]

Moreover, a diamond being one of the hardest bodies and one of the least to yield to pressure, he concludes that the vibrations of its parts must be '*exceeding* short', that is, of a very small amplitude.[19]

Hooke then goes on to consider how the motion of light is propagated through the interposed medium.

First, the medium 'must be a body *susceptible* and *impartible* of this motion that will deserve the name of a Transparent'.[20] He does not explain in physical terms what this 'susceptibility' consists in; it remained for Huygens to introduce elasticity as the requisite property. Hooke in fact preserved Descartes' conception of the light-bearing medium as a perfectly dense and incompressible body until at least 1680-2. In his lectures on optics dating from that period he defines light as

---

[18] Ibid., pp. 55–56.   [19] Ibid., p. 56   [20] Ibid.

nothing else but a peculiar Motion of the parts of the Luminous Body, which does affect a fluid Body that incompasses the Luminous Body, which is perfectly fluid and perfectly Dense, so as not to admit of any farther Condensation; but that the Parts next the Luminous Body being moved, the whole Expansum of that fluid is moved likewise.[21]

Second, the parts of the medium must be 'Homogeneous, or of the same kind'.[22] This was also a property of the Cartesian heaven, but Hooke makes use of it in arriving at an important result of his own conceptions, to be mentioned presently in the fifth remark.
    Third,

the constitution and motion of the parts must be such, that the appulse of the luminous body may be communicated or propagated through it to the greatest imaginable distance in the least imaginable time; though I see no reason to affirm, that it must be an instant.[23]

Hooke here questions Descartes' hypothesis of the instantaneous propagation of light. He does not actually assert that the velocity of light must be finite. But that he favoured such a view (at the time of writing the *Micrographia*) may be gathered from the following discussion of Descartes' arguments from the eclipses of the moon:

I know not any one Experiment or observation that does prove it [viz. instantaneous propagation]. And, whereas it may be objected, That we see the Sun risen at the very instant when it is above the sensible Horizon,[24]

---

[21] Hooke, *Posthumous Works* (1705), p. 113. (The whole passage is italicized in the text.) Quoted by A. Wolf, *op. cit.*, p. 258. Consistent with this conception of the medium, some passages of Hooke's lectures suggest that the propagation of light is instantaneous (see, for example, *Posthumous Works*, p. 77). It is strange that Hooke should revert to this Cartesian idea in 1680–2 after he had rejected in *Micrographia* (1665) Descartes' astronomical arguments to establish it (as will be seen shortly). Roemer's estimation of the velocity of light had been published before the date of the above-mentioned lectures. But Hooke, instead of accepting Roemer's calculations as a hypothesis, in fact maintained that Roemer's argument was not conclusive (cf. *Posthumous Works*, pp. 77–78, 130). It must be mentioned, however, that the passages relating to the velocity of light in those lectures are not always consistent (compare pp. 74, 130 with pp. 77–78 in *Posthumous Works*).
[22] *Micrographia*, p. 56.
[23] Ibid.
[24] This is not Descartes' argument yet. It seems to be one that was used by supporters of instantaneous propagation in general. In objecting against *this* argument in the above passage, Hooke was in fact repeating what had already been expressed by Galileo. Cf. Galileo, *Two New Sciences*, p. 42.

190

and that we see a Star hidden by the body of the Moon at the same instant, when the Star, the Moon, and our Eye are all in the same line; and the like Observations, or rather suppositions, may be urg'd. I have this to answer, That I can as easily deny as they affirm; for I would fain know by what means any one can be assured any more of the Affirmative, then I of the Negative. If indeed the propagation were very slow, 'tis possible something might be discovered by Eclypses of the Moon; but though we should grant the progress of the light from the Earth to the Moon, and from the Moon back to the Earth again to be full two Minutes in performing, I know not any possible means to discover it; nay, there may be some instances perhaps of Horizontal Eclypses that may seem very much to favour this supposition of the slower progression of Light than most imagine. And the like may be said of the Eclypses of the Sun, &c.[25]

Hooke remarks that the arguments from the eclipses presuppose what they propose to prove. He points out their inconclusiveness as they assume a sufficiently small velocity that would easily be detected by the observations described. He does not himself produce any positive arguments, experimental or theoretical, to support successive propagation (though he suggests that 'there may be some instances' in favour of non-instantaneous progression). But the *picture* which he gives in the fifth remark (below) clearly depicts the propagation of light as a process taking place at a finite speed.

Fourth, 'the motion is propagated every way through an *Homogeneous medium* by *direct* or *straight* lines extended every way like Rays from the center of a Sphere'.[26]

And fifth, the important result that

in an *Homogeneous medium* this motion is propagated every way with *equal velocity*, whence necessarily every *pulse* or *vibration* of the luminous body will generate a Sphere, which will continually increase, and grow bigger, just after the same manner (though indefinitely swifter) as the waves or rings on the surface of the water do swell into bigger and bigger circles about a point of it, where by the sinking of a Stone the motion was begun, whence it necessarily follows, that all the parts of these spheres undulated through an *Homogeneous medium* cut the Rays at right angles.[27]

---

[25] *Micrographia*, p. 56.
[26] Ibid, pp. 56–57. [27] *Micrographia*, p. 57.

These ideas represent a definite advance towards a wave theory. Hooke makes use here of the concept of wave-front which he illustrates by the propagation of water-waves. And he correctly remarks that in a 'homogeneous' (we would say *isotropic*) medium, the wave-front describes a sphere which, at every point of its surface, is perpendicular to the ray or direction of propagation from the centre of disturbance. But, in spite of the analogy with water-waves, there is no reason to suppose that he necessarily understood the vibrations in the light-bearing medium to be transverse, that is, at right angles to the direction of propagation. Nor does he explicitly say that his pulses or waves follow one another at regular intervals. Not that he was aware of any reasons for excluding one of these concepts. But, rather, he was not in a position to make specific use of them, and, in consequence, he left out the whole question of what type of waves should correspond to light. It will be seen later that Huygens went further than Hooke (but not forward) by expressly denying the periodicity of light waves.

We come now to Hooke's application of the preceding ideas in his treatment of refraction. This is of importance since he devised a construction for the refracted ray which has been described as a 'rather crude anticipation of that of Huygens'[28] and it will be worthwhile to try to define Hooke's contribution here. The following is his account of what happens to a 'pulse' or wave-front when it passes from one medium into another:

But because all transparent *mediums* are not *Homogeneous* to one another, therefore we will next examine how this pulse or motion will be propagated through differingly transparent *mediums*. And here, according to the most acute and excellent Philosopher *Des Cartes*, I suppose the sign [i.e. sine] of the angle of inclination in the first *medium* to be to the sign of refraction in the second, As the density of the first, to the density of the second, By density, I mean not the density in respect of gravity (with which the refractions or transparency of *mediums* hold no proportion) but in respect onely to the *trajection* of the Rays of light, in which respect they only differ in this; that the one propagates the pulse more easily and weakly, the other more slowly, but more strongly. But as for the pulses

---

[28] Cf. A. Wolf, op. cit., p. 258.

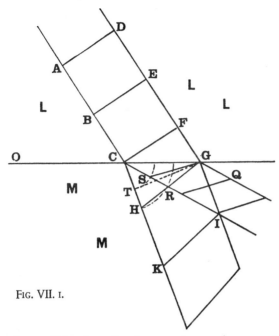

Fig. VII. I.

themselves, they will by the refraction acquire another propriety, which we shall now endeavour to explicate.

We will suppose therefore in the first Figure [Fig. VII. I] *ACFD* to be a physical Ray, or *ABC* and *DEF* to be two Mathematical Rays, *trajected* from a very remote point of a luminous body through an *Homogeneous* transparent *medium LLL*, and *DA, EB, FC*, to be small portions of the orbicular impulses which must therefore cut the Rays at right angles; these Rays meeting with the plain surface *NO* of a *medium* that yields an easier *transitus* to the propagation of light, and falling *obliquely* on it, they will in the medium *MMM* be refracted towards the perpendicular of the surface. And because this *medium* is more easily *trajected* then the former by a third, therefore the point *C* of the orbicular pulse *FC* will be mov'd to *H* four spaces in the same time that *F* the other end of it is mov'd to *G* three spaces, therefore the whole refracted pulse *GH* shall be *oblique* to the refracted Rays *CHK* and *GI*; and the angle *GHC* shall be an acute, and so much the more acute by how much the greater the refraction be, then which nothing is more evident, for the sign of the inclination is to be the sign of refraction as *GF* to *TC* the distance between the point *C* and the perpendicular from *G* to *CK*, which being as four to three, *HC* being longer then *GF* is longer also then *TC*, therefore the angle *GHC* is less than GTC. So that henceforth the parts of the pulses GH and IK are mov'd ascew, or cut the Rays at *oblique* angles.[29]

A clearer idea of how Hooke arrives at these results can be gathered from the dotted lines in his figure. Let the distance FG be equal to $v_i t$, where v_i is the velocity of incidence and t the time required for the side of the wave-front at F to cover that distance. It is supposed by Hooke that in the same time t, the side of the wave-front at C will have travelled in the refracting medium a distance $v_r t$ equal to $\frac{4}{3}FG$ (v_r being the velocity of refraction).

Accordingly, the larger dotted semi-circle in the figure is described with C as centre and a radius equal to $\frac{4}{3}FG$: the side of the wave-front at C will have reached some point on this semi-circle when the side at F arrives at G.

To find the *direction* of the refracted ray Hooke *assumes* with Descartes that

$$\frac{\sin i}{\sin r} = \frac{v_r}{v_i} = n = \frac{4}{3}.$$

Accordingly, with C as centre and a radius equal to $\frac{3}{4}FG$, he describes the smaller arc in the figure. He then draws the tangent to this arc from G; the line joining C and the point of tangency T gives the direction of refraction. For, FCG being equal to the angle of incidence and CGT being equal to the angle of refraction, we have

$$\frac{FG}{CT} = \frac{\sin i}{\sin r} = \frac{4}{3} = \frac{v_r}{v_i}.$$

To find the position of the wave-front after refraction, CT is extended to meet the larger arc at H. GH therefore represents the wave-front; it is oblique to the direction of propagation CH since, in the triangle GTH, the angle GTH is a right angle.

It is clear that according to Hooke's construction the velocity of light must be greater in denser media, since it is based on the Cartesian relation giving the sines in inverse ratio to the velocities.

We shall see that in Huygens' construction for refraction the wave-front must be perpendicular to the direction of propagation after

---

[29] *Micrographia*, p. 57.

refraction, and that this construction yields a law according to which one must adopt the opinion opposite to that of Descartes regarding the velocity of light in different media.[30] The crucial considerations on which Huygens relied in arriving at these results are lacking in Hooke's treatment. But, as Whittaker has remarked,[31] it was Hooke's merit to have introduced the concept of wave-front; and, by considering what the wave-front undergoes in passing from one medium into another, he has replaced Descartes' comparisons with a clear mechanical picture that was later more successfully used by Huygens.

Hooke attempted to explain the generation of colours by refraction as being due to the deflection of the wave-front. His explanation will be discussed later.[32]

3. Pardies' speculations on light were contained in an unfinished treatise on refraction which Huygens saw in manuscript before 24 June 1673.[33] This treatise has been lost;[34] but we can form some idea of its contents from Pardies' own reference to it in his published work, and from a book on optics published in 1682 by the Jesuit Father Pierre Ango who had also seen Pardies' manuscript and adopted some of its ideas.[35] As in the preceding discussion of Hooke we shall here concentrate on those conceptions which may have played a role in Huygens' theory.

Pardies intended his treatise to form part of a sixth discourse in a complete system of mechanics. Only the first two discourses have

---

[30] Had Hooke represented the wave-front by the perpendicular GT instead of the oblique GH, he would have obtained the following:

$$\frac{FG}{CT} = \frac{\sin i}{\sin r} = \frac{v_i t}{v_r t} = \frac{v_i}{v_r} = \frac{4}{3},$$

which is in accordance with Fermat's and Huygens' law. But this would of course imply that the velocity of light is *less* in denser media, and he would also have to give up the idea that the wave-front is deflected by refraction, that is, renounce the foundation of his theory of colours.

[31] Cf. Whittaker op. cit., pp. 15–16.

[32] See below, pp. 254 ff.

[33] Cf. Huygens to Oldenburg, 24 June 1673, H, VII, p. 316.

[34] Cf. H, XIX, p. 393 (*Avertissement*).

[35] Cf. Pierre Ango, *L'optique*, Paris, 1682.

been published (in 1670 and 1673 respectively).[36] We learn from the Preface to the second discourse that the sixth was to be devoted to wave motion 'on the example of these circles which are formed in the surface of water when a stone is dropped into it'.[37] The author added that he would consider how similar circles are formed, propagated, reflected and refracted—in air as well as in 'other more subtle substances'. He would also explain by considerations of wave motion 'all that concerns sounds';

and making a conjecture on the propagation of light, we examine if one could also suppose, that the vehicle of light were a similar movement in a more subtle air; and we show that in this hypothesis one would in fact explain all the properties of light and of colours in a very natural manner.

Pardies also promised that he would provide a demonstration of the law of refraction from his hypothesis. This in itself would constitute an advance over Hooke's treatment which, as we have seen, simply takes Descartes' formula for granted. We learn that Pardies did in fact attempt such a demonstration in his unpublished treatise from Huygens' testimony in the *Traité*.[38] But in order to get an idea of this demonstration we have to turn to Ango's *L'optique*.

Ango provides a construction for refraction which is not unlike that of Hooke[39]. Like Hooke, he considers what happens to an incident wave-front that is perpendicular to the direction of propagation when it obliquely strikes a refracting surface. The side of the wave-front which first meets the surface proceeds with the velocity proper to the refracting body while the other side travels with the velocity of the medium of incidence. But he departs from Hooke's construction in two important respects. First, he represents the wave-front as perpendicular to the direction of propagation *after refraction*. Second, he considers the velocity of light to be greater in rarer media. On these two points Ango's construction perfectly agrees with Huygens' theory. But we would search in vain for a

---

[36] Ignace Gaston Pardies, *Discours du mouvement local*, Paris, 1670; *La statique ou la science des forces mouvantes*, Paris, 1673.

[37] This and other quotations in the same paragraph are taken from the Preface to Pardies *La statique*. The pages of this Preface are not numbered.

[38] Cf. Huygens, *Treatise*, p. 20.   [39] Cf. Ango, op. cit., figure on p. 64; and pp. 59–67

satisfactory proof of the sine law in Ango's book. As Leibniz observed, he practically presupposes what is to be proved.[40] In view of Ango's opinion concerning the velocity of light in different media, the law which he adopts is the following

$$\frac{\sin i}{\sin r} = \frac{v_i}{v_r} = n.$$

This is the same as Huygens' (and Fermat's) law, as opposed to the law (giving the velocities in inverse ratio to the sines) proposed by Descartes and adopted by Hooke.

We may safely assume with Leibniz[41] that Ango's construction had been taken from Pardies' manuscript. Apart from Ango's own confession of his indebtedness to Pardies,[42] there is in his book a remark which may support this assumption. He informs us that Pardies had performed a demonstration of the principle of least time as applied to optical refraction and that Pardies had sent his demonstration to Fermat.[43] Neither Fermat nor Huygens mentions this demonstration. But, if correct, it would presuppose a law of refraction identical with that of Fermat and Huygens, and also the view (adopted by Ango and assumed in his construction) that the velocity of light is less in denser media.

We cannot of course be certain as to whether Pardies had provided in his manuscript a more satisfactory proof of the sine law than the one given in Ango's *L'optique*. At any rate, Huygens, who readily expressed his admiration for Pardies' treatise,[44] has stated in the *Traité* that the main foundation of his own theory (and of his deduction of the sine law) was lacking in Pardies' work.[45] And there is no reason to discredit this statement which is in fact confirmed by Ango's account. This foundation consists in the idea of secondary waves which Huygens made use of to explain the fact that the wave-front remains perpendicular to the direction of propagation after refraction. As it would seem from Ango's treatment, this fact Pardies had either simply assumed or attempted to explain in a different way from Huygens.

---

[40] Cf. Leibniz to Huygens, October 1690, H, IX, pp. 522–3. [41] Cf. ibid.
[42] Cf. Ango, op. cit., p. 14. [43] Cf. ibid., pp. 92–94.
[44] Cf. Huygens to Oldenburg, 24 June 1673, H, VII, p. 316.
[45] Cf. Huygens, *Treatise*, p. 20.

Chapter Eight

HUYGENS' WAVE THEORY

1. The fact that Huygens, like Hooke, occupied himself with problems in the Cartesian theory is clear enough from his *Traité de la Lumière*. Indeed, when he communicated this *Traité* to the *Académie des Sciences* in 1679, his task was understood as an attempt to solve 'difficulties' in Descartes' theory. The account of Huygens' communication, published in the *Histoire de l'Académie Royale des Sciences* in 1733, begins as follows:

> The inadequacies [*inconvenients*] of M. Descartes' system of light obliged M. Huygens to endeavour to conceive another more suitable for avoiding or resolving the difficulties. Such are the errors of M. Descartes that he often illuminates the other philosophers, either because, where he erred, he has not moved far from the goal and that the mistake can be easily corrected, or because he expresses deep views and produces ingenious ideas, even when he is mostly mistaken.[1]

There is in fact evidence to show that Huygens first arrived at his views regarding the nature of light and the mode of its propagation through an examination of Descartes' ideas.

Huygens started his work on dioptrics in 1652 when he was twenty-three years old[2] (about the same age as Newton when, in 1666, he arrived at his definitive theory of colours). He first conceived his wave theory in about 1672-3[3] (about three years after Newton had started lecturing on his theory of light in Cambridge). In (probably) 1673 Huygens was already planning a work under the title *Dioptrique* which was to contain his wave theory together with questions of dioptrics proper. We have in the *Projet*, which he wrote probably in that year, a brief but valuable indication of the problems

---

[1] Cited by Mouy in *Développement de la physique cartésienne*, p. 204. Cf. *Histoire de l'Académie Royale des Sciences*, I, *depuis son établissement en 1666 jusqu'à 1686*, Paris, 1733, pp. 283-93. This account was written in 1679 by Fontenelle in his capacity as secretary of the *Académie*.

[2] Cf. H, XIII, *fasc.* I, pp. v-x.

[3] Cf. ibid., pp. ix-x.

and applications which occupied Huygens' thought at that early stage in the formation of his theory. It will therefore be interesting first to examine this *Projet*, and to see the extent to which he was still under the influence of some of Descartes' views, while at the same time he realized, and tried to tackle, difficulties in the Cartesian theory. It was in this context that Huygens' wave theory was first conceived.

Owing to the particular character of the *Projet*, consisting mainly of phrases and incomplete sentences, I have adjoined the French original to my translations and, in some cases, inserted complement-ary words in brackets. To facilitate reference in my subsequent remarks I have provided Huygens' notes with numbers without changing their order in the published text.

From Huygens' '*Projet du Contenu de la Dioptrique*' (1673)[4]

(1) *Refraction comment expliquee par Pardies.*
    How refraction is explained by Pardies.

(2) *Comparée au son.*
    [Light] compared with sound.[5]

(3) *ondes en l'air.*
    waves in air.

(4) *comparées à celles de l'eau.*
    [Waves in air] compared to those of water.

(5) *la pesanteur est cause de celles cy comme le ressort des autres.*
    weight is the cause of these [water waves] as elasticity is the cause of the others.

(6) *transparence sans penetration.*
    transparency without penetration [of rays].[6]

(7) *corps capable de ce mouvement successif.*
    a body [or medium] capable of this successive movement [of light].

(8) *Propagation perpendiculaire aux cercles.*
    Propagation perpendicular to the circles [described by the expanding waves].

(9) *difficultez contre des Cartes.*
    difficulties against Descartes.

---

[4] H, XIII, *fasc.* II, p. 742.
[5] For the insertion here compare the following line from the same *Projet*: '*lumière comparée au son*', ibid., p. 739.
[6] Compare: '*transparence sans penetration de rayons*', ibid.

(10) *d'ou viendrait l'acceleration.*

whence would the acceleration come.

(11) *il fait la lumiere un conatus movendi, selon quoy il est malaisè d'entendre la refraction comme il l'explique, a mon avis au moins.*

he [Descartes] takes light to be a *conatus movendi* [or tendency to motion], in consequence of which it is difficult to understand refraction as he explains it, in my opinion at least.

(12) *Cause de la reflexion a angles egaux.*

Cause of reflection at equal angles.

(13) *lumiere s'estend circulairement et non dans l'instant, au moins dans les corps icy bas.*

light spreads in circles and not instantaneously, at least in the bodies here below.

(14) *car pour la lumiere des astres il n'est pas sans difficultè de dire qu'elle ne seroit pas instantanee.*

for as to the light from the stars, it is not without difficulty to say that it would not be instantaneous.

(15) *Cette explication convient avec les Experiences.*

This explanation agrees with the Experiments.

(16) *pour les sinus.*

for the sines.

(17) *pour le rayon entrant et sortant.*

for the ray passing into and out of [a medium].

(18) *pour celuy qui ne peut penetrer.*

for that [ray] which cannot penetrate [the interface].

(19) *pour le verre.*

for glass.

(20) *dans l'eau.*

in water.

(21) *maniere de Rohaut de faire voir les conveniences.*

the manner that Rohaut[7] shows the agreements.

(1) indicates that Huygens' projected work was to include an account of Pardies' explanation of refraction. The *Traité de la Lumière* does not give any such account. The analogy between light and sound, referred to in (2) must have been already developed by Pardies to some extent, as it appears from the discussion of his

---

[7] Reference to Jacques Rohault, author of *Traité de physique* which was first published in 1671. I have not been able to find the passage to which Huygens refers.

Statique in the previous chapter.[8] The comparison with water-waves did not lead Huygens to adopt transverse motion for light. Light waves are conceived in the *Traité* rather on the analogy of sound waves; that is, they are longitudinal. Consequently, elasticity of the ethereal matter, the vehicle of light, is assumed in accordance with the remark in (5). That Huygens was also thinking about Hooke's explanations at the time of writing the *Projet* is indicated by the following words which he wrote in the margin of the manuscript: 'vid. micrograph. Hookij'.[9]

(15–20) indicate applications of the theory. (16) refers to the sine law. (17) refers to the reciprocality law for refraction, namely the law stating that when a refracted ray is reversed, it retraces its original path in incidence after its emergence from the refracting body.[10] (18) refers to total reflection which takes place when a ray, falling at a sufficiently great angle of incidence on a separating surface from the side of the denser medium, does not penetrate the surface.[11]

(9–11), (13) and (14) directly deal with Descartes' theory. (10) points out the difficulty (noted in Chapter IV above) that is involved in Descartes' explanation of the increase in speed which, in his view, would take place when light passes from a rare into a dense body. We have seen that Descartes' model does not adequately provide for a plausible physical explanation of this hypothesis.[12] According to Huygens' theory the velocity of light must be less in denser media. On the assumption that light consists in waves in the ethereal

---

[8] Judging from Ango's *L'optique* which, as we have remarked, embodies Pardies' ideas, it would seem that Pardies had not developed the analogy between light and sound far enough. Ango observes (op. cit., pp. 38–39) that the loudness and pitch of sound respectively depend on the amplitude and the frequency of vibrations. But he does not attempt to relate the intensity and colour of light to corresponding properties of the light waves.

[9] H, XIII, *fasc.* II, p. 742, editors' note 1.

[10] Cf. Huygens, *Treatise*, pp. 35–39. Some historians have credited Ibn al-Haytham with the discovery of this law. He himself attributed it to Ptolemy, *Optics*, Bk. V. Cf. J. Baarmann, 'Abhandlung über das Licht von Ibn al-Haitam', *Zeitschrift der Deutschen Morgenländischen Gesellschaft*, XXXVI (1882) pp. 225–6 (Arabic text with German translation); M. Naẓīf, al-Ḥasan ibn al-Haytham' buḥūthuhu wa-kushūfuhu al-baṣariyya, I (Cairo, 1942), pp. 72–73. See Ptolemaei Optica (ed. Lejeune), V, 31–32, pp. 242–3 and editor's note 31.

[11] Cf. Huygens, *Treatise*, p. 39.

[12] See above, pp. 114 f, 133 ff.

matter which fills the spaces between the particles of transparent bodies, Huygens later suggested in the *Traité* that 'the progression of these waves ought to be a little slower in the interior of bodies, by reason of the small detours which the same particles cause'.[13]

(11) repeats the objection which had been raised by Fermat: how could light be a tendency to motion (rather than an actual motion) if, in the explanation of refraction, one has to argue (as Descartes did) in terms of greater and smaller velocities? The dilemma which Huygens seems to point out in Descartes' theory may be expressed thus: either light is an actual motion, or it is a tendency to motion existing simultaneously in the parts of the subtle matter. On the first assumption, the propagation of light must be successive and not instantaneous. On the second, it is 'difficult' to explain refraction. In (13) Huygens chooses the first alternative: 'light spreads in circles and not instantaneously. . . .'

At the same time (14) asserts that 'as to the light from the stars, it is not without difficulty to say that it would not be instantaneous'. (14) thus shows that Huygens had not completely shaken himself free from Descartes' eclipses arguments (which had been published in the second volume of Clerselier's edition of Descartes' letters in 1659). In contrast to Hooke, who questioned the conclusiveness of Descartes' arguments, Huygens thought (in 1673) that it would not be easy to assert the successive propagation of light in interplanetary space.[14]

But, at least 'in the bodies here below'—see (13)—the spreading of light must not be instantaneous; for here below, on earth, light is refracted by its passage through the atmosphere and the other transparent bodies; and to explain refraction, mechanically, one has to understand it as a process taking place in time.

Huygens therefore first adopted the finite velocity of light (in the bodies here below) about three years before Roemer's discovery,

---

[13] Huygens, *Treatise*, p. 32.
[14] In the communication of 1679 Huygens was still of the opinion that Descartes' belief in the instantaneous propagation of light as based on the arguments from the eclipses was 'not without reason', thus admitting their validity though, by that time, he was fully aware of their inconclusiveness and convinced of the contrary opinion to Descartes'. Cf. Huygens' *Treatise*, p. 5.

not because any terrestrial experiment had forced him to do so,[15] but simply because this hypothesis was required for a clear understanding of the properties of light, and particularly, of refraction. It is implied in the *Traité* that Huygens had changed his mind also about Descartes' eclipses arguments before Roemer announced his results. One of the reasons given by Huygens confirms the preceding account of the development of his thought:

I have then made no difficulty, in meditating on these things [Descartes' eclipses arguments], in supposing that the emanation of light is accomplished with time, *seeing that in this way all its phenomena can be explained, and that in following the contrary opinion everything is incomprehensible.* For it has always seemed to me that even Mr Des Cartes, whose aim has been to treat all the subjects of Physics intelligibly, and who assuredly has succeeded in this better than any one before him, has said nothing that is not full of difficulties, or even inconceivable, in dealing with Light and its properties.

But that which I employed only as a hypothesis, has recently received great seemingness [*vraisemblance*] as an established truth by the ingenious proof of Mr Römer . . . .[16]

We thus see what exactly Roemer's contribution was to Huygens' attitude towards Descartes' eclipses arguments. As Roemer had estimated that light would take eleven minutes to reach the earth from the sun, Huygens could now see precisely why the eclipses of the moon did not provide a sufficient argument for deciding the

---

[15] Galileo described in the *Two New Sciences* a terrestrial experiment designed to decide the question of the velocity of light. Two observers sent light messages to each other by uncovering a lantern; as soon as the second observer perceives the light sent off by the first, he uncovers his own lantern which is then seen by the first observer; the interval between sending a message by one observer and receiving an answer from the other would be the measure of the time required for the light to travel the distance between them twice. But, we are told by Salvatio, the result was negative: 'In fact I have tried the experiment only at a short distance, less than a mile, from which I have not been able to ascertain with certainty whether the appearance of the opposite light was instantaneous or not; but if not instantaneous it is extraordinarily rapid—I shall call it momentary.' Galileo's real reason for not accepting instantaneous propagation was theoretical, not experimental: 'What a sea we are gradually slipping into without knowing it! With vacua and infinities and indivisibles and instantaneous motions, shall we ever be able, even by means of a thousand discussions, to reach dry land?' *Two New Sciences*, pp. 43–44.
[16] *Treatise*, p. 7; italics added.

question of the velocity of light. This is Huygens' re-formulation of Descartes arguments:[17]

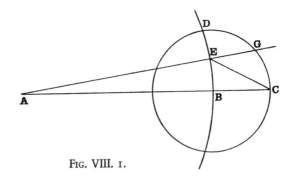

FIG. VIII. 1.

In the figure (Fig. VIII. 1) *A* is the place of the sun, supposed in the argument to be at rest. *BD* is part of the annual path of the earth round the sun, and the circle *CD*, the orbit of the moon. *AB* is a beam of light interrupted by the earth arriving at *B*. Assuming that light takes one hour to travel the distance *BC* (*AB* extended), a shadow will be cast on the moon if the moon happens to be at *C* one hour after the earth was at *B*. The reflection of the shadow will reach the earth at *E* another hour later, *BE* being the arc described by the earth in two hours. Therefore, two hours after the earth was at *B*, the moon will be seen from *E* eclipsed at *C* while the sun is still at *A*. According to Huygens' computations, the angle *GEC* (the supplementary of *AEC*) will be 'very sensible, and about 33 degrees'.[18] This was contrary to astronomical observations according to which this angle was unappreciable.

But it must be noted that the speed of light in this argument has been assumed such that it takes a time of one hour to make the passage from here to the Moon. If one supposes that for this it requires only one minute of time, then it is manifest that the angle *CEG* will only be 33 minutes; and if it requires only ten seconds of time, the angle will be less than six minutes. And then it will not be easy to perceive anything of it in

---

[17] Cf. ibid., pp. 5–6.
[18] Ibid., p. 6.

observations of the Eclipses; nor, consequently, will it be permissible to deduce from it that the movement of light is instantaneous.[19]

In the last supposition the velocity of light would be according to Huygens a hundred thousand times greater than that of sound. But as Roemer had in fact estimated the velocity of light to be 'at least six times greater'[20] than in this supposition, it was now understandable why the eclipses of the moon could not reveal the successive movement of light.

But it should now be observed that Roemer's discovery itself was not commonly accepted as definitive in Roemer's time. Roemer proposed the hypothesis that light travels with a finite velocity in connection with observations on the eclipses of the first (innermost) satellite of Jupiter which he had been making for several years with Cassini and other members of the *Académie des Sciences* in Paris. It was observed that the periods of revolutions of this satellite round Jupiter exhibited certain inequalities which were found to be related to the orbital motion of the earth: the eclipses of the satellite were delayed when the earth was farthest from Jupiter, and accelerated when the earth was in the contrary position. Roemer attributed these irregularities to the fact that the light had to travel a longer distance in the former case than in the latter and, on the basis of this hypothesis, he calculated the velocity of light to be such as to require eleven minutes to travel a distance equal to the radius of the earth's orbit.[21]

Roemer was not the first to propose the hypothesis of finite velocity of light to account for these observations. Cassini put forward the same hypothesis in August 1675 but withdrew it shortly afterwards.[22] Roemer adopted it and showed that it was in agree-

---

[19] Ibid., pp. 6–7

[20] Ibid., p. 7.

[21] Cf. '*Démonstration touchant le Mouvement de la Lumière trouvé Par M. Roemer*' in *Mémoires de l'Académie Royale des Sciences, depuis 1666 jusqu'à 1699*, X (Paris, 1730), pp. 575–7.

[22] This fact is generally ignored in recent histories of the subject. Cf. the account of Maraldi's communication to the *Académie des Sciences* on 9 February 1707 (entitled 'Sur la seconde inégalité des satellites de Jupiter') in *Histoire de l'Académie Royale des Sciences, année 1707. Avec les Mémoires de mathématique & de physique pour la même année. Tirés des Registres de cette Académie*. Paris, 1708, p. 78, Cf. also J.-E. Montucla, *Histoire des mathématiques, nouvelle édition*, II, Paris, AN VII, p. 579.

ment with all the observations on Jupiter's first satellite which had been recorded at the *Académie* for eight years. He also predicted that the emergence of the satellite from Jupiter's shadow which was to occur on 16 November 1676 (Roemer announced his hypothesis to the *Académie* in September 1676) should be later than would be expected on the ordinary calculations; and his prediction was confirmed. Nevertheless, Cassini 'who was better informed than any one else'[23] about the world of Jupiter, could not accept Roemer's explanation; and in this he was not alone. In fact there were serious difficulties. It was argued, for example, that if the hypothesis were true, then the periods of the other three satellites of Jupiter would exhibit the same inequalities as those of the first; but this was not observed.[24] We are told in the *compte-rendu* of Roemer's communication in the *Histoire de l'Académie des Sciences* that another astronomical hypothesis was put forward which would take 'all the Observations' into account without assuming the finite velocity of light, but that that hypothesis lacked the '*vrai-semblance*' which Roemer's hypothesis seemed to have.[25] '*Il falut donc admettre le Retardement de la Lumiere, si vrai-semblable selon la Physique, quand il ne seroit pas prouvé par l'Astronomie.*'[26] It would seem from the text of the *Histoire* that the discussion between members of the *Académie des Sciences* concerning Roemer's hypothesis gradually shifted from the purely astronomical arguments (which, to some of them, appeared inconclusive) to turn round the question whether that hypothesis was acceptable from a *physical* point of view. There is no doubt that Huygens would be foremost among those who answered this question in the affirmative. But, even after the publication of Huygens' *Traité de la Lumière* in 1690, Roemer's hypothesis was not universally accepted by members of the *Académie des Sciences*. Some of them still believed that the observed inequalities of periods might be due to the eccentricity of the satellite or to the irregularity of its

---

[23] Cf. *Histoire de l'Académie Royale des Sciences*, I, *depuis son établissement en 1666 jusqu'à 1686*, Paris, 1733, p. 214.

[24] On this and other difficulties, see Montucla, op. cit., II, pp. 580-1.

[25] Cf. *Histoire de l'Académie Royale des Sciences*, I, *depuis son établissement en 1666, jusqu'à 1686*, Paris, 1733, p. 214.

[26] Ibid., p. 215.

motion 'or to some other cause which may be revealed in time'.[27]
In view of all this one is inclined to conclude that Huygens accepted
Roemer's 'demonstration' not so much because he saw in it an
'impressive revelation of facts'[28] but, rather, because it was in agreement with what he had adopted as a physical hypothesis which, as
we have seen, he had required for a clear explanation of the properties of light.

Among the properties which Huygens proposed to explain in the
Traité is the fact that rays of light are not interrupted by crossing one
another.[29] His explanation was an attempt to solve another difficulty arising from the Cartesian conception of light and the nature of
the light-bearing medium. To sharpen the difficulty Huygens
considers the special case in which the rays are opposite and in the
same line, as when two torches illuminate each other or when two
eyes view one another: if, as Descartes supposes, light is a tendency
to movement and not a successive motion, how could one and the
same particle tend to move in two opposite directions at the same
instant? To overcome this difficulty Huygens finds it necessary to
endow the ethereal matter with the property of elasticity and,
consequently, he rejects the doctrine of instantaneous propagation.
This he illustrates by reference to the following experiment.

Imagine a row of contiguous spheres of equal size. If two similar
spheres strike the ends of the row in opposite directions, they will
both rebound with their velocity before impact, while the rest of the
spheres remain motionless. This indicates to Huygens that the
movements have passed through the length of the row in both
directions. 'And if these contrary movements happen to meet one

---

[27] Cf. the text quoted by the editors of Huygens' works from *Recueil d'Observations faites en
plusieurs voyages par ordre de Sa Majesté, pour perfectionner l'astronomie et la geographie, avec
divers traitez astronomiques, par Messieurs de l'Académie Royale des Sciences*, Paris, 1693, H, XIX,
pp. 400–1, *Avertissement*. For a detailed discussion of Roemer's investigations on the velocity
of light and a bibliography on this subject, see I. Bernard Cohen, 'Roemer and the first
determination of the velocity of light (1676)', *Isis*, XXXI (1940), pp. 327–9.

[28] Cf. Mouy, *Développement de la physique cartésienne*, p. 201.

[29] Cf. *Treatise*, pp. 21–22. Ibn al-Haytham described an experiment (the well-known
camera obscura experiment) to show that beams of coloured light do not mix, and therefore do
not affect one another, when they meet in space (at the opening of 'the dark place'). Alhazeni
Optica, I, 29, in *Opticae thesaurus*, p. 17: 'Et significatio, quod luces et colores non per-
misceantur in aere, neque in corporibus diaphanis, est: quod quando in uno loco fuerint

another at the middle sphere . . . that sphere will yield and act as a spring at both sides, and so will serve at the same instant to transmit these two movements.'[30] To understand the corresponding effects of light one must therefore suppose the ether particles to be capable of contraction and expansion, which must require time.

But in my judgement . . . [these effects] are not at all easy to explain according to the views of Mr Des Cartes, who makes Light to consist in a continuous pressure merely tending to movement. For this pressure not being able to act from two opposite sides at the same time, against bodies which have no inclination to approach one another, it is impossible so to understand what I have been saying about two persons mutually seeing one another's eyes, or how two torches can illuminate one another.[31]

On account of the extremely high velocity of light the particles of the ether are supposed to have perfect hardness and elasticity. Huygens is not willing to explain in detail in what the elasticity of the ether consists. But the explanation which he gives 'in passing' is interesting from the point of view of the present discussion.

. . . we may conceive that the particles of the ether, notwithstanding their smallness, are in turn composed of other parts and that their springiness

---

multae candelae in locis diversis et distinctis, et fuerint omnes oppositae uni foramin pertranseunti ad locum obscurum, et fuerit in oppositione illius foraminis in obscuro loco paries, aut corpus non diaphanum, luces illarum candelarum apparent super corpus vel super illum parietem, distinctae secundum numerum candelarum illarum, et quaelibet illarum apparet opposita uni candelae secundum lineam transeuntem per foramen. Et si cooperiatur una candela, destruetur lux opposita uni candelae tantum; et si auferatur coopertorium revertetur lux. Et hoc poterit omni hora probari: quod si luces admiscerentur cum aere admiscerentur cum aere foraminis, et deberent transire admixtae, et non distinguerentur postea. Et nos non invenimus ita. Luces ergo non admiscentur in aere, sed quaelibet illarum extenditur super verticationes rectas; et illae verticationes sunt aequidistantes, et secantes se, et diversi situs. Et forma cuiuslibet lucis extenditur super omnes verticationes, quae possunt extendi in illo aere ab illa hora; neque tamen admiscentur in aere, nec aer tingitur per eas, sed pertranseunt per ipsius diaphanitatem tantum, et aer non amittit suam formam. Et quod diximus de luce, et colore, et aere, intelligendum est de omnibus corporibus diaphanis, et tunicis visus diaphanis.' The sentence 'Et nos non invenimus ita.' has been wrongly interpreted to mean that Ibn al-Haytham had not himself discovered this (S. L. Polyak, *The Retina*, Chicago, 1941, p. 133; Colin Murray Turbayne, *The Myth of Metaphor*, Yale University Press, 1963, p. 155). In fact he is simply saying that we do not observe (find) what would have resulted had the lights been mixed at the opening; therefore, etc. Ibn al-Haytham did not attempt a mechanical explanation of the property he asserts here.

[30] *Treatise*, p. 18.   [31] Ibid, p. 22.

consists in the very rapid movement of a subtle matter which penetrates them from every side and constrains their structure to assume such a disposition as to give to this fluid matter the most overt and easy passage possible. This accords with the explanation which Mr Des Cartes gives for the spring [*ressort*: elasticity], though I do not, like him, suppose the pores to be in the form of round hollow canals. And it must not be thought that in this there is anything absurd or impossible, it being on the contrary quite credible that it is this infinite series of different sizes of corpuscles, having different degrees of velocity, of which Nature makes use to produce so many marvellous effects.[32]

Descartes had tried to explain the elasticity which 'generally exists in all the bodies whose parts are joined by the perfect contact of their small superficies' by the fluid matter which, being itself incapable of reduction in volume, was not elastic.[33] He assumed those bodies to have pores through which the subtle matter incessantly streams. The pores are normally disposed in such a way as to give the subtle matter the most free and easy passage and therefore are in the form of circular canals to accommodate the spherical shape of the parts of the streaming matter. When the body is deformed its pores take a different shape, and the subtle matter must consequently exert some pressure on the walls of the canals, thus endeavouring to make the body regain its original form. In Descartes' system the incompressibility of the fluid matter was suited to the supposed instantaneous transmission of light. Huygens, on the other hand, requires the elasticity of ether for the successive propagation of light. Proceeding on Cartesian lines he explains the elasticity of the ethereal particles by the action of smaller particles and these, in turn, he supposes to be penetrated by still smaller particles, and so on to infinity.

2. Huygens agreed with Descartes (against the views of Gassendi and Newton) that the motion of light cannot consist in the transport of bodies from the luminous object to the eye. Like Descartes, he

---

[32] Ibid., p. 14.
[33] Descartes, *Principles*, IV, 132.

thought it inconceivable that a body, however small, could travel with such a great velocity as that of light, and found the corpuscular conception of rays unsuited to the fact that light rays meet in space without affecting one another. But whereas Descartes thought the action of light in the intervening medium to be instantaneous, Huygens was led, through an examination of the various difficulties involved in Descartes' hypotheses, to adopt the view that light must be a process taking place at a finite rate. This in turn led him to introduce elasticity as a property of the light-bearing medium. Huygens thus replaced the Cartesian conception of light rays as mere geometrical lines whose points are all simultaneous with one another[34] by another picture, already developed by Hooke and Pardies, in which the straight lines are cut by spherical surfaces representing successive loci of a central disturbance. The reform thus introduced by seventeenth-century wave theorists consisted in replacing Descartes' geometristically static picture by another that was truly capable of clear mathematical treatment.[35] To Huygens belongs the merit of having taken the first step in the successful application of mathematics to it. This may be illustrated by Huygens' explanations of rectilinear propagation, ordinary and double refractions.

Like Pardies, Huygens draws the analogy between light and sound: both are propagated by spherical surfaces or waves. But Huygens does not extend the analogy any further than that. He first remarks that their modes of production are different: whereas sound is produced by the agitation of the sounding body as a whole, or of a considerable part of it, light originates 'as from each point of the luminous object'.[36] More important is the difference in their manner of transmission. For the medium serving for the transmission of sound is air; air can be very much compressed and reduced

---

[34] See above, p. 60.

[35] Descartes' conception of the transmission of light, and of light rays as 'nothing more' than geometrical lines, is another example of what Alexandre Koyré would have called 'géométrisation à outrance'; it was indeed geometrical, but, paradoxically, too geometrical to allow of anything much being done with it mathematically. To develop it mathematically the points on the rectilinear rays had to be made successive, not simultaneous.

[36] Huygens, Treatise, p. 10. Cf. Alhazeni Optica, I, 15, in Opticae thesaurus, p. 8.

in volume, and the more it is compressed the more it exerts an effort to regain its original volume. This proves to Huygens that air is made up of small bodies which are agitated by the smaller particles of the ethereal matter, and that sound spreads by the effort of the air particles to escape from the place where they are squeezed together at the regions of compression along the propagating waves. 'But the extreme velocity of Light, and other properties which it has, cannot admit of such a propagation of motion . . . .'[37] The picture which Huygens imagines for the propagation of light bears certain traces of the Cartesian picture. The particles of ether are supposed to act like the row of contiguous and equal spheres mentioned before. If a similar sphere strikes against one end of the row, it will communicate the whole of its motion to the sphere at the other end, while the rest of the spheres remain motionless. This shows that the movement has passed from one end to the other 'as in an instant',[38] but not instantaneously. For if the movement 'or the disposition to movement, if you will have it so',[39] were not transmitted successively, the spheres would all move together at the same time, which does not happen. The ether particles must touch one another in order to be able to transmit the action of light in the same way. This is to be contrasted with the picture which Newton envisaged and in which he supposed the ether to be 'exceedingly more rare . . . than Air'.[40] Huygens, unlike Descartes, sees no reason why the particles should be spherical in shape. But in order to render the propagation easier and to avoid considerable reflection of movement backwards, he finds it desirable that they should be of equal size.[41] Following Descartes, he also remarks that the various motions of the ether particles among themselves cannot hinder the movement of light, as the latter is not effected by the transport of these particles themselves. As has been noted before, every particle of

---

[37] *Treatise*, p. 12.

[38] Ibid., p. 13.

[39] Ibid., p. 13. These words do not occur in the 1678 version of the *Traité* but were added by Huygens in the edition of 1690. Their addition makes the picture even more Cartesian. Cf. H, XIX, p. 383 (*Avertissement*).

[40] Newton, *Opticks*, Query 18, p. 349.

[41] *Treatise*, p. 16.

the luminous body must be regarded as the centre of its own spherical wave. 'But', and this is where Huygens explicitly rejects the periodicity of light, 'as the percussions at the centres of these waves possess no regular succession, it must not be supposed that the waves themselves follow one another at equal distances'.[42] Thus, Huygens' 'waves' are in fact isolated pulses that are generated in the ether by a succession of impacts which follow one another at irregular intervals.

Into the midst of this picture Huygens introduces the idea of secondary waves which, in fact, forms the basis of his explanations of the properties of light. The first consideration in the principle governing these waves consists in the remark that

each particle of matter in which a wave spreads, ought not to communicate its motion only to the next particle which is in the straight line drawn from the luminous point, but that it also imparts some of it necessarily to all the others which touch it and which oppose themselves to its movement. So it arises that around each particle there is made a wave of which that particle is the centre.[43]

This may be stated briefly thus: on any wave surface, each point may be regarded as centre of a particular or secondary wave travelling with the same velocity as the initial principal wave. The condition about the velocity is implied in Huygens' exposition of the principle.

Thus if *DCF* [Fig. VIII. 2] is a wave emanating from the luminous point *A*, which is its centre, the particle *B*, one of those comprised within the sphere *DCF*, will have made its particular or partial wave, *KCL*, which will touch the wave *DCF* at *C* at the same moment that the principal wave emanating from the point *A* has arrived at *DCF*; and it is clear that it will be only the region *C* of the wave *KCL* which will touch the wave *DCF*, to wit, that which is in the straight line drawn through *AB*. Similarly the other particles of the sphere *DCF*, such as *bb*, *dd*, etc., will each make its own wave. But each of these waves can be infinitely feeble only as compared with the wave *DCF*, to the composition of which all the others contribute by the part of their surface which is most distant from the centre *A*.[44]

---

[42] *Treatise*, p. 17.   [43] Ibid., p. 19.   [44] Ibid.

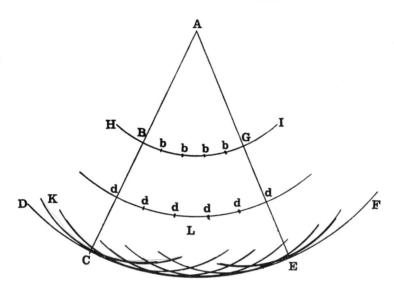

FIG. VIII. 2.

These remarks contain Huygens' fundamental contribution to the wave theory of light. They are not to be found in Hooke's treatment of the subject, and Huygens tells us that they were also lacking in Pardies' manuscript.[45]

To explain rectilinear propagation Huygens adds a further assumption: the particular waves are not effective except where they concur at the same instant, i.e. at their point of tangency with a principal wave. Thus, in Fig. VIII. 2, let BG be an opening limited by opaque bodies BH and GI. The portion BG of a principal wave-front emanating from A will in a certain time t spread out into the arc CE, where C and E are, respectively, in the lines AB and AG extended, and the distance $BC = GE = vt$ (v being the velocity of propagation). In the same time, the secondary waves which have originated (for example) at bb will have travelled, all in *forward* directions, a distance equal to vt.[46] Their common tangent (or

---

[46] Huygens excludes the possibility of back waves travelling towards the centre of disturbance A by supposing the ether particles to be equal. Accordingly, he does not consider the other envelope which can be obtained by completing the small circles in his figure. The mathematical construction is here limited by a physical condition.

envelope) will be the wave-front *CE*. And since it is assumed that the secondary waves are not effective except where they simultaneously concur to make up a single wave, it is concluded that their effect will be sensible only within the space *BGCE*; that is, the illuminated area will be limited by the straight lines *AC* and *AE*.

As opposed to the preceding explanation, Newton had maintained in the *Principia* (which appeared before the publication of Huygens' *Traité*) that any disturbance propagated in a fluid must diverge from a rectilinear progress.[47] On the basis of dynamical considerations he showed that when waves are partially allowed to pass through an opening in a barrier, their progress will not be terminated by straight lines drawn from the centre of disturbance, but will diverge beyond those lines and spread throughout the shaded portion of the medium. This was confirmed by familiar observations on the propagation of water waves and by the fact that sound can be heard round obstacles. As will be seen later,[48] these considerations led Newton to reject the wave hypothesis as applied to light, since in his view this hypothesis was founded on a false analogy between light and sound.

Huygens tried to answer Newton's objections in the *Addition* to his *Discours de la Cause de la Pesanteur* (published in the same volume with the *Traité* in 1690).[49] He recognized the lateral communication of movement into the geometrical shadow; but repeated his remark in the *Traité* that the secondary waves which do cross into that region are too feeble to produce any sensible effect, the reason being that they arrive there individually and not concurrently and, therefore, fail to strengthen one another. Newton was not convinced by this argument,[50] and quite understandably, since it was not clear why the same considerations could not apply to sound. In fact no satisfactory explanation of rectilinear propagation from the wave-theory point of view was given until the beginning of the nineteenth century. It was Fresnel who first showed that in places well beyond

---

[47] Newton, *Principia*, Bk. II, Sec. VIII, Prop. XLII, Theor. XXXIII, pp. 369–71.
[48] See below, pp. 282f.
[49] Cf. H, XXI, pp. 474–5.
[50] This we gather from the fact that he repeated the same objections in the *Opticks*. Cf. Query 28 (pp. 362–3) which first appeared as Query 20 in the Latin edition of 1706.

the limits of the geometrical shadow the secondary waves (which are, in Fresnel's theory, proper wave-trains) arrive with phase relations such that they neutralize one another to produce darkness. Fresnel's theory *simultaneously* explained the slight bending of light which does take place in the immediate vicinity of those limits. Some of the fundamental ideas on which Fresnel relied in his explanations were available neither to Huygens nor to Newton.

Ernst Mach in his *Principles of Physical Optics* considers the question of how Huygens arrived at the ideas involved in his principle of secondary waves and in his explanation of rectilinear propagation. He suggests that Huygens' ideas, in their essentials, 'evidently have a two-fold origin, his natural experience and his particular line of thought'.[51] That particular line of thought, according to Mach, derives from Huygens' earlier work on the percussion of elastic spheres which he took as a model for the transmission of light through ether. Regarding the contribution of Huygens' 'natural experience', Mach suggests the following:

It would be quite natural for the son of a seafaring people and the inhabitant of a town intersected by canals to make observations upon water waves. As a boy, no doubt, he would have thrown pebbles into water and observed the interaction of ripples generated simultaneously or successively at two different points. He could not fail to have observed the wedge-shaped waves produced by moving a stick rapidly through water in one direction, or the waves from the bow of a moving boat, and must have recognized that practically the *same* effect is produced by dropping several pebbles into the water in succession at intervals along a straight line. These observations contain the nucleus of his most important discoveries.[52]

The origin of these remarks is evidently Mach's particular line of thought, and they are in accordance with his view of scientific theories as being ultimately derived from observation and experiment. He actually deduces the sine relation from a situation similar to that in which waves are produced on a water surface by successively dropping pebbles into it along a straight line. But he assumes in his deduction that the velocity of the waves is different on either

---

[51] Mach, op. cit., p. 257. [52] Ibid.

side of the line joining their centres. This surely was not *observed* by Huygens in the canals of Amsterdam. It would be nearer to the truth to present Mach's deduction as an *application* of some already developed ideas to a hypothetical situation. In the absence of any evidence to the contrary, Mach's deduction cannot be regarded as indicating the 'origin' of Huygens' own discoveries. Indeed, as Newton pointed out, more obvious and less complex observations than those described by Mach clearly pronounced against Huygens' ideas as applied to rectilinear propagation.[53]

In any case Huygens himself does not cite observations in support of his principle of secondary waves. On the contrary, he was aware of its unfamiliar character. After the exposition of that principle he wrote: 'And all this ought not to seem fraught with too much minuteness or subtlety, since we shall see in the sequel that all the properties of Light, and everything pertaining to its reflexion and its refraction, can be explained in principle by this means.'[54] The only justification given here by Huygens for his new ideas is the fact that he could explain the properties of light by their means. Had he been familiar with observations supporting his principle, would he not have cited them instead of admitting that it might seem 'fraught with too much minuteness or subtlety'?

3. Huygens' construction for ordinary refraction may be obtained by application of his principle of secondary waves as follows.[55] In the figure (Fig. VIII. 3) let AC be a plane wave-front which obliquely strikes the separating surface AB at A. AC is perpendicular to the direction of incidence DA. Let the distance CB (parallel to DA) be equal to $v_i t$, where v_i is the velocity of the medium above and t the time required for C to arrive at B. In the same time, the secondary

---

[53] Historically more interesting than Mach's remarks is the similarity, noted by Dijksterhuis (*The Mechanization of the World Picture*, translated by C. Dikshoorn, Oxford, 1961, p. 149), between Huygens' principle and the medieval doctrine of the multiplication of species as applied to the propagation of light, in particular the idea which this doctrine entailed that every illuminated point becomes itself a further source of illumination by virtue of a certain agency being conferred upon it.

[54] Huygens, *Treatise*, p. 20.

[55] Cf. *Treatise*, pp. 35–39.

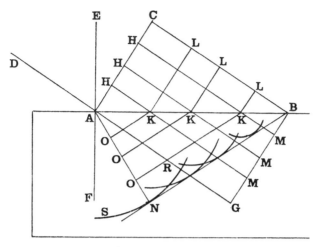

FIG. VIII, 3.

wave generated at A will have travelled in the medium of refraction a distance equal to $v_r t$, where v_r is the velocity in this medium.

Supposing, then, that

$$\frac{v_i}{v_r} = n,\ \text{a constant } greater\ or\ less\ \text{than unity,}$$

the circumference SNR, having A as centre and a radius equal to $\frac{1}{n}CB$, will give the position of the secondary wave at the time when the part of the incident wave-front at C arrives at B. The other arcs in the figure are similarly drawn (with points K as centres and radii that are equal to the distances LB each divided by n) by considering the wave-fronts KL in order. All these arcs will have as a common tangent the straight line BN which is the same as the tangent from B to the arc SNR. BN therefore gives the position which the wave-front has reached successively by taking the positions LKO in order. And the direction of propagation after refraction is represented by the perpendicular AN.

From this the sine relation is readily obtained. For since the angle of incidence EAD is equal to the angle CAB, and the angle of refraction FAN is equal to ABN, it follows that

$$\frac{\sin i}{\sin r} = \frac{CB}{AB} \times \frac{AB}{AN} = \frac{CB}{AN} = \frac{v_i t}{v_r t} = \frac{v_i}{v_r} = n.$$

217

This law implies that when the angle of refraction is smaller than the corresponding angle of incidence, the velocity must have been diminished by refraction. And since light in passing from a rare into a dense medium is deflected towards the normal, it must be concluded that the velocity of light is greater in rarer media. Huygens' law is the same as that deduced by Fermat (from the least-time principle) and maintained by Pardies and Ango. But whereas Ango and (perhaps) Pardies simply assumed the wave-front to be perpendicular to the direction of propagation after refraction, this is presented by Huygens as a *consequence* of regarding the wave-front as a resultant wave composed from the secondary waves generated successively at the surface of the refracting medium. Fermat's principle of least time then follows as a result of Huygens' law. Although this is already clear from Fermat's proof by synthesis (given in Chapter V above),[56] Huygens' simpler and shorter deduction may be summarized here.

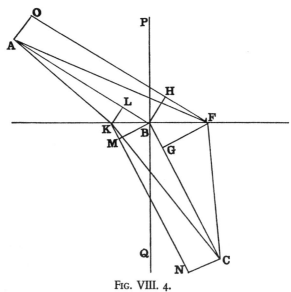

Fig. VIII. 4.

Let *A*, *C* (Fig. VIII. 4) be any two points on the incident and refracted rays *AB*, *BC* respectively—the direction of refraction being

---

[56] Above, pp. 150ff.

determined by the law stated above.[57] It is to be proved that the actual path ABC is that for which the time employed is a minimum. Let AFC be any other path on the right-hand side of B. Huygens proves first that the time along AFC is longer than the time along ABC, as follows.

He draws OA perpendicular to AB, FG perpendicular to BC, OF parallel to AB, and BH perpendicular to OF. BH may therefore represent an incident wave-front which takes the position FG after refraction, and we thus have:

$$\frac{\sin i}{\sin r} = \frac{HF}{BG} = \frac{v_i}{v_r} = n.$$

This means that the time along HF is equal to that along BG: from which it follows that the time along OF is equal to the time along ABG. But, AF being greater than OF, and FC greater than GC, it follows that the time along the hypothetical path AFC would be longer than that required along ABC. (In this case BM may be taken to represent the incident wave-front which occupies the position KL after refraction.) ABC is therefore the path for which the time required is a minimum.

The construction which Huygens gives for the reflection of light is obtained by a similar application of the principle of secondary waves as in the case of refraction.[58] On the supposition that the velocity is unaltered by reflection, he draws the arcs representing the secondary waves that are successively generated (by the incident wave-front) at the reflecting surface in the medium of incidence; their common tangent is drawn, and the direction of the resultant wave is then found to make an angle with the normal equal to the angle of incidence.

Huygens also shows[59] that when the light falls on a separating surface from the side of the denser medium (that is, the medium in which the velocity is less) at an angle whose sine is greater than the refractive index $n\left(\dfrac{v_i}{v_r}\right)$, no tangent can be found common to all the secondary waves which are supposed to be excited in the rarer

---

[57] Huygens, *Treatise*, pp. 42–45.　　　　[58] Ibid., pp. 23–25.
[59] Ibid., pp. 39–41.

medium on the other side of the interface. This means that no resultant wave can be constructed in the medium of refraction, and it is therefore inferred that, in this case, the refracted light will not be perceived. At the same time a common envelope can be found for the secondary waves in the medium of incidence. Huygens' principle is thus shown to be in agreement with the phenomenon of total reflection. At the critical angle of incidence whose sine is exactly equal to *n*, the secondary waves strengthen one another in the direction of the separating surface; that is, the refracted ray will travel along the surface, also in accordance with observation.

It is known from experience that, in general, light is *partially* reflected at refracting surfaces. For example, a beam of light travelling through a block of glass is in general split up into two beams upon meeting the surface of the glass: the one beam is refracted into the outer medium (say air) while the other is internally reflected within the block. From a geometrical point of view this raises no difficulties for Huygens' theory: the refracted beam is determined by the construction for ordinary refraction, and the reflected beam is determined by the construction for reflection. But how should we understand this phenomenon mechanically; how does one and the same surface reflect some of the incident light and refract the rest? Huygens' suggestion was that (in this case) the interior reflection is produced by the particles of the air outside, while the refracted light is transmitted through the ethereal matter. But he himself realized the 'difficulty' involved in such an explanation: 'It is true that there remains here some difficulty in those experiments in which this interior reflection occurs without the particles of air being able to contribute to it, as in vessels or tubes from which the air has been extracted.'[60]

---

[60] *Treatise*, p. 42. The 14th-century Persian scholar Kamāl al-Dīn noted the problem of partial reflection in the following comment on Ibn al-Haytham's explanation of reflection by analogy with the mechanical reflection of bodies from smooth surfaces: 'This is questionable. For if smoothness prevents the transmission of light and forces it to be reflected, then how is it refracted into bodies whose transparency differs from that of the bodies in which it exists; and if smoothness does not prevent its transmission, then how is light reflected from the surfaces of liquids while it still goes through them? And it cannot be said that one and the same light is both transmitted and reflected, the one thus becoming two.' Kamāl al-Dīn concluded by rejecting the analogy with moving spheres and preferred a wave interpretation

Newton interpreted these experiments (and other related experiments which he described in the *Opticks*)[61] to indicate that a purely mechanical picture of reflection was fundamentally unsuitable for explaining the observed behaviour of light. In this he has been confirmed by the later development of optics. Thus, while Huygens' geometrical construction remains classical, the mechanical picture which he used has long been discarded. Duhem would find in this an illustration of his view that only the representative (geometrical or experimental) parts of scientific theories are to be counted as valuable, since the explanatory parts (such as the mechanical considerations involved in Huygens' theory) are always doomed to be abandoned in time.[62] Yet it is the ephemeral nature itself of explanatory theories that indicates their important, indeed essential role in the development of science. For it is only in the light of previous theories that certain experiments, which might otherwise be taken for granted, acquire a problematic character which calls for new and more adequate explanations. For example, to account for the experiments referred to by Huygens above, Newton appealed to his theory of fits which simultaneously served him as an explanation of the colours of thin transparent bodies. Newton's theory itself was not free from difficulties, but it brought out the significance of new experiments which, in their turn, called for new theories. As Popper has insisted, it is through the falsification of preceding conjectural explanatory theories that important progress is often made in science.

4. Huygens was already aware of the phenomenon of double refraction when, in 1673, he drafted the *Projet du Contenu de la Dioptrique*. We learn in fact from the *Projet* that he intended to include a chapter in his *Dioptrique* on 'Cristal d'Island' which exhibited that phenomenon. But, as the editors of Huygens' works observe, this chapter would have contained only a description of

---

of light; he suggested that the movement of light 'is like the movement of sounds, not the movement of bodies, and therefore the repulsion mentioned by [Ibn al-Haytham] cannot be conceived in it [the movement of light]' (*Tanqīḥ al-manāzir*, I, 1928, p. 374). Quoted by M. Nazif, *al-Ḥasan ibn al-Haytham*, I (Cairo, 1942), pp. 136–37. See above, pp. 72ff.
[61] See below, pp. 319f.   [62] Cf. P. Duhem, *Théorie physique*, Pt. I, Ch. III, Sec. I.

Iceland crystal and its properties;[63] the only words relating to that subject in the *Projet* are the following: '*difficultè du cristal ou talc de Islande.sa description. figure. proprietez.*'[64] Huygens first succeeded in explaining some of the phenomena connected with double refraction in 1677;[65] hence his idea of writing the *Traité de la Lumière* as a book wholly devoted to the wave theory and its applications without going into the questions of dioptrics which he had planned to include in the *Dioptrique*. Huygens' explanations of the phenomena of double refraction were thus the result of an extended application of a theory which he had previously devised mainly to account for ordinary refraction. He himself has made this quite clear in the fifth chapter of his *Traité*: 'It was after having explained the refraction of ordinary transparent bodies by means of the spherical emanations of light, as above, that I resumed my examination of the nature of this Crystal, wherein I had previously been unable to discover anything.'[66] This resumed examination did not result in deducing the anomalous phenomena of the crystal from his already formed theory as it stood. In fact they persistently appeared to threaten the validity of its basic assumptions. But in this lay their chief interest. They constituted a challenge which could not be ignored by Huygens; and to solve the difficulties he had to invent new hypotheses. He was not completely successful in his attempt. But some of the principles which he introduced are still valid; and he was certainly entitled to regard his partial success as a confirmation of his fundamental principle of secondary waves.

As has been remarked above, a description of Iceland crystal and of its chief phenomena with respect to optical refraction was first given by Erasmus Bartholinus in 1669. Pieces of that crystal were found in the shape of an oblique parallelepiped; each of its faces being a parallelogram. It was observed by Bartholinus that when a small

---

[63] Cf. H, XIII, *fasc.* II, p. 743, note 7.

[64] Cf. H, XIII, *fasc.* II, p. 739.

[65] Cf. Huygens to Colbert, 14 October 1677, H, VIII, pp. 36 and 37.

[66] *Treatise*, p. 61. On 22 November 1679 (the year in which Huygens communicated his theory to the *Académie des Sciences*) he wrote to Leibniz: 'I worked a great deal all last summer on my refractions, especially regarding the Iceland Crystal, which has very strange phenomena the reasons of all of which I have not yet unravelled. But what I have found of them greatly confirms my theory of light and of ordinary refractions' (H, VIII, p. 244).

object was viewed through two opposite faces it appeared double. He attributed this to the fact that when the incident ray entered the crystal it was divided into two refracted rays. One of these (called the *ordinary* ray) followed the usual rules of refraction while the other (called the *extraordinary* ray) did not. For example, a ray falling perpendicularly on one of the parallelogram faces was split up at the point of incidence into two: the one continued in the normal direction, as would be expected, while the other proceeded at an angle with the first. Also (and this was observed by Huygens) there was a certain angle of incidence at which the ray falling obliquely on the surface of the crystal would partially continue in the same oblique direction (thus giving rise to an extraordinary ray), while its other part would be deflected at an angle of refraction bearing a certain constant ratio of the sines to the angle of incidence.[67]

How did Huygens attempt to explain these phenomena on the basis of his original ideas?

As there were two different refractions, I conceived that there were also two different emanations of waves of light, and that one could occur in the ethereal matter extending through the body of the Crystal. Which matter, being present in much larger quantity than is that of the particles which compose it, was alone capable of causing transparency . . . I attributed to this emanation of waves the regular refraction which is observed in this stone, by supposing these waves to be ordinarily of spherical form, and having a slower progression within the Crystal than they have outside it; whence proceeds refraction as I have demonstrated.[68]

This accounts for the ordinary ray. But what about the extraordinary ray? What form of waves correspond to it, and how are we to picture the manner of propagation of these waves in the same crystal while the other, ordinary ray is proceeding by spherical surfaces as was assumed in the explanation of ordinary refraction?

As to the other emanation which should produce the irregular refraction, I wished to try what Elliptical waves, or rather spheroidal waves, would

---

[67] Cf. Huygens, *Treatise*, pp. 56–57. The two refracted rays would continue to be in the plane of incidence only if that plane was parallel to a certain plane of the crystal called by Huygens the *principal section*.

[68] Ibid., p. 61.

do; and these I supposed would spread indifferently both in the ethereal matter diffused throughout the crystal and in the particles of which it is composed, according to the last mode in which I have explained transparency. It seemed to me that the disposition or regular arrangement of these particles could contribute to form spheroidal waves (nothing more being required for this than that the successive movement of light should spread a little more quickly in one direction than in the other) and I scarcely doubted that there were in this crystal such an arrangement of equal and similar particles, because of its figure and of its angles with their determinate and invariable measure.[69]

Thus, according to Huygens, when a wave-front impinges on the surface of Iceland crystal it generates two series of waves which simultaneously traverse the crystal. The waves of the first series are carried through the particles of the ethereal matter alone, and they proceed by spherical surfaces. The waves of the second series travel through both the ether particles and the particles of the crystal, and they are transmitted by spheroidal surfaces with different velocities in different directions. From these assumptions the directions of refractions within the crystal could be determined by application of the principle of secondary waves. The direction of the ordinary ray is determined by the construction for ordinary refraction; and the direction of the extraordinary ray is found by a similar construction in which spheroidal secondary waves take the place of the spherical secondary waves in the ordinary construction.

It is to be noted that Huygens does not tell us that his choice of the spheroidal shape, rather than any other, was determined by a prior quantitative analysis of the phenomena in question. Rather, he adopted the assumption that the extraordinary ray was produced by *spheroidal* waves as a hypothesis which he later set out to examine experimentally. Thus he writes: 'Supposing then these spheroidal waves besides the spherical ones, I began to examine whether they could serve to explain the phenomena of the irregular refraction, and how by these same phenomena I could determine the figure and position of the spheroids . . . .'[70]

---

[69] Cf. Huygens, *Treatise*, pp. 56–57.
[70] Ibid., p. 63.

Huygens' analysis of the wave surface for the extraordinary ray as a spheroid still stands in geometrical optics.[71] But there were certain phenomena which he could not explain by his theory. He discovered these phenomena himself after he had written the greater portion of the *Traité* and of the chapter on Iceland crystal.

Before finishing the treatise on this Crystal, I will add one more marvellous phenomenon which I discovered after having written all the foregoing. For though I have not been able till now to find its cause, I do not for that reason wish to desist from describing it, in order to give opportunity to others to investigate it. It seems that it will be necessary to make still further suppositions besides those which I have made; but these will not for all that cease to keep their probability [*vraisemblance*] after having been confirmed by so many tests.[72]

The phenomenon discovered by Huygens was an unexpected one. For he expected a ray emerging from a piece of Iceland crystal to behave in relation to a second piece as the light coming directly from the light source had behaved in relation to the first. But he found that this was not so. He first placed two crystals at a distance above one another such that the faces of the one crystal were parallel to those of the other. A beam of light was allowed to fall perpendicularly on the surface of the first (higher) crystal. This gave rise to two rays within the crystal, the one continuing perpendicularly in the line of incidence (the ordinary ray); and the other (extraordinary) ray, being deflected from the normal direction. At their emergence from the first crystal, the ordinary ray persisted in the same perpendicular direction; and the extraordinary ray was again deflected in such a way as to be parallel to the ordinary ray but displaced from it. Upon their entrance into the second crystal none of the two rays was divided; the ordinary ray gave rise to one ordinary ray (thus again continuing in the same line), and the extraordinary ray was refracted into an extraordinary ray. Naturally, Huygens found it strange that the rays incident on the lower crystal

---

[71] For a detailed account of Huygens' analysis, see Mach, *Principles*, pp. 260–4; Bell, *Christian Huygens*, pp. 186–9.
[72] *Treatise*, p. 92 (H, XIX, p. 517).

should not behave in the same way as the original beam did through the higher one. Could that mean, he asked, that the ordinary ray emerging from the first crystal has lost something which is necessary to move the matter that serves for the transmission of extraordinary refraction; and that likewise the extraordinary ray has lost that which is necessary to move the matter that serves for the transmission of ordinary refraction?[73] If that hypothesis were true, then, it would seem, the ordinary ray would in all circumstances excite only one ordinary ray in the second crystal, and the extraordinary ray would always give rise to only one extraordinary ray. But this was not the case.

When he rotated the lower crystal about the original beam as axis through an angle equal to 90°, the two rays incident on that crystal were still not divided by refraction; but the ordinary ray gave rise to an extraordinary ray, and vice versa. When the lower crystal was held in an intermediate position between the last position and the first, *each* of the two rays incident on it gave rise to *two* rays, one ordinary and the other extraordinary.

When one considers here how, while the rays CE, DG [these are the rays incident on the second crystal], remain the same, it depends on the position that one gives to the lower piece, whether it divides them both in two, or whether it does not divide them, and yet how the ray AB above [this is the original beam incident on the higher crystal] is always divided, it seems that one is obliged to conclude that the waves of light, after having passed through the first crystal, acquire a certain form or disposition in virtue of which, when meeting the texture of the second crystal, in certain positions, they can move the two different kinds of matter which serve for the two species of refraction; and when meeting the second crystal in another position are able to move only one of these kinds of matter. But to tell how this occurs, I have hitherto found nothing which satisfies me.[74]

The problem was taken up by Newton in Queries 17, 18, 20, and 21 of the Latin edition (1706) of the *Opticks*.[75] To him, Huygens'

---

[73] Cf. *Treatise*, pp. 93–4.　　　　　　　　　[74] Ibid., p. 94.

[75] These re-appeared in the English edition of 1718, and in all subsequent editions, as Queries 25, 26, 28 and 29 respectively.

confession of failure to explain the experiment just described was a confirming sign of defeat of the wave hypothesis itself.[76] He argued[77] that if light consisted in a motion propagated through a uniform medium its properties would be the same in all directions, and the two beams emerging from the first crystal would not change their behaviour with respect to the position of the second. The real explanation must, therefore, be sought not in some 'new modifications'[78] which the rays might undergo in their passage through the first crystal (as Huygens had thought), but rather in the fact that they have certain *original* and *unchangeable* properties on which their fortunes through each one of the crystals depend. Accordingly, he suggested that a ray of light should be conceived as having two opposite 'sides' endowed with an original property on which the extraordinary refraction depends, and two other opposite sides which do not possess this property. Whether or not the ray will be refracted according to the ordinary rules will depend on the position of these sides in relation to the crystal or, rather, to its principal plane.

Thus Newton interprets Huygens' two-crystal experiment in the following manner.[79] The rays contained in the beam of ordinary light that is incident on the surface of the first crystal arrive there with their sides in various (random) positions relative to the principal plane of the crystal. The exact position of the sides of each ray determines whether it will be refracted according to the ordinary rules, or whether it will take the direction of the extraordinary beam. Two beams will thus traverse the first crystal, all the rays contained in the one beam having their sides oriented in a contrary way to those of the rays contained in the other.

Suppose now that the principal plane of the second crystal is parallel to that of the first. Then, since the sides of *all* rays contained in the *two* beams that are incident on the second crystal have the same position with respect to its principal plane as they had with

---

[76] Cf. *Opticks*, Query 28, p. 364.
[77] Cf. ibid., Query 25, p. 358; Query 28, p. 363.
[78] Cf. ibid., Query 28, p. 363.
[79] Cf. ibid., Query 26.

respect to the principal plane of the first, the ordinary beam will again be refracted as an ordinary beam and the extraordinary beam will give rise to one extraordinary beam. If the second crystal is rotated through a right angle, then the sides of the rays in the *ordinary* beam will have the same position to its principal plane as the sides of the rays in the *extraordinary* beam had with respect to the principal plane of the *first* crystal. Consequently, the ordinary beam will in this case traverse the second crystal as an extraordinary beam and, for the same reasons, the extraordinary beam will be refracted as an ordinary beam. By holding the second crystal in an intermediate position we are, according to Newton, back to the case of the ordinary light falling on the first crystal; each beam will therefore give rise to two beams, one ordinary and the other extraordinary.

The basic intuition in these explanations, namely Newton's realization that the rays emerging from the first crystal behave in such a way as to indicate that they do not have the same properties with respect to all directions perpendicular to their direction of propagation, contains in fact the discovery of what is known as the *polarization* of light.[80] This discovery could not be expressed in terms of the wave picture used by Huygens, and to that extent Newton was justified in regarding the two-crystal experiment as a stumbling block to Huygens' wave hypothesis. But he was mistaken in regarding it as a crucial test against the wave hypothesis as such. The idea of the *sidedness* of rays was in fact assimilated into the wave theory when, at the beginning of the nineteenth century, transverse wave motion was adopted for light. It was then shown by Fresnel that ordinary (unpolarized) light could be represented by a transverse wave motion in which the particles of the medium vibrate in all conceivable planes that are perpendicular to the direction of propagation. And polarized light (such as that which has passed through a crystal of the Iceland spar) was characterized by the

---

[80] Cf. Whittaker's Introduction to Newton's *Opticks*, p. lxxvi. The term *polarization* is derived from Newton's reference in this connection to the 'Poles of two Magnets' which, he suggests, act on one another as the crystal acts on the sides of the ray through a 'virtue' which he does not consider to be magnetic. Cf. *Opticks*, Query 29. pp. 373–4.

fact that the vibrations occur in precisely one of these planes. On this representation it could be significantly asserted in wave language that the two rays emerging from the first crystal in Huygens' experiment are polarized, and that their difference consists in the fact that their planes of polarization are perpendicular to one another.

5. Huygens' investigations in his *Traité de la Lumière* do not show any influence of those Baconian suggestions which he submitted to the *Académie des Sciences* in 1668 regarding the proper method of conducting physical research. Unlike Boyle who, as a follower of the Baconian programme, devoted his work on colours to collecting a natural history of those phenomena, Huygens does not attempt to make a complete enumeration of the properties of light, nor does he seem to think that such an enumeration would be necessary for his purpose. His only concern is to give a clear mechanical explanation of a few properties which, he believed, other philosophers had failed to explain before him. True, what he attempts to explain are 'facts of experience' which he takes for granted. But he nowhere claims to *argue from* these facts to the theory which he puts forward. Rather, his arguments consist in showing how these facts agree better with his conceptions than with those which had been proposed by other writers before him. And thus his starting point was the *problems* arising from previous theories rather than the facts themselves. By examining the difficulties involved in already existing hypotheses he was led to modify those hypotheses or to suggest new ones; he then set out to show how they could account for the experiments. His procedure was always from the hypothesis to the experiment, not vice versa, and accordingly, he did not claim for his explanations more than a certain degree of 'vraisemblance'. In this the character of Huygens' *Traité* greatly contrasts with Newton's *Opticks*. Newton, in faithful observance of Baconian teaching, always sets out the Observations and Experiments before the Propositions which are claimed to have been simply derived from what precedes. The 'hypotheses', those fictions which we might be tempted to invent in the absence of sufficient experiments, are relegated to the end of the *Opticks* where they appear as problems or Queries. Underlying this mode of exposition there is a radical distinction between discoveries or theories on the one hand, and hypotheses on the other. Such a distinction does not exist for Huygens. Newton would explain this by the fact that Huygens was committed to a mechanistic view of nature which forced him to formulate his explanations in terms of hypothetical entities such as the ether. This of course is true and Huygens would readily admit it. But does this mean that by innocent and careful observation of what nature does or may be made to do by experimentation, one would be able to arrive at theories that are free from all hypothetical elements? Newton thought that would be possible and he actually claimed that there was nothing in his *asserted* theories which had not been deduced from the experiments. An examination of that claim will constitute an important part of the chapters that follow.

Chapter Nine

NEWTON'S THEORY OF LIGHT AND COLOURS, 1672

1. Newton first declared his views on scientific method during the famous controversy which took place immediately after the publication of his theory of light and colours in 1672. As is well known, the prominent and most important scientists against whose objections Newton had to defend his theory were Robert Hooke, Ignatius Pardies, and Christian Huygens.[1] All three of them were Cartesians of sorts and all had been engaged in developing a wave theory of light. Perceiving an atomistic tendency in Newton's ideas, they regarded his theory as a hypothesis which indeed agreed with the experiments, but which was no more than one among several possible interpretations of them. Newton rejected this characterization of his doctrine, claiming that there was nothing in the propounded properties of light that was not positively and directly concluded from the experiments. Therefore, in his view, to propose alternative hypotheses about the nature of light was beside the business at hand, which was to deduce the properties of light from the observed facts. This he defended as the proper method of scientific inquiry as opposed to the method of his critics—which, it appeared to him, sought to predetermine the properties of things from the mere possibility of hypotheses. Not that, according to Newton, hypotheses had no function whatsoever in natural philosophy; he expressly allowed that hypotheses might be employed, tentatively, to explain the phenomena after they had been discovered

---

[1] The far less important and not very revealing discussion which started later (in 1674) between Newton and Francis Hall (or Linus), S. J., professor of mathematics at the English Jesuit College at Liège, and which was continued after Hall's death in 1675 by his student Gascoigne and afterwards by Anthony Lucas, will not be of interest to us here. For a bibliography and analysis of this discussion, see: L. Rosenfeld, 'La théorie des couleurs de Newton et ses adversaires,' *Isis*, IX (1927), pp. 47–51. See also: Michael Roberts and E. R. Thomas, *Newton and the Origin of Colours*, London, 1934, Ch. XIII; David Brewster, *Memoirs of the Life, Writings and Discoveries of Sir Isaac Newton*, Edinburgh, 1855, vol. I, Ch. IV.

and ascertained, and even that they might be used to suggest new experiments that had not yet been performed. But he advised that they should not be held against the evidence of experiments; and he believed that when inquiry regarding a certain matter has been successfully accomplished, all hypotheses will have completely vanished. That is to say, hypotheses may be employed in the course of scientific inquiry, but they may not form part of asserted scientific doctrine.

It is a notorious fact that Newton's declarations about hypotheses are often torn out of their context and given a meaning that was never intended by their author. It is, therefore, not a misplaced hope to try first, as is partly the aim of this and the two following chapters, to understand his methodological views in the context in which they were first expressed and argued; this will prepare the way for a discussion of further examples in Chapters XII and XIII.

For this purpose it will be first necessary to analyse Newton's theory of colours as it was presented to the scientific world in 1672. I shall try to show, in terms as close as possible to Newton's, what his theory propounded, and which if any of his expressions justified the corpuscular interpretation which his contemporaries readily placed upon it. It will be also necessary, in Chapter X, to examine closely the objections raised by Newton's critics and see whether they took the experimental results into account, and to what extent they were determined by theoretical preferences. A sufficient understanding of the standpoint of Newton's opponents should naturally help in appreciating his reaction. Finally, in Chapter XI, an examination of Newton's answers will involve further analysis of his theory, and an assessment of his position will be attempted.

Various aspects of this controversy have been studied more than once, sometimes even with a view to clarifying Newton's methodological position. The judgement generally has been that Newton's critics failed to understand his theory; their fault was that while they were faced with an experimental discovery, they preferred to wrangle over hypotheses.[2] Now it is certainly true, particularly

---

[2] R. S. Westfall in an interesting recent article, 'Newton and his critics on the nature of colors', *Archives internationales d'histoire des sciences*, XV (1962), pp. 47–58, put forward the view

at the beginning of the discussion, that the opponents of Newton's theory did not fully appreciate the experimental facts with which they were presented; this is an important factor which will have to be borne in mind in trying to understand Newton's reaction. But we should not be too much influenced by Newton's own view of himself as the man who had been grossly misunderstood. Not only was he partly responsible for the initial misunderstanding, but he also completely and consistently failed to distinguish in his doctrine of colours between the strictly experimental propositions and the particular interpretation of white light which he attached to them. This interpretation, which the experiments certainly did not prove, was to the effect that white light should be viewed as a mixture of differentiated elements. The most important among Newton's critics, viz. Hooke, Pardies and Huygens, eventually conceded the only thing that Newton could have justifiably claimed to have proved experimentally, namely the fact that to every colour there is attached a constant of refrangibility which is not the same for any other colour. But they correctly refused to grant an equal status to Newton's doctrine of white light as a heterogeneous aggregate. Hooke in particular envisaged a new formulation of his pulse hypothesis which would have fully taken the experimental results into account without conceding the original heterogeneity of white light. According to this formulation the pulse of white light could be imagined as the resultant of a large number of 'vibrations' each of which when differentiated would produce a given colour.

---

that the basis of the controversy between Newton and his critics was an ironical misunderstanding: while Newton presented his discovery as further support of the mechanical philosophy, his critics interpreted his position as a reversion to the rejected peripatetic philosophy. Thus he writes: 'Not the nature of light but the nature of qualities, not undulatory vs. corpuscular but mechanical vs. peripatetic – these were the primary concerns of the critics' (ibid., p. 47). This seems to me to lay the emphasis on the wrong place. The suspicion of peripatetic conceptions in Newton's doctrine of colours was only a false start, one, moreover, for which Newton himself was partly responsible. The discussion was not, even from the beginning, solely concerned with this initial misunderstanding, nor did it come to a stop when Newton reassured Hooke on this score. Professor Westfall states that 'Like Hooke, Huygens saw in Newton's theory something that appeared incompatible with mechanical philosophy' (ibid., p. 54). The passages quoted from Huygens, however, show no more than a reserved attitude. He believed that colours should be explained mechanically, but that Newton had not yet provided such an explanation. As we have seen, Huygens himself later advanced a mechanical wave-theory of light which admitted its incapability of explaining colours.

This would imply that white light is compounded only in a mathematical sense, and prismatic analysis would be understood as a process in which colours are manufactured out of the physically simple and undifferentiated pulse. For reasons which will be discussed in the two following chapters Newton rejected this suggestion as unintelligible. And historians of science, confusing a generalized concept of composition with Newton's narrower interpretation, have continued to view this controversy with his own eyes. In this, however, they forget that Hooke's idea was reintroduced in the nineteenth century in a mathematically developed form as a plausible representation of white light from the wave-theory point of view.

2. Newton's first published account of his theory of the prismatic colours was contained in a letter to the Secretary of the Royal Society, Henry Oldenburg, dated 6 February 1671/2. The letter was read before the Society (in Newton's absence) on the eighth of the same month[3] and subsequently printed in No. 80 of the *Philosophical Transactions* of 19 February 1671/2.[4] Newton seems to imply in his letter that his experimental research on the prismatic colours dated from the beginning of 1666 when he applied himself to the grinding of lenses of non-spherical shapes.[5] It was generally believed at that time that the perfection of telescopes, and in particular, the elimination of chromatic aberration, would depend on the shape in

---

[3] Cf. Thomas Birch, *The History of the Royal Society of London*, London, 1756–7, III, p. 9.

[4] Cf. *Correspondence of Isaac Newton*, ed. Turnbull, I (1959), pp. 92–102.

[5] '. . . in the beginning of the Year 1666 (at which time I applyed my self to the grinding of Optick glasses of other figures than *Spherical*,) I procured me a Triangular glass-Prisme, to try therewith the celebrated *Phaenomena of Colours*' (*Correspondence*, I, p. 92). A. R. Hall has pointed out reasons 'for thinking that, in his letter, Newton wrote "1666" for 1665, and that his note of 1699 (verified with his own records by himself) is more accurate, the definitive theory of colours being obtained in January 1666, as a result of experiments on the refraction of a beam of light, carried out either with a new prism or one bought in 1664' ('Sir Isaac Newton's Note-Book, 1661–65', *Cambridge Historical Journal*, IX (1948), p. 246). For Newton's note of 1699 referred to here, see ibid., p. 240, note 6. For the 'period about the middle of 1665' as the more plausible date at which 'Newton's interest in Descartes's suggestion for the use of lenses of non-spherical curvatures was greatest', see Hall, 'Further optical experiments of Isaac Newton', *Annals of Science*, XI (1955), p. 36. Turnbull pushes the date back to 1664 (*Correspondence*, I, p. 59, note 11). But 1665 seems to be the more cautious date.

in which the lenses were formed. In particular, it was hoped that lenses having one of the shapes of the conic sections might serve the purpose. Newton had himself embarked on that task when, as he tells us, he obtained a triangular glass prism to experiment on the phenomena of colours. Having darkened his chamber and made a small (circular) hole in the window-shutters,[6] to let in the sun's light, he placed the prism close to the hole so that the light might be refracted to the opposite wall. 'It was at first a very pleasing divertisement, to view the vivid and intense colours produced thereby....'[7] But that was only a passing divertisement. The thing that Newton found puzzling was the shape rather than the colours of the solar spectrum on the wall: 'I became surprised to see them in an *oblong* form; which, according to the received laws of Refraction, I expected should have been *circular*.'[8] The colours were terminated at the sides with straight lines but faded gradually at the ends in semicircular shapes. But why did Newton find the oblong shape surprising? For, in fact, except for one definite position of the prism, namely that of minimum deviation, a certain elongation of the image should have been expected. As we go on reading Newton's paper, however, we soon discover that the prism was fixed at precisely that position. We are therefore here presented with a carefully planned experiment and not, as the opening sentence of Newton's paper might convey,[9] a chance observation. Newton's expectation that the image should be circular rather than oblong was

---

[6] Newton does not state in his paper that the hole was circular. This, however, is understood from his subsequent reasonings. The shape of the hole is essential for appreciating the problematic character of the image projected on the opposite wall. Compare Newton's *Opticks*, Bk. I, Pt. I, Prop. II, Theor. II, p. 26. See below, n. 9.

[7] *Correspondence*, I, p. 92.

[8] Ibid.

[9] 'The original letter in Newton's handwriting has not been found, but a transcript, written by his copyist Wickins and bearing a few verbal corrections in Newton's own hand, is preserved in the Portsmouth Collection . . .' (*Correspondence*, I, p. 102, editor's note 1). Variants between the two versions are recorded in the *Correspondence* and will be indicated here when relevant. The letter seems, in certain respects, to have been written rather informally and even without much care. It lacked the detailed clarifications and illustrations that were necessary for securing a full appreciation of his new and revolutionary discovery. Newton was thus, to a certain extent, responsible for his paper being misunderstood by his contemporaries. He did not make it sufficiently clear and explicit that he was considering a special case in which the image should be circular; and Pardies' first objections were based on a misunder-

based on calculations involving a great deal of information drawn from geometrical optics.[10]

Comparing the dimensions of the image, Newton found that its length was about five times greater than its breadth, 'a disproportion so extravagant, that it excited me to a more then ordinary curiosity of examining, from whence it might proceed'.[11] To determine the cause of this disproportion, he began to consider several hypotheses or, as he called them, suspicions which suggested themselves to his mind. These were all hypotheses conceived on the implicit assumption that the sine law, in its then accepted form, was not to be modified. Could the shape of the spectrum be due, for example, to the thickness of the prism, or the bigness of the hole in the window-shutters? But observing that, by allowing the light to pass through different parts of the prism, or by varying the magnitude of the hole, the oblong shape was always the same. Or, perhaps, it was caused by some unevenness or another irregularity in the glass? To examine this possibility he placed a second prism in a contrary position behind the first. In this arrangement the second prism should destroy the regular effects of the first and, at the same time, augment the supposed irregular ones. He found, however, that the image became round, as if the light had not been refracted at all. Newton then concluded that the cause of the phenomenon could not be 'any contingent irregularity'[12] of the prism.

A third hypothesis was that the shape might be explained by the different inclinations to the prism of rays coming from different

---

standing of this point. The *experimentum crucis* (which will be soon described) was not illustrated with diagrams; and both Pardies and Huygens could not at first understand it. As we shall see, Newton later admitted that if he had meant his letter for publication, some additions would have been introduced (see below, p. 267, n. 44). In fact, it was not without reluctance that he allowed its publication in the first place (cf. Brewster, *Memoirs*, I, pp. 72–73). Had he adopted in that letter the strictly geometrical and systematic approach which is to be found in his *Optical Lectures* (which had been written, but not published, before his letter appeared in the *Philosophical Transactions*) a considerable part of the controversies with his adversaries would no doubt have been spared. It is partly to these controversies, however, that we should be grateful for the wealth of beautiful experiments and illustrations which Newton adduced in later papers and in the *Opticks* to avoid further misunderstandings.

[10] Cf. Newton, *Optical Lectures*, London, 1728, Sec. I, Arts. 4 and 5, pp. 8–11. See below, p. 237, n. 14.

[11] *Correspondence*, I, p. 92.   [12] *Correspondence*, I, p. 93.

parts of the sun. To decide whether this was the case, Newton made the following measurements and calculations. The diameter of the hole was $\frac{1}{4}$ of an inch, its distance from the image 22 feet, the length of the image was $13\frac{1}{4}$ inches, its breadth $2\frac{5}{8}$ inches; and 'the angle, which the Rays, tending towards the middle of the image, made with those lines, in which they would have proceeded without refraction, 44 deg. 56′. And the vertical Angle of the Prisme, 63 deg. 12′. Also the Refractions on both sides the Prisme, that is, of the Incident, and Emergent Rays, were as near, as I could make them, equal, and consequently about 54 deg. 4″'.[13]

The preceding statements imply that the prism was set in the position of minimum deviation for the middle colour of the spectrum. Newton was thus simply assuming it to be known that the equality of the refractions on both sides of the prism (i.e. the symmetrical arrangement of the incident and emergent rays with respect to those sides) is a condition for that position in which the image should be circular and not oblong. By not providing any explicit geometrical explanation of this assumption he was certainly expecting too much from his readers.[14]

Having subtracted the diameter of the hole from the breadth and from the length of the image, its breadth was found to subtend an angle of 31 minutes, corresponding to the diameter of the sun,

---

[13] *Correspondence*, I, p. 93. An earlier description of this experiment occurs in a Note-Book which Newton probably began in 1665–6; see A. R. Hall, 'Further optical experiments of Isaac Newton', *Annals of Science*, XI (1955), pp. 28–29. Some of the measurements in this description approach those of the 1672 paper: diameter of the hole=$\frac{1}{8}$ of an inch [?]; distance of wall from prism=260 inches; breadth of image=$2\frac{1}{4}$ inches; vertical angle of prism= 'about 60^{gr}'. But the length of the image is much smaller than that reported to the Royal Society, 'about 7 or 8 inches' (ibid.).

[14] For the necessary geometrical demonstrations relating to this case, see *Optical Lectures*, London, 1728, Sec. I, Art. 10 pp. 19–20; Sec. III, Prop. 25, pp. 164–6. Compare *Opticks*, Bk. I, Pt. I, Prop. II, Theor. II, p. 28. The *Optical Lectures* were not published until 1728. These were a translation of *Pars* I of the Latin original (*Lectiones opticae*) which appeared in the following year. The title-page of this edition stated that the *Lectiones* were delivered at Cambridge in the years 1669–71. According to W. W. Rouse Ball, *An Essay on Newton's Principia*, London, 1893, pp. 27–28, Newton started his lectures in the Lent Term 1670 and continued them in Michaelmas 1670, 1671 and 1672 (referred to by A. R. Hall, 'Newton's First Book (I)', *Archives internationales d'histoire des sciences*, XIII (1960), p. 44 and note 23). Hall notes that Part I of the manuscript of the *Lectiones* in the Cambridge University Library bears the date 'Jan. 1669'. See *Correspondence*, I, p. 103, note 1. The *Opticks* (1704) does not include the demonstrations in question.

whereas its length subtended an angle equal to 2°49′. Newton then computed the refractive index of the glass (obtaining the ratio of sines 20 : 31) and, using this result, he computed the refractions of two rays coming from opposite parts of the sun so as to subtend 31′ between them. He found that the emergent rays should have subtended about the same angle. Therefore, he had to account for a difference of 2°18′. 'But because this computation was founded on the Hypothesis of the proportionality of the *sines* of Incidence, and Refraction, which though by my own & others Experience I could not imagine to be so erroneous, as to make that Angle but 31′, which in reality was 2 deg. 49′; yet my curiosity caused me again to take my Prisme.'[15]

By rotating the prism about its axis to and fro through a large angle (more than 4 or 5 degrees), the colours were not sensibly displaced from their place on the wall. This further confirmed that a difference in incidence of such a small angle as 31′ could not be responsible for the large angle between the emergent rays.

The next hypothesis to be examined by Newton is vividly reminiscent of Descartes' explanation of the prismatic colours. It will be remembered that Descartes ascribed those colours to the rotary motion acquired by the globules of the subtle matter when they obliquely strike the side of the prism.[16] Now Newton suspected that the rays might move in curved lines after their emergence—for the following reason:

it increased my suspition, when I remembred that I had often seen a Tennis-ball, struck with an oblique Racket, describe such a curve line. For, a circular as well as a progressive motion being communicated to it by that stroak, its parts on that side, where the motions conspire, must press and beat the contiguous Air more violently than on the other, and there excite a reluctancy and reaction of the Air proportionably greater. And for the same reason, if the Rays of light should possibly be globular bodies, and by their oblique passage out of one medium into another acquire a circulating motion, they ought to feel the greater resistance

---

[15] *Correspondence*, I, p. 93; '& others' omitted in the text of the *Transactions*.
[16] See above, pp. 64ff.

from the ambient Æther, on that side, where the motions conspire, and thence be continually bowed to the other.[17]

The hypothesis involved in this passage differs from the other 'suspicions' in that it assumes the idea of a hypothetical entity, the ether. It should therefore be observed that Newton was not reluctant to *examine* a hypothesis of the Cartesian type in the course of arriving at his theory.[18] On the contrary, by drawing the analogy between the light corpuscle and the tennis ball, he was trying to relate that hypothesis to observation in order to see whether it could stand the experimental test. The analogy suggested to him that the rays would travel in curved lines after their emergence from the prism. But by receiving the image on a board at various distances from the prism, he found that the variation of the magnitude of the image with the distance was as it should be on the assumption that the rays proceeded in straight lines after refraction.[19] If, therefore, the considered hypothesis was finally rejected, that was not because, *being a hypothesis*, it was unworthy of serious examination, but because it was found to be false. That was in fact Newton's consistent attitude to rival hypotheses which he could not accept.

Having removed all these 'suspicions', Newton was finally led to devise what he called an *experimentum crucis*. That was an arrangement involving two prisms.[20] The first was placed, as before, near the hole in the window-shutter; the emerging rays passed through a small opening in a board set up close behind the prism. At a distance of about 12 feet, he fixed the second prism behind another board with a small hole in it to allow only some of the incident light

---

[17] *Correspondence*, I, p. 94. While Descartes also allowed the 'ambient ether' to play a part in the production of the rotary motion, he did not envisage any curvature of the rays' path. Newton seems to have been the first to explain a ball's swerve correctly; see ibid., pp. 103–4, editor's note 8.

[18] Compare Cajori's remarks to the same effect in the Appendix to his edition of Newton's *Principia*, note 55, pp. 671–2.

[19] *Correspondence*, I, p. 94.

[20] Note that it is only in the course of Newton's 'historicall narration' (*Correspondence*, I, p. 97) that we learn that he had *two* prisms, not one, at his disposal. The narration starts: '. . . I procured me a Triangular glass-Prisme . . .' (ibid., p. 92). A second prism was already required to remove the suspicion that the colours might have been produced by the unevenness in the glass of the first. See above, p. 237, n. 13.

to reach the prism. By rotating the first prism about its axis, the colours cast on the second board were made to pass successively through the opening in it. In this manner he could observe the several positions on the wall where the various colours were refracted by the second prism. Noting that the lower colours of the image projected on the second board were more refracted by the second prism than the higher ones, he arrived at the following conclusion: that

... The true cause of the length of that Image was detected to be no other, then that *Light* consists of *Rays differently refrangible* which, without any respect to a difference in their incidence, were, according to their degrees of refrangibility, transmitted towards divers parts of the wall.[21]

The import of this experiment and the significance which Newton attributed to it will be examined later. But it must be noted here that the fundamental proposition which Newton immediately based on his *experimentum crucis* concerns the refrangibility of the rays rather than their colour. For, in fact, the real problem from which he started in this account was a problem of refraction rather than colour. The appearance of colours by the refraction of light through a prism had been observed long before Newton. Several explanations of these colours had been proposed. But these explanations were mainly concerned with the qualitative aspect of the phenomenon, that is, the appearance of the colours as such. That was one reason why they all missed the problem which engaged Newton's attention. Descartes, for example, was content to ascribe the prismatic colours to the rotation of the ethereal globules; the ratio of the velocity of rotation to the velocity of the globules in the direction of propagation determined the colour. Alternatively, Hooke would explain the same phenomenon by the obliquation of

---

[21] *Correspondence*, I, p. 95. Newton then tells us that it was this discovery, namely his realization that 'Light it self is a *Heterogeneous mixture of differently refrangible Rays*', that made him despair of perfecting refracting telescopes. It appeared to him that however exactly figured the lenses might be, and whatever the shape given them, they would not collect the rays of white light in one focus. This led him to the consideration which finally resulted in the construction of his reflecting telescope. Cf. A. R. Hall, *Annals of Science*, XI (1955), p. 37.

the ethereal pulse to the direction of propagation: the manner in which this oblique pulse struck the eye determined the colour.

But neither Descartes nor Hooke (nor, indeed, anyone before Newton) had paid attention to the particular *geometrical problem of* refraction that was involved in Newton's experiment. To Newton, the appearance of colours on the wall was, as we have seen, more of a 'divertisement' than a problem. The real problem was one of refraction; it consisted in explaining the *shape* of the spectrum *for a given position of the prism,* and this on the supposition that the rays from the sun were all equally refrangible. That supposition was *implicitly* accepted by all writers on optics, or, at least, it had not been challenged by anyone. It was on the basis of that supposition that Newton had expected to see the solar image in a round, and not as he found, in an oblong shape. After having dismissed some of the most obvious hypotheses that might be thought to account for the elongation of the image, he performed the *experimentum crucis.* This experiment showed him that the rays were more or less refracted through the second prism, according as they were more or less refracted by the first. He was thus led to discard the initial supposition itself and to declare that the sun's light is 'not similar, or homogeneal, but consists of *difform* Rays, some of which are more refrangible than others'.[22]

But the same experiment showed that the rays preserved their colours as well as their degrees of refrangibility upon their being refracted through the second prism; the same index of refraction was attached to the same colour. It was thus apparent that with the difference in refrangibility of the rays went the difference 'in their disposition to exhibit this or that particular colour'.[23] Therefore, Newton concluded, colours were not 'qualifications', or modifications suffered by light upon reflection or refraction, as it was generally assumed; they were '*Original* and *connate properties*' of the rays just as their respective degrees of refrangibility were.[24]

Newton's doctrine of colour further stated the unalterability of the rays' colours (and, of course, the degrees of refrangibility associated

---

[22] *Correspondence,* I, p. 96.   [23] Ibid., p. 97.   [24] Ibid.

with them) either by reflection, or refraction, or by any other circumstance that he had tried. Seeming alterations of colours might take place when several sorts of rays were mixed together. But the component colours could be made to reappear when the difform rays were separated by refraction. There are thus two sorts of colours: 'The one original and simple, the other compounded of these.'[25] The original or 'primary' ones are all those that are exhibited in the 'indefinite variety' of colours of the solar spectrum.[26] White is the most compounded of all composite colours since its analysis gives all the spectrum colours, and for its composition all the primary colours should be mixed in due proportion. To illustrate the latter point Newton described an experiment in which the rays of the sun's light were made to converge after they had been dispersed, and a white light just like the light coming directly from the sun was thereby produced. Interpreting this experiment Newton asserted that white light is 'a confused aggregate of Rays indued with all sorts of Colors, as they are promiscuously darted from the various parts of luminous bodies'.[27]

As every sort of ray has a different degree of refrangibility, the production of colours from white light by the prism was thus understood in the following way. The rays which are all originally endowed with colour are merely 'severed and dispersed'[28] by the prism according to their various degrees of refrangibility and consequently the primary colours are distinguished.

It was this interpretation of the constitution of white light and the role which Newton assigned to the prism in the production of colours that occupied the greatest and most important part of the controversy between Newton and the critics of his theory, Hooke, Pardies and Huygens. They all suspected from Newton's expressions that he was inclined towards an atomistic interpretation of light which they found disagreeable. Their feeling was certainly justified by Newton's stand in the same paper on a scholastic issue concerning whether light was a substance or quality. Newton explicitly subscribed (though with some caution) to the view that light was a

---

[25] *Correspondence*, I, p. 98.   [26] Ibid.   [27] Ibid.   [28] Ibid., p. 99.

substance. This opinion he maintained in connection with his explanation of the colours of natural bodies. Those colours, he asserted, have their origin in the fact that opaque bodies reflect one sort of rays in greater quantity than others. He observed that in the dark any body appeared with any colour with which it was illuminated, though in various degrees of vividness. When natural bodies are exposed to the sun's light they put on the colour which they reflect most copiously. Then he wrote:

These things being so, it can be no longer disputed, whether there be colours in the dark, nor whether they be the qualities of the objects we see, no nor perhaps, whether Light be a Body. For, since Colours are the *qualities* of Light, having its Rays for their intire and immediate subject, how can we think those Rays *qualities* also, unless one quality may be the subject of and sustain another; which in effect is to call it *substance*. We should not know Bodies for substances, were it not for their sensible qualities, and the Principal of those being now found due to something else, we have as good reason to believe that to be a substance also.[29]

Newton later drew attention to the word 'perhaps' in the preceding passage when he tried to point out that his doctrine was not necessarily committed to a substantial view of light. But we read, immediately after that passage, the following emphatic words at the head of a new paragraph: 'Besides, whoever thought any quality to be a *heterogeneous* aggregate, such as Light is discovered to be.'[30] That white light is a heterogeneous mixture is a thesis which Newton never abandoned; it was for him an experimental fact proved beyond any possible doubt. But since he expressed himself in terms of this dichotomy, substance or quality, was it not only natural for his readers to assume that he favoured a corpuscular hypothesis about light? Nevertheless, we read again in Newton's paper: 'But, to determine more absolutely, what Light is, after what manner refracted, and by what modes or actions it produceth in our minds the Phantasms of Colours, is not so easie. And I shall not mingle conjectures with certainties.'[31] What was certain for Newton?

---

[29] Ibid., p. 100.   [30] Ibid.   [31] Ibid.

That light is a heterogeneous aggregate. But if, as he himself argued, it was not possible to conceive of quality as such an aggregate, did he leave much doubt as to the conclusion he wanted to draw?

3. In a later printed version[32] of the 1672 paper Newton tried to extricate himself from the position he appeared to be taking in the passage suggesting the corporeal nature of light. He wrote in a footnote to this passage:

Through an improper distinction which some make of mechanical Hypotheses, into those where light is put a body and those where it is put the action of a body, understanding the first of bodies trajected through a medium, the last of motion or pression propagated through it, this place may be by some unwarily understood of the former: Whereas light is equally a body or the action of a body in both cases. If you call its rays the bodies trajected in the former case, then in the latter case they are the bodies which propagate motion from one to another in right lines till the last strike the sense. The only difference is, that in one case a ray is but one body, in the other many. So in the latter case, if you call the rays motion propagated through bodies, in the former it will be motion continued in the same bodies. The bodies in both cases must cause vision by their motion.[33]

Newton seems to be equivocating here. Is it not odd that he should describe the distinction between the corpuscular and wave views of light as an '*improper* distinction which *some* make of mechanical Hypotheses'? Should we take this to be a sincere judgement when we realize (as we shall see) that he always considered all forms of the wave hypothesis to be false, and that he had already decided in favour of the corpuscular interpretation?
The footnote continues:

---

[32] Only four sheets (pp. 9–16) of this version have survived in two copies. Discovered in a decayed binding by Derek J. de Solla Price, these Waste Sheets (as A. R. Hall calls them) were first published in facsimile and analysed by I. Bernard Cohen, 'Versions of Isaac Newton's first published paper', *Archives internationales d'histoire des sciences*, XI, (1958), pp. 357–75. The circumstances surrounding their first abortive publication (which may have taken place in 1676–7) are not known. See A. R. Hall, 'Newton's First Book (I)', ibid., XIII (1960), pp. 51–54; *Correspondence*, I, pp. 105–7, where variants of the Waste Sheets (here called 'the Print') with the 1672 version are recorded.

[33] *Correspondence*, I, pp. 105–6.

Now in this place my design being to touch upon the notion of the Peripateticks, I took not body in opposition to motion as in the said distinction, but in opposition to a Peripatetick quality, stating the question between the Peripateticks and Mechanick Philosophy by inquiring whether light be a quality or body. Which that it was my meaning may appear by my joyning this question with others hitherto disputed between the two Philosophies; and using in respect of the one side the Peripatetick terms *Quality, Subject, Substance, Sensible qualities*; in respect of the other the Mechanick ones *Body, Modes, Actions*; and leaving undetermined the kinds of those actions (suppose whether they be pressions, strokes, or other dashings,) by which light may produce in our minds the phantasms of colours.[34]

This is not quite exact; for Newton did *not* draw the line of demarcation where he claims here to have drawn it. By arguing that light was a body on the basis of having shown it to be a substance of which colours were qualities, he was in fact arguing *in* peripatetic terms for his own mechanical view. Again one is impelled to ask: Was it really his intention to argue for a generalized mechanical hypothesis which did not distinguish between the transport of body and the transmission of motion? Could he really have expected a man like Hooke (whom he certainly had in mind when he wrote his first paper)[35] to hear the description of light as a substance and still consider the new doctrine as compatible with his own wave-like interpretation?

The first part of Newton's paper, moving with increasing suspense to the final denouement in the crucial experiment, reads like a narrative of events which took place in a rather quick and logical succession over a short period of time: '. . . having darkened my chamber, and made a small hole in my window-shuts, . . . It was at first a very pleasing divertisement, . . . but after a while . . . I became surprised . . . Then I suspected . . . I then proceeded to examine more critically . . . Then I began to suspect . . . The gradual removal of these suspitions at length led me to the *Experimentum Crucis.* . . .' As

---

[34] Ibid., p. 106.
[35] Cf. R S. Westfall, 'The development of Newton's theory of color', *Isis*, LIII (1962), p. 354. Newton's arguments in the Waste Sheets are in fact in contradiction with what he stated in his reply to Hooke's 'Considerations'; see below, p. 276, n. 6.

a compelling tale of discovery, it has captured the imagination of readers since its first publication in the *Philosophical Transactions*. As history, however, it has the mark of implausibility. The progression is sustained throughout; there are no obstacles or hesitations; suspicions are formulated to be immediately removed; the final result is clear and definitive. Not even the 'fortunate Newton'[36] could have been fortunate enough to have achieved this result in such a smooth manner.

Recent studies of Newton's manuscripts have provided a more plausible picture of how he arrived at the position expressed in 1672.[37] In the earliest Note-Book of Newton, one which he started as an undergraduate at Cambridge in 1661 and continued until at least 1665, he asserts the varying refrangibility of different colours, with reference to an experiment which he later incorporated in his published work:

That ye rays which make blew are refracted more y^n ye rays which make red appears from this experiment. If one halfe of ye thred *abc* be blew & ye other red & a shade or black body be put behind it then looking on ye thred through a prism one halfe of ye thred shall appear higher y^n ye other, not both in one direct line, by reason of unequal refractions in ye differing colours.[38]

All experiments in the 1661–5 Note-Book are, like this one, performed with light reflected from coloured objects. No experi-

---

[36] 'Fortunate Newton, happy childhood of science!... Nature to him was an open book, whose letters he could read without effort. The conceptions which he used to reduce the material of experience to order seemed to flow spontaneously from experience itself, from the beautiful experiments which he ranged in order like playthings and describes with an affectionate wealth of detail.' Albert Einstein, Foreword to the 1931 edition of Newton's *Opticks*. Einstein is not of course suggesting that the conceptions did in fact flow spontaneously from experience. But that Newton thought they did is quite certain.

[37] Cf. A. R. Hall, *Cambridge Historical Journal*, IX (1948), pp. 239–50; *Annals of science*, XI (1955), pp. 27–43; and Richard S. Westfall, *Isis*, LIII (1962), pp. 339–58; 'The foundations of Newton's philosophy of nature', *The British Journal for the History of Science*, I (1962), pp. 171–82. For a different view of the evidence discussed by Hall, see T. S. Kuhn, 'Newton's optical papers', in *Isaac Newton's Papers & Letters* (ed. I. Bernard Cohen), Cambridge, Mass., 1958, pp. 27–45.

[38] Quoted by A. R. Hall, *Cambridge Historical Journal*, IX (1948), pp. 247–8. In a later Note-Book Newton described a similar experiment in which he viewed through the prism a parti-coloured line drawn on a black piece of paper. Here he also mentioned the experiment with the parti-coloured thread. See A. R. Hall, *Annals of Science*, XI (1955), p. 28.

ment is described in which, as in the 1672 paper, a direct beam from the sun is refracted through the prism. Each colour is further associated with a physical property, the speed of the ray producing that colour in the sensorium. Thus he suggests that the red-end colours are produced by more swiftly moving rays, and the blue-end colours are produced by more slowly moving rays.[39] But the speeds with which the rays reach the sensorium depend on the particular effect they receive from the reflecting body:

Hence redness, yellowness &c are made in bodys by stoping ye slowly moved rays w^{th}out much hindering of ye motion of ye swifter rays, & blew, greene & purple by diminishing ye motion of ye swifter rays & not of ye slower.[40]

But this means that there is no one speed-value that is always associated with the ray, right from the beginning. In other words, the physical property responsible for producing a given colour is not original or connate with the ray, as was insisted upon in the 1672 doctrine (already stated in the *Optical Lectures*). Thus Newton had not, at the time of the Note-Book, clearly abandoned the modification theory which he positively renounced in his paper to the Royal Society. It should be pointed out that already in this early Note-Book Newton readily operated with the notion of the rays as 'globules' which may vary in size as well as in speed.[41] His preference

---

[39] Newton also considers that 'slowly moved rays are refracted more then swift ones' (*Cambridge Historical Journal*, IX, 1948, p. 247).

[40] Ibid., p. 248.

[41] For example: 'Though 2 rays be equally swift yet if one ray be lesse y^n ye other that ray shall have so much lesse effect on ye sensorium as it has lesse motion y^n ye others &c.

Whence supposing y^t there are loose particles in ye pores of a body bearing proportion to ye greater rays, as 9:12 & ye less globules is in proportion to y^e greater as 2:9, y^e greater globulus by impinging on such a particle will loose 6/7 parts of its motion y^e less glob. will loose 2/7 parts of its motion & y^e remaining motion of y^e glob. will have almost such a proportion to one another as their quantity have viz. $5/7: 1/7:: 9: 1\frac{4}{5}$ wch is almost 2 y^e lesse glob. & such a body may produce blews and purples. But if y^e particles on wch y^e globuli reflect are equal to y^e lesse globulus it shall loose its motion & y^e greater glob. shall loose 2/11 parts of its motion and such a body may be red or yellow' (A. R. Hall, *Cambridge Historical Journal*, IX, 1948, p. 248). On one page of the Note-Book Newton drew a sketch of the 'Globulus of light' surrounded by 'a cone of subtile matter wch it carrys before it the better to cut ye ether, wch serves also to reflect it from other bodys' (Cambridge University Library MS, Add. 3996, fol. 104^v).

was, right from the start, for a corpuscular interpretation of light.

As was first suggested by A. R. Hall, it would thus appear that the theory of colours, announced to the world in 1672, was not completed until winter 1666 after an extended investigation which had occupied Newton's mind for about one year and a half. In spite of its autobiographical character, what the 1672 paper presents us with is the final result of a long series of inquiries which had passed through several stages, rather than the actual way in which that result had been achieved. More positively and more exactly, it presents us with an *argument* designed to establish the final doctrine beyond all possible objections. The form of this argument was clearly indicated by Newton himself in the same paper:

A naturalist would scearce expect to see ye science of those [colours] become mathematicall, & yet I dare affirm that there is as much certainty in it as in any other part of Opticks. For what I shall tell concerning them is not an Hypothesis but most rigid consequence, not conjectured by barely inferring 'tis thus because not otherwise or because it satisfies all phaenomena (the Philosophers universall Topick,) but evinced by ye mediation of experiments concluding directly & wthout any suspicion of doubt.[42]

This passage is of great importance for understanding the controversy which followed the publication of Newton's paper. He already repudiates a method of hypothetico-deductive explanation which he finds to be common among philosophers ('the Philosophers universall Topick'), a method in which hypotheses are merely conjectured to account for all the relevant phenomena. In contrast to this objectionable method, he claims to have directly deduced his theory of colours from experiments; in other words, his theory is a (proved) conclusion, not a (conjectured) assumption. Now all this sounds very Baconian. Newton proceeds in his paper as Bacon and

---

[42] *Correspondence*, I, pp. 96–97. This passage was omitted in the text of the *Philosophical Transactions*; it was first published by I. Bernard Cohen, *Archives internationales d'histoire des sciences*, XI (1958), p. 367. Both Newton and Hooke alluded to the suppressed passage in the course of their discussion; cf. *Correspondence*, I, p. 105, editor's note 19.

members of the Royal Society would have required him to do,[43] from the experiments through a series of negatives—but only to reach finally an affirmative proposition.[44] This end is achieved with the aid of an *experimentum crucis* which has in Newton's paper exactly the same role as the *instantiae crucis* of the *Novum Organum*: namely, it does not only refute the false doctrine but also positively establishes the true one. Newton's argument may thus be summarized as follows: the appearance of colours when a beam of the sun's light passes through a prism must be due to one of two and only two causes; the colours are either manufactured by the prism, or they have been with the rays from their origin; on the first supposition white light would be simple or homogeneous and colours would be modifications or confusions or disturbances of white light; on the second, white light would be a heterogeneous aggregate of difform rays which the prism would merely separate in accordance with their original degrees of refrangibility; experiment proves that an isolated coloured beam does not suffer any modification in respect of either colour or refrangibility on being refracted through a second prism; therefore, white light is demonstrated to be, not simple or homogeneous, but a heterogeneous aggregate of difform rays.

The effect of this 'demonstration' is, it must be admitted, almost hypnotic. Nevertheless, it is certainly inconclusive. Why should the

---

[43] See passage quoted above from Hook's *Micrographia*, p. 187.

[44] Bacon's advice was not to venture anything in the affirmative (conjectures, hypotheses) but to work for eliciting the affirmative from the experiments. This eliciting consisted in rejections which finally gave way to affirmatives necessarily implied by the experiments. Thus he maintained that to man 'it is granted only to proceed at first by negatives, and at last to end in affirmatives after exclusion has been exhausted' (*Novum Organum*, II, 15). In the Plan of the *Great Instauration* he described induction as a method which will 'analyse experience and take it to pieces, and by a due process of exclusion and rejection lead to an inevitable conclusion [*necessario concludat*]' (B, IV, p. 25). And again: 'But the induction which is to be available for the discovery and demonstration of sciences and arts, must analyse nature by proper rejections and exclusions; and then, after a sufficient number of negatives, come to a conclusion on the affirmative instances . . .' (*Novum Organum*, I, 105). In an exactly parallel fashion Newton first describes an experiment revealing a certain anomaly (the length of the image) which is to be accounted for. Various possibilities (suspicions) are then successively considered; they suggest new experiments by which they are, in turn, rejected. Finally a doctrine of light and colours is asserted, not simply because other doctrines have been proved false ('inferring 'tis thus because not otherwise'), but because it is *necessarily entailed* by affirmative instances, in particular the *experimentum crucis*, itself representing a decisive stage in the Baconian scheme.

second prism have the same effect on a *coloured* beam as the first had on a *white* beam? And why should not a property acquired by white light through a first refraction remain unaltered on undergoing further refractions?

Some of Newton's contemporaries saw the inconclusiveness of his argument; it is therefore curious that most historians of science have fallen under its spell.[45]

---

[45] The following more recent statement by R. S. Westfall reflects a view shared by almost all writers on this controversy: 'Whereas the modification theory held ordinary sunlight to be simple and homogeneous, Newton *demonstrated that it is a heterogeneous mixture* of what he called difform rays, rays differing in refrangibility, in reflexibility, and in the colour they exhibit' (*Isis*, LIII, 1962, p. 354; italics added).

Chapter Ten

THREE CRITICS OF NEWTON'S THEORY: HOOKE, PARDIES, HUYGENS

1. Hooke was present at the meeting of the Royal Society when Newton's letter was read. As requested by the Society he formulated what came to be known as his Considerations upon Newton's theory and he read them before the Society on 15 February 1671/2, that is, only seven days after Newton's paper had been delivered. Hooke was thus the first to raise objections against Newton's doctrine. His Considerations were not published in the *Philosophical Transactions*; only Newton's Answer to them was printed (in No. 88, 18 November 1672) without revealing the identity of the considerer to whom the Answer was addressed. They did not appear in print until 1757 when Thomas Birch published them in his *History of the Royal Society*.[1]

In general, Hooke's attitude to Newton's theory was the following: he was willing to grant the experimental results reported by Newton, but was unable to accept the 'hypothesis' which the latter had proposed to explain them;[2] he further expressed his

---

[1] Birch, op. cit., III (1757), pp. 10–15; references will be to the later edition from Hooke's autograph, in *Correspondence of Isaac Newton*, I, pp. 110–14.

The immediate effect of Newton's letter on the Society and the manner in which it was received by those of its members who were present (who included Hooke) is described by Oldenburg in the opening passage of a letter to Newton, dated 8 February 1671/2: 'The effect of your promise came to my hands this day, just as I was going to attend the publick meeting of the R. Society, where the reading of your discourse concerning Light and Colours was almost their only entertainement for that time. I can assure you, Sir, that it there mett both with a singular attention and an uncommon applause, insomuch that after they had order'd me to returne you very solemne and ample thankes in their name (which herewith I doe most cheerfully) they voted unanimously, that if you contradicted it not, this discourse should without delay be printed, there being cause to apprehend that the ingenuous & surprising notion therein contain'd (for such they were taken to be) may easily be snatched from you, and the Honor of it be assumed by forainers, some of them, as I formerly told you being apt enough to make shew of and to vend, what is not of the growth of their Country' (*Correspondence*, I, pp. 107–8).

[2] *Correspondence*, I, pp. 110–11.

belief that all of Newton's experiments, including the *experimentum crucis*, could be equally well explained by his own hypothesis about the nature of light and colours.

In order to be able to appreciate Newton's reaction to Hooke's objections, it will be necessary to understand the way Hooke interpreted Newton's position and the extent to which he really understood the import of his experiments. In the course of the following analysis of Hooke's objections, it will also be of interest to watch his attempts to envisage possibilities of adjusting his own conception of light, as he had already formulated it in the *Micrographia* (1665), with a view to accommodating Newton's observations. These attempts constitute the first trial to translate the Newtonian account of dispersion, which appeared to Hooke fundamentally (and undesirably) atomistic, into wave terms. The obvious limitations of Hooke's ideas seem to have discouraged historians from paying enough attention to his attempts. For the purpose of the present discussion, however, they will be of importance in bringing out the real point of difference between Hooke and Newton over the proper representation of white light. It will be seen[3] that Hooke's ideas, in a polished and mathematical form developed, no doubt, beyond Hooke's own expectations, were later introduced into optics to replace the Newtonian conception of white light.

Hooke understood Newton's 'first supposition' to be a corpuscular hypothesis about the nature of light, namely,

> that light is a body, and that as many colours or degrees thereof as there may be, soe many severall sorts of bodys there may be, all wch compounded together would make white, and . . . further, that all luminous bodys are compounded of such substances condensed, and that, whilst they shine, they do continually send out an indefinite quantity thereof, every way *in orbem*, which in a moment of time doth disperse itself to the outmost and most indefinite bounds of the universe.[4]

The rest of Newton's 'curious Theory', Hooke readily granted, could be demonstrated from this supposition without any difficulty.[5]

---

[3] See below, pp. 280ff.
[4] *Correspondence*, I, pp. 113–14.   [5] *Correspondence*, I, p. 114.

He objected to this fundamental hypothesis, however, on the grounds that 'all the coloured bodys in the world compounded together should not make a white body; and I should be glad to see an expt of the kind'.[6]

Whittaker has rightly observed[7] that, in making this and similar remarks, Newton's opponents were confusing the physiological or subjective aspect of colours with the physical problem with which alone Newton's experiments were concerned. But Newton himself had spoken of light as a substance, and of colours as *qualities* of light existing, as it were, without us. This ambiguous way of speaking was at least partly responsible for the misunderstanding of which his adversaries were victims. It was only natural, in the absence of any explicit remarks to the contrary, that they should understand the *heterogeneous aggregate*, supposed by Newton to constitute white light, on the analogy of the mixture of variously coloured powders. We shall come back to this important point when Newton's answers to Huygens are discussed.[8]

Hooke called Newton's theory a *hypothesis* because he believed that the same phenomena of colours would equally agree with various other views on the nature of light.[9] To him, Newton's supposition was by no means the only possible one. In his Considerations, therefore, he proposed his own hypothesis and tried to show that it would achieve the same purpose. This hypothesis expounded the view, just opposite to Newton's, about the constitution of white light and of colour; it declared that

---

[6] Ibid.

[7] Cf. Whittaker's Introduction to Newton's *Opticks*, New York, 1952, p. lxix.

[8] See below, pp. 290ff.

[9] 'I doe not therefore see any absolute necessity to beleive his theory demonstrated, since I can assure Mr. Newton, I cannot only salve all the Phænomena of Light and colours by the Hypothesis, I have formerly printed and now explicate yt by, but by two or three other, very differing from it, and from this, which he hath described in his Ingenious Discourse. Nor would I be understood to have said all this against his theory as it is an hypothesis, for I doe most Readily agree wth him in every part thereof, and esteem it very subtill and ingenious, and capable of salving all the phænomena of colours; but I cannot think it to be the only hypothesis; not soe certain as mathematicall Demonstrations' (*Correspondence*, I, p. 113). Hooke added in a following paragraph: 'If Mr Newton hath any argument, that he supposeth an absolute Demonstration of his theory, I should by very glad to be convinced by it . . .' (ibid., p. 114).

light is nothing but a pulse or motion propagated through an homogeneous, uniform and transparent medium: And that Colour is nothing but the Disturbance of yt light by the communication of that pulse to other transparent mediums, that is by the refraction thereof: that whiteness and blackness are nothing but the plenty or scarcity of the undisturbd Rayes of light; and that the two colours [blue and red] (then which there are noe more uncompounded in Nature) are nothing but the effects of a compounded pulse or disturbed propagation of motion caused by Refraction.[10]

It appears from the preceding passage that, for Hooke, white is the effect of an undisturbed and uncompounded motion or pulse; and that all colours are the result of the disturbance of light by refraction. In spite of the clause between parentheses, Hooke should not be understood to regard the colours blue and red as simple or primary in Newton's sense. They are themselves 'the effects of a compounded . . . motion caused by refraction' and not *original* as Newton had asserted of all colours. According to Hooke, however, red and blue are simple or uncompounded relative to the other colours, since he regarded the latter as due to the composition in different variations of these two. In order to make these ideas of Hooke clear, we should first look at his explanation of colours in the *Micrographia* where it is given in more detail, and then come back to his new suggestions which Newton's theory had provoked.

In the *Micrographia*,[11] Hooke considers a 'physical ray' falling obliquely on a refracting surface. What he calls physical ray is a portion of the ever expanding sphere described by the propagating pulse or wave-front. In the simpler case of a plane pulse, as in the figure (Fig. X. 1), the physical ray, incident on the refracting surface, is defined, for example, by the parallel lines *EB*, *FA*, which he calls 'mathematical rays', while the perpendiculars *EF*, and *BL*, represent the pulse itself in two successive positions. The mathematical rays thus represent the direction of propagation and they are made

---

[10] *Correspondence*, I, p. 110. The text published by Birch from the Register book of the Royal Society substitutes 'white' for 'light' in its first occurrence at the beginning of this quotation. See Birch, *History*, III, p. 11; *Correspondence*, I, p. 115, editor's note 3.

[11] Cf. *Micrographia*, Observation IX, especially pp. 62–64.

parallel on the assumption that the centre of propagation is infinitely remote.

According to Hooke, when the pulse is perpendicular to the direction of propagation, as in the ray *EFLB*, the light is said to be simple, uniform and undisturbed; it generates the sensation of white. Colours appear when the uniform (or undifferentiated) pulse is confused by refraction—in the following way. Since the ray is falling obliquely on the surface of separation, the side of the pulse

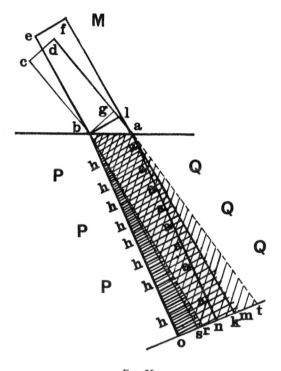

Fig. X.1

at *B* arrives at the surface and starts penetrating into the refracting medium while the other side is still travelling in the medium of incidence. We have seen[12] how Hooke, using Descartes' formula, arrives at the result that, in the medium of refraction, the pulse becomes oblique, instead of perpendicular, to the direction of

---

[12] See above, pp. 192ff.

propagation. This obliquity of the pulse to the mathematical rays constitutes the disturbance or confusion which Hooke asserts to be responsible for the appearance of colours. The refracted mathematical rays *BO, AN* remain parallel; and, of course, the sides of the refracted pulse, at *H* and at *A*, proceed with the same speed. (Otherwise, the inclination of the pulse to the direction of propagation would be continually changing while the ray is travelling in one and the same medium, contrary to what the theory assumes.)

But how are the colours produced as a result of the obliqueness of the pulse? Hooke considers that the preceding part of the pulse which first strikes the refracting surface 'must necessarily be somewhat *obtunded*, or *impeded* by the resistance of the transparent medium'.[13] When the second end of the pulse arrives at the surface, the way will have been, as it were, prepared for it by the first, and will accordingly meet with less resistance from the medium. There will thus proceed in the refracting medium a pulse whose weaker part precedes and whose stronger part follows. The dark or undisturbed medium *P* will further have a 'deadening'[14] effect on the part of the pulse that is adjacent to it and which has already been weakened at the surface; and this effect will penetrate deeper and deeper into the ray as it goes away from the surface. 'Whence all the parts of the triangle, *RBHO* will be of a dead *Blue* colour, and so much the deeper, by how much the nearer they lie to the line *BHH*, which is most deaded or impeded, and so much the more *dilute*, by how much the nearer it approaches the line *BR*.'[15]

To account for the generation of red, Hooke explains that the other side of the pulse at *AA*, being stronger than the part at *HH*, will induce a faint motion into the dark medium *Q*. As the ray goes forward this faint motion will be propagated farther and farther into the medium, that is, as far as the line *AM*, '. . . whence all the triangle *MAN* will be ting'd with a *Red*, and that *Red* will be the deeper the nearer it approaches the line *MA*, and the *paler* or *yellower* the nearer it is the line *NA*'.[16] Thus, on this account, while blue is due

[13] *Micrographia*, pp. 62–63.
[14] Ibid., p. 63.
[15] Ibid.
[16] Ibid.

to the deadening effect of the medium *P* on the weak side of the pulse at *HH*, red is the result of the disturbing effect of the strong side of the pulse at *AA* on the medium *Q*.

Though Hooke does not explicitly mention the spreading out of light by refraction in the *Micrographia*, it should be noted that his preceding explanation implicitly takes that fact into account, at least in a qualitative manner. For the refracted ray is limited in his figure by the *diverging* lines *BO* and *AM*. That is, one can infer from his figure that the incident ray, defined by the parallel lines *EB* and *FA*, has been dilated, or dispersed by refraction. It should be further noted that his figure represents blue as the most refracted, and red as the least refracted, these two colours being defined respectively by the lines *BO* and *AM*.

These observations may enable us to understand why Hooke did not feel that there was anything new in Newton's experiments.[17]

Hooke goes on to explain the appearance of the other colours: As the refracted ray continues, the lines *BR* and *AN*, representing the innermost bounds of the diluted blue and red (that is, pale blue and yellow), will intersect; beyond that intersection 'all kinds of Green' will be generated, owing to the overlapping of the regions occupied by blue and yellow.[18] If we further imagine another incident ray like *BCDA*, its corresponding refracted ray will be confined between the lines *BS* and *AT*. Again the blue will be adjacent to *BS*, and the red on the side of *AT*. All the rays falling between *BCDA* and *BEFA* will be refracted between *BO* and *AT*, and will be partially superimposed on one another. The disturbance at any point in the medium, say on the line *OT*, will thus depend on the resultant effect of the various pulses at that point; and every point in the disturbed medium will give rise to a particular colour according to its state of motion.

---

[17] Hooke in fact readily wrote in his Considerations: 'That the Ray of Light is as twere split or Rarifyd by Refraction, is most certaine' (*Correspondence*, I, p. 111); but he insisted that that splitting of light can be *explained* by what the pulse, in his theory, undergoes in refraction: 'But why there is a necessity, that all these motions, or whatever els it be that makes colours, should be originally in the simple rayes of light, I doe not yet understand the necessity . . .' (ibid.). Newton, on the other hand, was convinced that dispersion should be interpreted as the separation of heterogeneous elements.

[18] *Micrographia*, p. 63.

To complete Hooke's account, it will be sufficient to explain how, in his view, the two extreme colours, blue and red, are produced in vision. He imagines two rays falling at different inclinations on the cornea. The rays are focused by the eye-lens on two points on the retina. He shows that, at one of these points, the weaker side of the pulses will strike the retina before the stronger part—which excites the sensation of blue; whereas, at the other point, the stronger part precedes the weaker, and a sensation of red is generated. Thus, he concludes, 'That *Blue is an impression on the Retina of an oblique and confus'd pulse of light, whose weakest part precedes, and whose strongest follows. And, that Red is an impression on the Retina of an oblique and confus'd pulse of light, whose strongest part precedes, and whose weakest follows.*'[19] The other colours will be generated as a result of the combined effect on the retina of the various pulses that are focused between the extreme points of blue and red.

That Hooke's explanations are unsatisfactory is only too obvious. Yet his theory is much more subtle than might appear at first consideration. In particular, the sense in which it is dualistic, since it admits only two principal colours, is very sophisticated. For though Hooke repeatedly asserts that all colours are but the various mixtures or dilutions of blue and red, his meaning can only be appreciated by reference to his explanation of what the refracting medium undergoes when a refracted pulse passes through. From this explanation it appears that there corresponds to every colour a certain *physical* property of the medium depending on the degree of 'strength' or 'weakness' of the pulses at any given point. The real deficiency in Hooke's theory, however, lies in the fact that this physical property is not clearly defined: it is not easy to understand what the relative strength or weakness of the various parts of the pulse consist in, since, as has been remarked before, those parts proceed with the same speed. Nevertheless, had Hooke tried to develop long enough the analogy between sound and light that was current in his time, he might have said, for example, that the strength of the pulse at any given point in the medium will depend on the degree of promptness of the vibration at that point. *Promptness* is in

---

[19] *Micrographia*, p. 64.

fact the word that was used by Malebranche about eighteen years later (i.e. in 1699) to denote frequency.[20] It will be seen later that Newton also suggested to Hooke a similar idea in order to bring their respective theories closer together.

We may now proceed to examine Hooke's approach to Newton's experiments and doctrine from the point of view of the theory expounded in the *Micrographia*. He felt there was no difficulty in accounting for the Newtonian experiment in which white light was obtained by allowing the colours to be refracted through a second prism in the opposite direction. Before light suffers any refraction at all, he explained, it consists in a simple and undifferentiated motion; at the first refraction by the first prism, a multiplicity of motions or vibrations are generated and colours appear; when these are refracted again in the opposite direction, the acquired vibrations are destroyed and the motion of light is restored to its original simplicity. There was thus, for Hooke, no need to postulate the original heterogeneity of white light.[21]

From Hooke's point of view, to assert the doctrine of original heterogeneity would mean that the motions (which according to him are created by refraction) 'should be originally in the simple rays of [white] light'.[22] But he found that as unnecessary to maintain as to say that sound exists in a string before it is plucked. Nevertheless, he himself proceeded to develop further this analogy:

which string (by the way) is a pretty representation of the shape of a refracted ray to ye eye; and the manner of it may be somewt imagined by the similitude there; for the ray is like the string, strained between the luminous object & the eye, & the stop or finger is like the refracting surface, on ye one side of wch the string hath noe motion; on the other, a vibrating one. Now we may say indeed or imagine that the rest or streightness of the string is caused by the cessation of motions or coalition of all vibrations, and that all the vibrations are dormant in it.[23]

---

[20] Cf. P. Duhem, 'L'optique de Malebranche', in *Revue de métaphysique et de morale*, 23ᵉ année (1916), pp. 37ff.

[21] *Correspondence*, I, pp. 111, 112–13.

[22] Ibid., p. 111.

[23] Ibid. The text continues: 'but yet it seems more naturall to me, to imagine it the other way.' And yet Hooke returns to this idea no less than four times in the same letter containing

That is to say, we may imagine that the various vibrations (into which the simple motion of white light is differentiated by a first refraction) are made to coalesce or add up together, and in so doing destroy each other. But, again, Hooke saw no need to adopt this suggestion. This was unfortunate. For the preceding remarks contain in fact an expression of the idea that was later known as the principle of the superposition of waves. Hooke, it seems, was the first to conceive of this idea as applied to light. His gropings are the more interesting for the fact that he went as far as to consider the application of that idea to the problem of representing white light from the point of view of the wave theory. In doing this he was envisaging a possibility that was only realized by M. Gouy in 1886.[24] The degree of clarity with which Hooke conceived that interesting idea is revealed in the following words which come at the end of his Considerations:

Tis true I can in my supposition conceive the white or uniforme motion of light to be compounded of the compound motions of all the other

his Considerations. Thus he wrote: 'the motion of Light in an uniform medium, in wch it is generated, is propagated by simple & uniform pulses or waves, which are at Right angles with the line of Direction; but falling obliquely on the Refracting medium, it Receives an other impression or motion, which disturbes the former motion:—somewt like the vibration of a string; and that, which was before a line, now becomes a triangular superficies, in which the pulse is not propagated at Right angles with its line of Direction, but askew; as I have more at large explained in my *micrographia*' (ibid., p. 112). Note the expression 'pulses or waves' and the recognition of dispersion in the consideration that 'that, which was before a line, now becomes a triangular superficies'. Again: '. . . the compound motions are made to coalesse into one simple, where they meet, but keep their disturbd and compounded [motions], when they begin againe to Diverge and Separate . . .' (ibid., p. 113). And again, with reference to superposition of water waves: 'How the split Ray by being made to coalesse does produce a cleer or uniforme light, I have before shewd, that is, by being united thereby from a super-ficiall motion, which is susceptible of two, to a lineary, which is susceptible of one only motion: And tis as easy to conceive, how all these motions again appear after ye Rayes are again split or Rarifyd: He that shall but a little consider the undulations on the surface of a small rill of water in a gutter, or the like, will easily see the whole manner curiously exem-pliffyd' (ibid., p. 113). The fourth return, with reference to Descartes' use of the composition and resolution of linear motion, is quoted above. Can it be doubted that Hooke was attracted to this idea, even though he obviously conceived it under pressure of Newton's experiments? As will be seen, Newton certainly understood him to be proposing the coalescence of pulses (or vibrations or waves) as a possible representation that would not concede the conception of white light as a heterogeneous mixture, and he consequently rejected it. See below, pp. 281f.

[24] See below, pp. 28of.

colours, as any one strait and uniform motion may be compounded of thousands of compound motions, in the same manner as Descartes explicates the Reason of the Refraction, but I see noe necessity of it.[25]

Here Hooke is pointing out that the idea of the composition of motion that was applied by Descartes in the *Dioptric*[26] to linear motion *could* also be applied to the vibrating motion (or waves) which, according to him, constitutes light. If this *were* to be done, the simple, uniform (that is, undifferentiated) *pulse* which constitutes white light would be conceived as the resultant of all the various vibrations corresponding to the various colours. On this supposition, the motion of white light would be compounded in the sense that the variety of vibrations *potentially* exist in it before the pulse is decomposed by refraction. This was fundamentally Gouy's suggestion at the end of the nineteenth century. Unfortunately Hooke saw no necessity for it.

In any case, no theory of light could have profitably digested that suggestion before the analogy between light and sound was fully developed and the necessary mathematical apparatus (i.e. Fourier's theorem) for carrying out that suggestion was available. The important thing to notice, however, is that Hooke was able to conceive of a plausible (though qualitative) representation of white light which, had it been properly developed, would have taken Newton's experiments into full account. He was hindered from developing his ideas because he was content with his dualistic theory in its vague form. It was Newton, an enemy of the wave theory of light, who first clearly saw the necessity of introducing new elements into Hooke's wave picture if it were to give a proper representation of the discovered properties of colours. But we shall see that Newton determinedly rejected Hooke's idea of the 'coalition' or composition of vibrations for the reason that it would do away with his (Newton's) doctrine of the original heterogeneity of white light as he understood it.

It was remarked at the beginning of this account that Hooke

---

[25] *Correspondence*, I, p. 114.
[26] See also: Descartes, *Principles*, II, 32.

claimed to accept the experimental propositions contained in Newton's paper as long as they did not go beyond the actual observations. But since he maintained throughout that there were no more than two principal colours, he could not in fact do justice to Newton's experiments. To support his view, Hooke recalled an experiment, which had been related in the *Micrographia*[27], in which one of two wedge-like glass boxes was filled with a red tincture of aloes and the other with a blue copper solution. The first box exhibited all degrees of colours from deep red near the base to pale yellow near the edge. The colours exhibited by the other box varied from indigo to pale blue. All the other colours were produced by viewing the two boxes together through various thicknesses. Pale green, for example, was produced by passing white light through the boxes near the edges; and so on.

Newton objected in his Answer[28] to Hooke's Considerations that in the production of colours other than those exhibited by the boxes individually, more than two colours were in fact made use of, since neither of the boxes was of one uniform colour. To produce different sorts of green, for example, various degrees of yellow and blue were necessary. And to say, as Hooke had asserted, that yellow was only a diluted red was simply begging the question. But Hooke had already argued in the *Micrographia* that since red and yellow are produced by the same solution according to the various thicknesses of the box, they must be degrees of the same colour; or, in other words, yellow is a diluted red, and red is a deepened yellow. The difficulty, as Michael Roberts and E. R. Thomas have observed, is a real one; and to meet this difficulty, Newton himself was led to propose a theory of selective absorption:[29]

The Tincture of *Aloes* is qualified to transmit most easily the rays indued with red, most difficultly the rays indued with Violet, & with intermediate degrees of facility the rays indued with intermediate colours. So that

---

[27] *Micrographia*, Observation X, pp. 73–75.
[28] Cf. *Philosophical Transactions*, No. 88, 18 November 1672, pp. 5093–7; *Correspondence*, I, pp. 171–88.
[29] Cf. Michael Roberts and E. R. Thomas, *Newton and the Origin of Colours*, London, 1934, pp. 99–100.

where the liquor is very thin it may suffice to intercept most of the Violet, & yet transmit most of the other colours; all wch together must compound a middle colour, that is, a faint Yellow. And where it is so much thicker as also to intercept most of the Blew & Green, the remaining Green Yellow & Red must compound an Orang. And where the thicknesse is so great that scarce any rays can passe through it besides those indued with Red, it must appear of that colour, & yt so much ye deeper & obscurer by how much ye liquor is thicker.[30]

There is also a certain difficulty involved in interpreting the term 'dilution' in Hooke's account. For he had radically distinguished in the *Micrographia* between diluting and deepening a colour on the one hand, and whitening and darkening it on the other.[31] The second operation consists in varying the intensity of the colour by increasing or decreasing the quantity of light. But '*diluting* a colour, is to make the colour'd parts more thin, so that the ting'd light, which is made by trajecting those ting'd bodies, does not receive so deep a tincture'; and 'deepning a colour, is to make the light pass through a greater quantity of the same tinging body'.[32] From this it appears that Hooke does not want to assert that the difference between yellow and red, for example, is simply a difference in intensity. His remarks in fact cannot be understood without reference to his elaborate theory of the colours of thin plates where there is an imaginative attempt to explain how colour may depend on the thickness of the body through which the light has been trajected. Hooke's attempt will be discussed in Chapter XIII.

In spite of all these subtleties and elaborations, there is one thing which did escape Hooke's careful appreciation when he wrote his Considerations upon Newton's theory. That was the fact, discovered by Newton and of essential importance in his doctrine, that a middle colour, like orange or green, can be produced either from several primary colours or from only one such colour. In the first case the colour can be resolved into its components; whereas, in the second, the colour cannot be altered. This is what Hooke failed to make

---

[30] *Correspondence*, I, p. 180.
[31] Cf. *Micrographia*, Observation X, p. 75.
[32] Ibid.

room for in his dualistic theory. We should not therefore expect Newton to be satisfied with Hooke's claim that he granted the totality of Newton's experimental propositions. It also naturally appeared to Newton that Hooke's inability to appreciate the full import of the experiments was due to his (Hooke's) inclination towards a particular hypothesis about the nature of light. It will be shown later how this interpretation of Hooke's attitude, and the attitude of Newton's adversaries in general, influenced Newton's views on hypotheses.

2. It is clear from Pardies' first two letters[33] to Oldenburg concerning Newton's theory of colours that, like Hooke, he understood Newton's fundamental assertion of the unequal refrangibility of rays to be essentially bound up with a corpuscular view about the nature of light. This view, this 'very ingenious hypothesis', as he referred to it in the opening sentence of his first letter, he found 'very extraordinary' in that it viewed (white) light as an aggregate of an 'almost infinite' number of rays originally endowed with different colours and different degrees of refrangibility.[34] Pardies was naturally alarmed at a theory which obviously upset the very foundations of dioptrics as they were then universally accepted. He applied the word 'hypothesis' to Newton's theory, however, in all innocence; he did not mean that word to carry with it any derogatory implications. Thus, when he found that Newton had taken offence at this term,[35] he hastened to apologize, explaining that he had only used

---

[33] Dated, respectively, 30 March 1672, and 11 May 1672. Cf. *Philosophical Transactions*, No. 84, 17 June 1672, pp. 4087-90; and No. 85, 15 July 1672, pp. 5012-3. Re-edited in *Correspondence*, I, pp. 130-3, 156-8. English translation reproduced from *Phil. Trans., Abridged* (London, 1809) in *Isaac Newton's Papers & Letters on Natural Philosophy*, ed. I Bernard Cohen, Cambridge, Mass., 1958., pp 86-89, 104-5. The dates given in *Phil. Trans.* (viz. 9 April 1672 and 21 May 1672) are those at which Oldenburg forwarded the letters to Newton. The original Latin texts are followed in Turnbull's edition of the *Correspondence* with English paraphrases.

[34] *Isaac Newton's Papers & Letters*, p. 87.

[35] In his reply to Pardies' first letter Newton wrote to Oldenburg (13 April 1672): 'I do not take it amiss that the Rev. Father calls my theory an hypothesis, inasmuch as he was not acquainted with it [*siquidem ipsi nondum constet*]. But my design was quite different, for it seems to contain only certain properties of light, which, now discovered, I think easy to be proved, and which if I had not considered them as true, I would rather have them rejected

the word which first came to his mind.[36]   And we find him, in his second letter, equally applying it with the same readiness to alternative views which he himself favoured about the nature of light. The fact that he referred to Newton's view of the original compositeness of white light as a hypothesis only reflected his understanding that this view was the fundamental postulate of Newton's doctrine, believing of course at the same time that there was no compelling reason for maintaining it.

But his reasons for this belief were not altogether fortunate. Thus he thought, for instance, that the elongated shape of the solar spectrum could be explained by the accepted rules of refraction.[37] To demonstrate this he considered a particular case in which two rays coming from opposite parts of the sun fell at one point on the side of the triangular glass prism whose refracting angle was 60°. The two rays had angles of incidence equal to 30° and 29°30' respectively, and, therefore, their inclination to one another, measuring the apparent diameter of the sun, was 30'. Assuming the same refractive index for all incident rays, he found by calculation that the emerging rays from the other side of the prism subtended an angle of 2°23' (the ray incident on the prism at 30° emerging at an angle of 76°22' with the normal, while the other emerged at an angle of 78°49'). Observing that the angle contained between the emergent rays in his case was smaller by only 26' than Newton's estimation of the corresponding angle, Pardies concluded that there was no need to postulate the varying refrangibility of rays: the slight discrepancy could have been due to an error of observation.

Pardies had of course overlooked in these calculations that his prism was placed in a different position from that required in Newton's experiment. He had obviously overlooked the sentence in Newton's paper, stating that the refractions were made equal on both sides of the prism.[38] Not, however, because he was prejudiced

---

as vain and empty speculation, than acknowledged even as an hypothesis' (*Isaac Newton's Papers & Letters*, p. 92; cf. *Correspondence*, I, pp. 142, 144).

[36] *Isaac Newton's Papers & Letters*, p. 105: 'And as to my calling the author's theory an hypothesis, that was done without any design, having only used that word as first occured to me; and therefore request it may not be thought as done out of any disrespect.'

[37] *Isaac Newton's Papers & Letters*, pp. 87–88.

[38] See above, p. 237.

against Newton's theory: Pardies' sincerity and his readiness to understand are clearly apparent in his second letter to Oldenburg. After having received Newton's further explanations,[39] he recognized that, in the particular case considered by Newton, the image should be circular, not oblong; and, therefore, that was no 'difficulty' (objection) left for him regarding that point.[40]

But only regarding that point. For Pardies still could not accept the varying refrangibility of rays, although he admitted that the oblong shape of the spectrum was something problematic. It was here that he appealed to the hypotheses of Grimaldi and Hooke in order to account for the observed elongation. In Grimaldi's hypothesis light was conceived as a fluid substance capable of a very rapid motion. And Pardies briefly suggested that, on this supposition, dispersion could be attributed to a certain 'diffusion'[41] of the light at the farther side of the hole through which it passed (that is, at the light's entrance into the prism). Alternatively, Pardies went on, Hooke's hypothesis asserted that light was propagated by means of undulatory motion taking place in a very subtle matter. On this hypothesis, dispersion would be attributed to a certain 'diffusion' or

---

[39] For Newton's reply to Pardies' first letter, see *Phil. Trans.*, No. 84, 17 June 1672, pp. 4091–3; *Correspondence*, I, pp. 140–2; English translation, *Isaac Newton's Papers & Letters*, pp. 90–92. Oldenburg was not at all fair in what he wrote to Newton (9 April 1672) referring to Pardies' first letter: 'You see by the enclosed written to me by the Jesuite Pardies . . . how nimble yt sort of men is to animadvert upon new Theories' (*Correspondence*, I, p. 135). As we shall see, Huygens equally failed at first to see the essential point in Newton's experiment. To avoid these misunderstandings, something like the full explanations already elaborated in the *Lectiones Opticae* should have been provided in the first place. See above, p. 235, n. 9.

[40] Pardies wrote in his second letter to Oldenburg (11 May 1672): '. . . with respect to that experiment of the greater breadth of the colours than what is required by the common theory of refractions; I confess that I supposed the refractions at the opposite sides of the prism unequal, till informed by the letter[s] [*in leteris*] in the Transactions, that the greater breadth was observed by Newton in that case in which the refractions are supposed reciprocally equal, in the manner mentioned in these observations [Sed nec ab eo tempore in iisdem *Transactionibus* videre licuit cum eas non potuerim recuperare]. But since I now see that it was in that case that the greater breadth of the colours was observed, on that head I find no difficulty' (*Isaac Newton's Papers & Letters*, p. 104; cf. *Phil. Trans.*, No. 85, 15 July 1672, p. 5012; *Correspondence*, I, pp. 156–7). The 'letters' here referred to are (1) Newton's first letter to the Royal Society on light and colours, and (2) his reply to some experimental proposals (offered by Sir Robert Moray). The latter contained no clarifications concerning the position of the prism. Cf. *Phil. Trans.*, No. 83, 20 May 1672, pp. 4059–62: *Correspondence*, I, pp. 136–9 and note 3.

[41] *Isaac Newton's Papers & Letters*, p. 104.

'expansion' of the undulations at the farther side of the hole 'by virtue of the very contiguity and simple continuity of matter'.[42] Pardies then declared that he had himself adopted the latter hypothesis in a treatise which he had written on undulatory motion, in which colours were explained by lateral spreading of the motion propagated in straight lines.[43]

Besides these brief suggestions, Pardies expressed his feeling of certain difficulties in the Newtonian account of colours, especially in connection with the *experimentum crucis*. He thought that by rotating the first prism near the window in order to allow the colours to fall successively on the second prism, the incidence of the rays on the second prism was thereby continually changed. If this were the case, then obviously Newton's experiment would appear to be inconclusive: the unequal refractions of the various colours by the second prism might be due to their unequal incidence. But, again, Pardies had overlooked the fact that the incidence on the second prism was kept constant by the fixity of the two boards set up between the two prisms at a sufficiently large distance from one another.[44] Pardies was willing to assume that the 'sagacious Newton' had taken all necessary precautions to ensure that the experiment was properly performed.[45] But a second reply from Newton,[46] containing a sufficient clarification of the difficulty in question and

---

[42] 'quae [undulationes] fiat ad latera radiorum ultra foramen, ipsa contagio ipsàque materiae continuatione' (Correspondence, I, p. 157).

[43] See above, pp. 195f.

[44] Newton had written in his first paper: 'I took two boards, and placed one of them close behind the Prisme at the window, so that the light might pass through a small hole, made in it for that purpose, and fall on the other board, which I placed at about 12 foot distance, having first made a small hole in it also, for some of that Incident light to pass through. Then I placed another Prisme behind this second board . . .' (Correspondence, I, p. 94; compare the drawing opposite p. 107). Pardies' difficulty in understanding Newton's arrangement for the *experimentum crucis* should not be surprising. Huygens also found it difficult to understand, simply 'because it is somewhat obscurely written' (see Huygens' letter to Oldenburg, 1 July 1672, H, VII, p. 186; quoted by Oldenburg in letter to Newton, 2 July 1672, Correspondence, I, p. 207). Newton himself wrote the following in reply to Huygens' complaint: 'As to ye Theory of Light and Colors, I am apt to believe that some of the experiments may seem obscure by reason of the brevity wherewith I writ them wch should have been described more largely & explained with schemes if they had been intended for the publick' (Newton to Oldenburg, 8 July 1672, Correspondence, I, p. 212; cf. H, VII, p. 207).

[45] Isaac Newton's Papers & Letters, p. 105.

[46] Cf. Correspondence, I, pp. 163–8; Phil. Trans., No. 85, 15 July 1672, especially pp. 5016–17.

providing for the first time a diagram of the *experimentum crucis*, stimulated Pardies to write a last and highly appreciative and extremely courteous letter in which he declared that he was 'entirely satisfied'.[47]

3. On 11 March 1672, Oldenburg sent to Huygens a copy of No. 80 (19 February 1671/2) of the *Philosophical Transactions* which contained Newton's first paper to the Royal Society on light and colours. In a covering letter,[48] Oldenburg particularly referred to Newton's paper and requested Huygens to give his opinion of it. Strangely enough, it would seem from Huygens' reply (9 April 1672)[49] that he completely missed the point in Newton's discovery; and this, in spite of the fact that Oldenburg had explicitly referred to Newton's doctrine as a '*theorie nouuelle*' in which the author '*maintient, que la lumiere n'est pas vne chose similaire, mais vn meslange de rayons refrangibles differemment*',[50] thus directing Huygens' attention to the fundamental thesis of Newton's doctrine. The following is what Huygens wrote in answer to Oldenburg's request:

I was pleased to find in the last [*Transactions*] what Monsieur Newton has written regarding the effect of glasses and mirrors in the matter of telescopes, *where I see that he has noticed, like myself, the defect of refraction of convex object glasses due to the inclination of their surfaces.* As for his new Theory about colours, it appears to me very ingenious; but it will be necessary to see whether it agrees with all the experiments.[51]

Now Newton had indeed pointed out the defectiveness of refraction in the construction of telescopes. But he had not ascribed that

---

[47] Pardies to Oldenburg, 30 June 1672: 'Je suis tres satisfait de la réponse que M. Newton a bien voulu faire à mes instances le dernier scrupule qui me restoit touchant *l'experience de la croix*, a esté parfaitement levé. Et je conçois tres-bien par sa figure ce que je n'avois pas compris auparavant. L'experience ayant esté faite de cette façon je n'ay rien á dire' (*Correspondence*, I, pp. 205–6). A Latin translation was published in the *Philosophical Transactions*, No. 85, p. 5018, following Newton's reply.

[48] Cf. H, VII, p. 156; *Correspondence*, I, p. 117.

[49] H, VII, pp. 165–7.

[50] *Correspondence*, I, p. 117.

[51] H, VII, p. 165; italics added. Compare the extract in Oldenburg's Letter to Newton, 9 April 1672, *Correspondence*, I, p. 135.

defectiveness, as Huygens wrote, to anything regarding the shapes of lenses. On the contrary, Newton's point was that the fault lay in the constitution of light itself, namely the varying refrangibility of its rays.[52] The italicized words in Huygens' passage thus show that he still viewed the problem of perfecting telescopes as if Newton's paper had not thrown any new light on it. There is no mention in Huygens' letter of the doctrine of the unequal refrangibility of rays. He obviously understood Newton's contribution to constitute a new theory about *colours* rather than *refractions*.

After a repeated request from Oldenburg, [53] Huygens returned to Newton's doctrine of colours, referring to it as a *hypothesis*:

So far, I find the hypothesis of Monsieur Newton about colours very probable [*fort probable*]. The *experimentum crucis* is presented a little obscurely, but if I understand it well it very much confirms his new opinion ... As to his new hypothesis about colours ... I confess that, so far, it appears to me very plausible [*tres vraysemblable*], and the *experimentum crucis* (if I understand it well, for it is written somewhat obscurely) very much confirms it.[54]

Again, we find no mention of the different refrangibility of rays with which Newton's *experimentum crucis* was in fact directly concerned.

On 8 July 1672, Oldenburg forwarded to Huygens a copy of the *Transactions* No. 84 (17 June 1672) which contained Pardies' first letter and Newton's answer to it.[55] And on 29 July 1672, he sent him a copy of No. 85 (15 July 1672).[56] This contained the remaining letters exchanged through Oldenburg between Newton and Pardies, and a further article by Newton on his theory of colours.[57]

---

[52] See above, p. 240.
[53] Cf. Oldenburg to Huygens, 8 April 1672, H, VII, p. 168.
[54] Huygens to Oldenburg, 1 July, 1672, H, VII, pp. 185 and 186. Compare extract in Oldenburg's letter to Newton, 2 July 1672, in *Correspondence*, I, p. 207. Newton's reply has been quoted, above, p. 267, n. 44.
[55] Cf. Oldenburg to Huygens, 8 July 1672, H, VII, p. 196; *Correspondence*, I, p. 213–14.
[56] Cf. Oldenburg to Huygens, 29 July 1672, H, VII, p. 215; *Correspondence*, I, p. 220.
[57] Cf. *Phil. Trans.*, No. 85, 15 July 1672, pp. 5004–5. Newton's article had the following title: 'A Serie's of Quere's propounded by Mr. Isaac Newton, to be determin'd by Experiments, positively and directly concluding his new Theory of Light and Colours; and recommended to the Industry of the Lovers of Experimental Philosophy, as they were generously imparted to the Publisher in a Letter of the said Mr. Newtons of July 8. 1672.' The

We may therefore assume that Huygens was (or should have been) sufficiently informed about the Newtonian theory when, on 27 September 1672, he wrote the following to Oldenburg:

What you have presented of Monsieur Newton's in one of your last journals [i.e. No. 85 of the *Transactions*] confirms even more his doctrine of colours. Nevertheless, the matter could well be otherwise; and it seems to me that he should be content to propose it [i.e. Newton's doctrine] as a very plausible [*fort vrai semblable*] hypothesis. Besides, if it were true that the rays of light were, from their origin, some red, some blue etc., there would still remain the great difficulty of explaining by the mechanical physics wherein this diversity of colours consists.[58]

Still Huygens does not pronounce a word about the unequal refrangibility of rays. Instead, he expresses his concern about a mechanical explanation of colours. As a mechanist and a Cartesian, he was not satisfied with Newton's doctrine as long as it did not make it clear in what sense colours are original properties of the rays. His remark that Newton should be content to propose his theory of colours only as a very plausible hypothesis was no doubt an answer to Newton's claim, in the article published in No. 85 of the *Transactions*, that the experiments *positively* and *directly* concluded his theory.

The sense in which Huygens demanded a mechanical explanation of colours, and the extent to which this primary requirement influenced his understanding of and attitude towards Newton's theory, are clearly revealed in a letter to Oldenburg of 14 January 1672/3.[59] He wrote this letter after he had read Newton's Answer[60] to Hooke's

original bears the date 6 July 1672; cf. *Correspondence*, I, pp. 208–11. Although the title was apparently provided by Oldenburg, the claim that the experiments 'positively and directly' concluded Newton's theory was made by Newton in the article itself. This claim will be examined in the following chapter.

[58] H, VII, pp. 228–9; cf. extract from this letter in *Correspondence*, I, pp. 235–6.

[59] Cf. H, VII, pp. 242–4; quoted in Oldenburg to] Newton, 18 January 1672/3, *Correspondence*, I, pp. 255–6 (see p. 257, editor's note 2). Part of this letter was translated from French into English by Oldenburg and published in the *Philosophical Transactions* (No. 96, 21 July 1673, pp. 6086–7) under the following title: 'An Extract of a Letter lately written by an ingenious person from Paris, containing some Considerations upon Mr. Newtons doctrine of Colors, as also upon the effects of the different Refractions of the Rays in Telescopical Glasses.' Quotations from this letter will be taken from Oldenburg's translation.

[60] Cf. *Phil. Trans.*, No. 88, 18 November 1672, pp. 5084–5103; *Correspondence*, I, pp. 171–88.

Considerations. Following Hooke's suggestion, Huygens wrote that 'the most important Objection, which is made against him [Newton] by way of *Quaere*, is that, Whether there be more than two sorts of Colours'.[61] He believed that 'an *Hypothesis*, that should explain mechanically and by the nature of motion the Colors *Yellow* and *Blew*, would be sufficient for all the rest'.[62] But why should Huygens prefer to allow no more than two primary colors: 'because it will be much more easy to find an *Hypothesis* by Motion, that may explicate these two differences, than for so many diversities as there are of other Colors?'[63] Then at last we read the following sentence in which Huygens, for the first time, recognized the unequal refrangibility of rays:

And till he hath found this *Hypothesis*, he hath not taught us, what it is wherein consists the nature and difference of Colours, but only this accident (which certainly is very considerable,) of their *different Refrangibility*.[64]

As to the composition of white, he further suggests, in spite of Newton's detailed explanations to Hooke,[65] that 'it may possibly be, that *Yellow* and *Blew* might also be sufficient for that [difference in refrangibility]'.[66]

Finally this discussion ended in an unfortunate misunderstanding over Huygens' last suggestions. In his answer to Huygens, Newton wrote the following sentence:

If therefore M Hugens would conclude any thing, he must show how white may be produced out of two uncompounded colours; wch when he hath done, I will further tell him, why he can conclude nothing from that.[67]

---

[61] *Phil. Trans.*, No. 96, 21 July 1673, p. 6086.
[32] Ibid.
[63] Ibid.
[64] Ibid.
[65] Cf. *Phil. Trans.*, No. 88 (1672), pp. 5097–5102; *Correspondence*, I, pp. 181–6.
[66] *Phil. Trans.*, No. 96 (1673), p. 6086.
[67] Newton to Oldenburg, 3 April 1673, *Correspondence*, I, p. 265; in the *Philosophical Transactions* (No. 97, 6 October 1673, p. 6110) Huygens' name was replaced by 'N'.

What Newton meant by this sentence was that a white compounded from two primary colours, if it could be produced, would not have the same (physical) properties as the white light of the sun. For the analysis of the former would yield only the two component colours, whereas the analysis of the sun's light gives all the colours of the spectrum.[68] But Huygens misinterpreted Newton's sentence which seems to have sounded rather unreasonable to him. He wrote back in a letter to Oldenburg (10 June 1673, N.S.):

... seeing that he [Newton] maintains his opinion with so much concern [*avec tant de chaleur*] I list not to dispute. But what means it, I pray, that he saith; *Though I should shew him, that the White could be produced of only two Un-compounded colors, yet I could conclude nothing from that.* And yet he hath affirm'd in p. 3083, of the *Transactions* [i.e. No. 80, 19 February 1671/2,] that to compose the White, all primitive colors are necessary.[69]

This caused Newton to write a further letter to Oldenburg in which he clearly explained his meaning and, at the same time, expressed his impatience with the Dutch physicist: 'As for M. Hugens [his] expression, that I maintain my doctrine wth some concern, I confess it was a little ungrateful to me to meet wth objections wch had been answered before, without having the least reason given me why those answers were insufficient.'[70]

To think that a discussion on light and colours between the two greatest writers on optics in the seventeenth century should end in such an anticlimax!

---

[68] Cf. *Phil. Trans.*, No. 96, 21 July 1673, pp. 6087–92; *Correspondence*, I, pp. 291–5.

[69] *Phil. Trans.*, No. 97, 6 October 1673, p. 6112; H, VII, p. 302; quoted in Oldenburg to Newton, 7 June 1673, *Correspondence*, I, p. 285. The translation is Oldenburg's; in the quotation sent to Newton he altered '*tant de chaleur*' to '*quelque chaleur*'.

[70] Newton to Oldenburg, 23 June 1673, *Correspondence*, I, p. 292; cf. *Phil. Trans.*, No. 96, 21 July 1673, p. 6089. It should be noted that Newton's letter printed in No. 96 of the *Transactions* was a reply to Huygens' letter (10 June 1673 N.S.) which was published *afterwards* in No. 97; see *Phil. Trans.*, No. 97, 6 October 1673, p. 6112, editor's note at bottom of page.

Chapter Eleven

NEWTON'S DOGMATISM AND THE REPRESENTATION OF WHITE LIGHT

1. We are now, I hope, in a position to appreciate Newton's reaction to the critics of his theory. It will have been noticed that the objections of Hooke, Pardies and Huygens exhibited somewhat the same pattern. They all understood Newton to be propounding a corpuscular view of light from which, they all granted, the rest of Newton's doctrine of colours could be deduced. They referred to Newton's theory as a hypothesis, thereby implying that the experiments could be interpreted in other ways. Being themselves inclined towards a view according to which light consisted in the transmission of motion through a medium rather than the transport of body, they either proposed alternative hypotheses (Hooke and Pardies) or expressed reservations regarding the Newtonian theory (Huygens). At the same time their proposals failed to show a sufficient appreciation of Newton's experimental findings. Thus Hooke held to his dualistic theory of colours; whether or not his theory was capable of successful development, he appeared to be satisfied with it as it stood (at least in his first reaction to Newton's first paper).[1] Pardies thought that the shape of the solar image could be explained by the diffusion of light on its passing through the hole near the face of the prism. But this opinion was contradicted by the fact that difference between the diameter of the hole and the *breadth* of the image was proportioned to the distance between the hole and the opposite wall; no spreading of the light was observed along the axis of the prism. Pardies' suggestion was more suited to the phenomenon for which it had in fact been originally proposed by Grimaldi, namely diffraction.[2] Huygens expressed the view that two primary colours might be sufficient for mechanically explaining

---

[1] See below, pp. 327f.
[2] Cf. Brewster, *Memoirs*, I, p. 78.

all the others together with their various degrees of refrangibility. This could in no way be reconciled with the experimental results as Newton had reported them.

In the face of such opposition, Newton adopted the following strategy. Seeing that his experimental propositions were being confused with his suggested hypothesis about the corporeity of light, he hastened to deny that this hypothesis was an essential part of his theory. On the other hand he tried to convince his critics (Hooke, in particular) that their own (wave) hypothesis was in no way opposed to his fundamental doctrine of the original heterogeneity of light. Further, suspecting that understanding of his experiments had been hampered by preconceived ideas, he appealed for the separation of the experimental propositions from the task of explaining them by hypotheses; he advised that the properties of light should be investigated independently of any view about its nature. Finally, he claimed that it was not at all necessary to explain his doctrine of colours by any hypothesis regarding the nature of light; that his doctrine, being sufficiently and firmly founded on the experiments, was absolutely infallible. It was in connection with this claim that Newton made his first pronouncements on the proper method of scientific procedure. According to him, this method consisted in deducing the properties of things from experiments; not in arguing from hypotheses.

Before the controversy about colours, Newton had applied the term 'hypothesis' to propositions which were only *sufficiently* but not *accurately* true.[3] For example, the proposition adopted before the discovery of the sine law that for sufficiently small incidence the angles of incidence and of refraction were in a given ratio, was, for Newton, a hypothesis; it was only an approximately true proposition that was not well established by experiments. In this sense he also applied the word 'hypothesis' (in his first paper to the Royal Society

---

[3] Cf. Newton, *Optical Lectures*, Sec. II, 25, pp. 46–47. For detailed studies of the historical development of Newton's views on hypotheses, see: I. Bernard Cohen, *Franklin and Newton* (1956), pp. 129ff and Appendix I, pp. 575ff; 'The first English version of Newton's *Hypotheses non fingo*', *Isis*, LII (1962), pp. 379–88; A Koyré, 'L'hypothèse et l'expérience chez Newton', *Bulletin de la Société Française de la Philosophie*, L (1956), pp. 59–89; 'Les *Regulae philosophandi*', *Archives internationales d'histoire des sciences*, XIII (1960), pp. 3–14.

on colours) to the sine law itself as it had been accepted before his discovery of dispersion.[4] For, disregarding the qualification which he introduced in the same paper (namely that the index of refraction varies with colour), the sine law could only have been an approximation. As he explained in other places,[5] the successful application of the law before he had made his discovery, must have been due to the tacit assumption that, in the determination of refractive indices, the *middle* colour determined the position of *the* refracted ray. Thus, as the old law of refraction was not accurately true for all angles of incidence, the sine law in its unqualified form was not accurately true for all colours. But a proposition (or hypothesis) that is only approximately true, is, strictly speaking, false. This is not altered by the fact that the old law *worked* for small angles of incidence, or that the first formulation of the sine law served some practical purposes. They were sufficiently accurate for the purpose for which they were intended by their authors, but not to serve as a foundation for the science of dioptrics. The fact that Newton has used the word 'hypothesis' in this sense may help to explain his resentment at this word being applied to his theory of colours by his critics.

In any case, Newton was already using the term 'hypothesis' in the 1672 paper itself, and during the discussion about colours, in the sense adopted by all his opponents. According to this usage, a hypothesis is a proposition involving an entity (like the ether) which cannot be perceived. Since a proposition of this kind could not be directly related to observation, it was generally recognized that a hypothesis could be false even if the consequences derived from it agreed with the observations. Only a greater or smaller degree of practical certainty (Descartes) or verisimilitude (Huygens) could be claimed for it. Never after the controversy about colours did Newton apply the word 'hypothesis' to a proposition of the same level as, for example, the sine law. This word came to mean a proposition (concerning a hidden entity) that was only *probably* true, in contrast to a verified experimental law that was *certainly*

---

[4] See above, p. 238.
[5] Cf. Newton, *Optical Lectures*, Sec. II, Art. 27, pp. 50–51; *Opticks*, Bk. I, Pt. I, Prop. VI, Theor. V, p. 76.

true. The probability or lack of certainty of a hypothesis was, according to Newton, due to the fact that, instead of being *deduced* from experiments, it was simply *supposed* in order to *explain* experiments.

Judging from his own experience with his Cartesian opponents, Newton felt that a preconceived supposition had prejudiced their attitude towards his new discoveries. His advice to abstain from contemplating hypotheses, in this context, therefore meant that all views about the nature of light should be suspended until all the relevant experiments had been examined and ascertained. After this difficult task had been fulfilled, hypotheses might be considered to explain the experimentally established facts; but that latter task was, for Newton, a relatively easy matter. In all these comments, Newton took it for granted that his own doctrine of the heterogeneity of white light in respect of both refrangibility and colour was one of these experimentally established facts. On this understanding, he ascribed to his *experimentum crucis* a role which was not accepted by his opponents. They rightly perceived that what he presented as a neutral doctrine was, even after withdrawing his suggestions as to the corporeity of light, identifiable with a corpuscular view and, therefore, no more than a plausible interpretation of the experiments.

In what follows I shall try to illustrate and develop these remarks with a view to clarifying Newton's specific representations of white light as well as his methodological position in this controversy.

2. In his Answer to Hooke's Considerations, Newton denied that the corporeal nature of light was part of his doctrine:

Tis true that from my Theory I argue the corporeity of light, but I doe it without any absolute positiveness, as the word *perhaps* intimates, & make it at most but a very plausible consequence of the Doctrine, and not a fundamentall supposition, nor so much as any part of it.[6]

---

[6] *Correspondence*, I, p. 173. See above, p. 243.

As we have seen, Newton stated in the Waste Sheets that by suggesting the corporeity of light in the 1672 paper his intention had been to exclude the peripatetic conception of light as a quality, but not the mechanical conception of light as a motion transmitted through a corporeal medium (see above, pp. 244f). In other words, he had meant his suggestion to be

The whole of the doctrine of colours, Newton maintained, was contained in the experimental propositions.

He then went on to indicate that the same doctrine could be explained *to a certain extent* by various other hypotheses:

Had I intended any such Hypothesis [as that of the corporeity of light] I should somewhere have explained it. But I knew that the Properties, wch I declared of light were in some measure capable of being explicated not onely by that, but by many other Mechanicall Hypotheses.[7]

Similar remarks were made by Newton in his reply to Pardies' suggestions.[8] But the qualification in the above passage 'in some measure' is indicative of his real attitude. Newton in fact never admitted the possibility of giving an adequate explanation of the properties of light in terms of the wave picture alone. This belief was partly due to his realization of the inherent difficulties which the wave theory had to face and which at that time had not been solved. These difficulties, which he believed to be insurmountable, he himself pointed out at the same time as he offered his suggestions to reconcile his doctrine of colours with Hooke's hypothesis. Newton further maintained that the wave theory could not explain the rectilinear propagation of light; that is to say, it seemed to him to be falsified by experiment. His only purpose in making these

---

in opposition to *peripateticism*, not to *mechanism*. Here, in his Answer to Hooke, Newton states that the doctrine expounded in the 1672 paper did not commit him to a corporeal view of light—this view being only a very plausible consequence of the doctrine. This statement, if taken in conjunction with what is said in the Waste Sheets, would mean that the doctrine was open to a peripatetic interpretation! I would suggest that Newton's real position is expressed in his Answer to Hooke, and not in what he said later in the Waste Sheets. What he wished to argue 'without any absolute positiveness' could not have been a generalized mechanical view of light, but a specific view within mechanism—namely the view that light consisted of corpuscles.

[7] Ibid., p. 174.

[8] Newton wrote in reply to Pardies' second letter: 'F. Pardies says, that the length of the coloured image can be explained, without having recourse to the divers refrangibility of the rays of light; as supposed by the hypothesis of F. Grimaldi, viz. by a diffusion of light . . . or by Mr. Hooke's hypothesis [*ex Hypothesi Hookij nostri*] . . . . To which may be added the hypothesis of Descartes, in which a similar diffusion of *conatus*, or pression of the globules, may be conceived, like as is supposed in accounting for the tails of comets. And the same diffusion or expansion may be devised according to any other hypotheses, in which light is supposed to be a power, action, quality, or certain substance emitted every way from luminous bodies' (*Isaac Newton's Papers & Letters*, p. 106; see *Correspondence*, I, p. 164).

suggestions, therefore, was to show that Hooke (and the others) had no reason to reject his doctrine because of their theoretical commitments.

Newton's attempt to develop Hooke's concepts, however, is very interesting and should be quoted in full:

> ... the most free & naturall application of this Hypothesis [of Hooke] to the Solution of *Phænomena* I take to be this: That the agitated parts of the bodies according to their severall sizes, figures, & motions, excite vibrations in the Æther of various depths or bignesses, wch being promiscuously propagated through that Medium to our eyes, effect in us a sensation of light of a white colour; but if by any meanes those of unequall bignesses be separated from one another, the largest beget a sensation of a Red colour, ye least or shortest of a deep Violet, & the intermediate ones of intermediate colors: Much after the manner that bodies according to their severall sizes shapes & motions, excite vibrations in the air of various bignesses, wch according to those bignesses make severall tones in sound. That the largest Vibrations are best able to overcome the resistance of refracting superficies, & so break through it with least refraction: Whence ye vibrations of severall bignesses, that is, the rays of severall colors, wch are blended together in light, must be parted from one another by refraction, & so cause the Phœnomena of Prisms & other refracting substances.[9]

Newton introduced in this passage a new idea which marked a great improvement upon Hooke's treatment of light. Led by the analogy between light and sound, he realized that in an account of the properties of light in wave terms, the vibrations (or waves) corresponding to the various colours must differ from one another in some definite property, their bigness or wave-length. The heterogeneity of white light would then consist in the infinite plurality of the wave-forms corresponding to the component colours. But it must be realized that, for Newton, the component wave-forms (the 'unequal vibrations') must exist with their respective characteristic regularities *before* they are separated by refraction. In other words, the *original* compositeness of white light (whatever be

---

[9] *Correspondence*, I, p. 175. I have argued elsewhere that by 'bignesses of vibrations' Newton meant wave-lengths; see *Isis*, LIV (1963), pp. 267–8.

the terms, wave or otherwise, in which it is conceived) must have a definite *physical* meaning. The blending of the various vibrations to compose white light did not mean, in Newton's suggestion, that the waves would lose their individual characteristics in such a mixture; these waves, as they are sent out by the various parts of the luminous body (which according to their different sizes, figures and motions excite different vibrations in the ether) might *cross* one another, but they might not combine, or alter one another;[10] they must exist as differentiated elements of a *heterogeneous* mixture.

That this was for Newton a condition that would have to be satisfied in any wave picture is manifest from the following words which come shortly after the preceding passage in his Answer to Hooke:

But yet if any man can think it possible [to maintain that a pressure or motion in a medium may be propagated in straight lines], he must at least allow that those motions or endeavours to motion caused in ye Æther by the severall parts of any lucid body wch differ in size figure & agitation, must necessarily be unequall. Which is enough to denominate light an aggregate of difform rays according to any of those Hypotheses. And if those originall inequalities may suffice to difference the rays in colour & refrangibility, I see no reason why they that adhere to any of those Hypotheses, should seek for other causes of these effects, unlesse (to use Mr Hooks' argument) they will multiply entities without necessity.[11]

Thus, for Newton, the original difformity of white light should be preserved in any case. Hooke, on the other hand, was inclined towards a different view. Being convinced, as we have seen, of the

---

[10] See Newton's passage quoted below, p. 281.

[11] *Correspondence*, I, p. 176. The last sentence in this quotation is a reference to a passage in Hooke's Considerations (ibid., p. 113) where Newton was accused of multiplying entities without necessity, for admitting an indefinite number of colours when two were sufficient. Newton here repays the compliment by accusing Hooke of committing the same error, by seeking superfluous hypotheses. It would seem from this exchange that such tools (so-called methodological rules) as Occam's razor only serve as *machines de guerre* which one can always direct against the adherents of rival theories. Occam's razor does not inform us, for example, when and under what circumstances it is *not necessary* to multiply entities. Such a question is always left to the choice of the scientist who bases his decision on the requirements of his own theory. These requirements are not themselves methodological rules. The problem under discussion between Newton and Hooke, for example, could be formulated, investigated and solved without reference to any methodological rules.

uniformity (homogeneity, simplicity) of white light, he would rather have the various waves unite and compose one undifferentiated motion, a pulse.[12] The regularities suggested by Newton as necessary characteristics of the waves corresponding to the various colours would, according to Hooke's conception, disappear in the compounded white light. Hooke's reference to Descartes in this connection makes it clear that the compoundedness of white light would only have a mathematical sense, in contrast to Newton's conception. Hooke would thus argue, somewhat like Descartes, in the following manner: the motion constituting white light can be *imagined* to be composed of an infinite number of component motions; when the light passes through a prism, the motion is resolved into those components of which it is imagined to be composed. There would be no meaning in ascribing a *real* compositeness to the initial motion; and the prism could then be said to *manufacture* the colours from the simple motion of white light.

Now Hooke did not in 1672 endow his vibrations or waves with the property of periodicity. Nevertheless, his fundamental idea of the 'coalescence' or composition of vibrations which he envisaged in his Considerations upon Newton's theory came very near to M. Gouy's representation of white light in wave terms. Gouy showed[13] that the motion of white light can be represented as the sum or superposition of an infinite number of waves each corresponding to one of the spectrum colours. As it happens, the composed motion in this representation turns out to be a single *pulse*, such as that produced by a pistol shot or an electric spark. Using Fourier's theorem it could be shown that the Fourier analysis of such a pulse yields a continuous spectrum such as that produced by a dispersive apparatus. *Fourier analysis* is, of course, a mathematical concept; and the regularities produced by the dispersive apparatus (for example, the prism) would not exist in a real or physical sense in white light. What would exist in that sense in white light would be the pulses

---

[12] See above, pp. 255ff.
[13] Cf. M. Gouy, 'Sur le mouvement lumineux', *Journal de physique*, 2ᵉ série, V (1886), pp. 354–62; R. W. Wood, *Physical Optics*, New York, 1911 (second edition), Ch. XXIII: 'The Nature of White Light'. This chapter is not reprinted in the third edition.

sent out by the luminous source at irregular intervals.[14] The point in common between Gouy's interpretation and Hooke's idea is that, in both, the colours are considered to be generated by the prism; and in both, Newton's doctrine of the original difformity of white light has no place.

But as Hooke dismissed his own idea as 'unnecessary', Newton rejected it as 'unintelligible'; he wrote:

. . . though I can easily imagin how unlike motions may crosse one another, yet I cannot well conceive how they should coalesce into one *uniforme* motion [the pulse], & then part again [by refraction] & recover their former unlikenesse; notwithstanding that I conjecture the ways by wch Mr Hook may endeavour to explain it.[15]

This passage should leave no doubt as to Newton's conception of white light: even if it were possible to explain the properties of light in wave terms (as he was willing to assume for a while), it must be

---

[14] R. W. Wood (op. cit., p. 649) writes: 'A pulse or single irregular wave can be represented by Fourier's theorem, as the resultant of a large number of sine waves which extend to infinity on either side of the pulse. The spectroscope will spread this disturbance out into a spectrum, and at every point of the spectrum we shall have a periodic disturbance. In other words, the spectroscope will sort out the Fourier components into periodic trains of waves, just as if these wave-trains were really present in the incident light.'

See also F. A. Jenkins and H. E. White, *Fundamentals of Optics*, 2nd edition, London, 1951, pp. 214–16. Following the publication of Gouy's paper the scientific world witnessed a curious re-enactment of the dispute between Newton and his critics. Poincaré maintained that a high degree of regularity must be attributed to white light, while Schuster and Lord Rayleigh argued in support of Gouy's representation. (See R. W. Wood, op cit., p. 650.) Needless to say, this repetition of the seventeenth-century controversy occurred on a definitely higher level of mathematical and experimental sophistication. It should, however, indicate that Newton's opponents were not simply motivated by their pigheadedness or their addiction to hypotheses. And, in any case, it clearly shows that Newton's experiments had not been sufficient to 'prove' his doctrine of white light. A parallel study of these two disputes, with due attention to their different contexts, would be illuminating.

[15] *Correspondence*, I, p. 177; see p. 176. The text continues: "So that the direct uniform & undisturbed pulses should be split & disturbed by refraction, & yet the oblique & disturbed pulses persist without splitting or further disturbance by following refractions, is as unintelligible.' This objection derives its apparent plausibility only from interpreting the change produced by refraction as a disturbance (or confusion) rather than a *regularization*. But Newton was himself proposing to Hooke a hypothesis according to which the vibrations corresponding to the various colours were distinguished by a *definite* property, their 'bigness'. It should be mentioned here that, according to Newton's testimony, Hooke eventually (i.e before December 1675 and possibly in March of the same year) adopted Newton's suggestion —thereby abandoning both his dualistic theory of colours and the idea that the effect of the prism was of the nature of a 'disturbance' or 'confusion'. See below, pp. 327f.

allowed that the unlike motions, the waves corresponding to the various colours, should keep their individual characteristics unaltered when they mix and compose white light. The function of the prism then consisted in simply *separating* what was already differentiated though blended, not in *generating* new characteristics.

Whittaker has remarked that the word 'mixture', in Newton's assertion that white light is a mixture of rays of every variety of colour,

must not be taken to imply that the rays of different colours, when compounded together, preserve their separate existence and identity unaltered within the compound, like the constituents of a mechanical mixture. On the contrary, as was shown by Gouy in 1886 (*Journal de Physique*,[(2)] V, p. 354), natural white light is to be pictured, in the undulatory representation, as a succession of short *pulses*, out of which any spectroscopic apparatus such as a prism *manufactures* the different monochromatic rays, by a process which is physically equivalent to the mathematical resolution of an arbitrary function into periodic terms by Fourier's integral theorem.[16]

Since it is my purpose to differentiate between Gouy's interpretation (qualitatively conjectured in Hooke's idea of the coalescence of vibrations) and Newton's conception of white light, it must be observed that, from Newton's last passage just quoted, it is quite clear that he understood white light as a mixture in which the rays of various colours preserve, in Whittaker's words, 'their separate existence and identity unaltered within the compound'. Not to appreciate this would be to miss the point at issue between Newton and Hooke. To adopt Gouy's representation implies a rejection of Newton's particular understanding of the compositeness of white light.

Apart from his formal reservations, Newton further expressed the opinion that Hooke's hypothesis, and indeed any other hypothesis proposing to explain the properties of light by the propagation of pressure or motion in a medium, is fundamentally 'impossible'.[17]

---

[16] E. W. Whittaker, *A History of the Theories of Æther and Electricity, the classical theories*, Edinburgh, 1951, p. 17, note 1.

[17] *Correspondence*, I, p. 176.

In his view, both '*Experiment & Demonstration*' proved that pressures, waves or vibrations in any fluid bend round obstacles into the geometrical shadow.[18] Consequently, a theory of the same type as Descartes' or Hooke's was proved false by the rectilinear propagation of light, that is, by the formation of shadows in the ordinary way. By '*Experiment*' he was referring to observations on the propagation of water waves and the fact that sound is heard round corners; the '*Demonstration*', which was to the effect that the motion will be diffused throughout the space behind the obstacle, came later in the *Principia*.[19] These objections were perfectly valid in Newton's time; and it is not quite fair to answer them bluntly, as Duhem did,[20] by saying that light does in fact bend into the geometrical shadow (as Grimaldi's experiments on diffraction had shown). For Newton maintained that any wave motion will be diffused *throughout* the whole region that is screened off by the obstacle, and he was therefore asking how any shadow at all could be formed. In fact no satisfactory explanation of rectilinear propagation in wave terms was offered until Fresnel supplied such an explanation in the beginning of the nineteenth century. Only then was it shown that *both* diffraction *and* the apparent rectilinear propagation can be understood on the same principles. Using Young's law of superposition and Huygens' principle of secondary waves Fresnel could show that whether or not the light will be completely diffused throughout the space behind a screen (as required by Newton's demonstration) will depend on the dimensions of the aperture and the relation of these to the wave-length. The formation of shadows in the ordinary way (when the aperture is sufficiently large) was then explained as a result of the destructive interference of the secondary waves pre-

---

[18] Ibid., p. 175.

[19] Cf. Newton, *Principia*, Bk. II, Section VIII (the motion propagated through fluids); especially Prop. XLII. Theor. XXXIII, asserting that 'All motion propagated through a fluid diverges from a rectilinear progress into the unmoved spaces.' Newton's demonstration covers three cases: first, he is concerned with the motion of 'waves in the surface of standing water', in which case the motion is transverse; second, he considers pulses 'propagated by successive condensations and rarefactions of the [elastic] medium', as in the propagation of sound; and last, the proposition is asserted of 'motion of any kind'.

[20] Cf. P. Duhem, 'L'optique de Malebranche', in *Revue de métaphysique et de morale*, 23 e année (1916), p. 71.

sumed to proceed from every point at the aperture in *all* directions.[21]
All these explanations involved ideas that were either unknown or
not sufficiently developed in Newton's time.

3. So far we have witnessed Newton repudiating his initial corporeal
hypothesis as a part of his doctrine of light; and we have acquainted
ourselves with his suggestion that the rival hypotheses favoured by
his opponents were not against his theory. We have also seen with
what conditions this suggestion was qualified; these were to the
effect that the original heterogeneity of light had to be preserved.
At this point it might seem as if Newton would have accepted a
wave picture of light, had it not been for his belief that such a picture
would not make room for rectilinear propagation. It will, however,
be shown in what follows that, owing to his conception of what
constituted a ray of light, Newton was in fact committed to a
corpuscular view. This Newton was not aware of, since he presented
his conception as a doctrine completely neutral to any possible
explanatory hypothesis regarding the nature of light.

That Newton made the contention that his *theory* of light was free
from hypothetical elements is perfectly clear from his first paper,
from his replies to his critics, from his letter to Oldenburg on the
experimental method (which was published during the controversy
about colours), [22] and from his later writings on optics. His declarations on the role of hypotheses, in this context, were not only connected with that contention but obviously assumed it: it was only
because he believed that his own doctrine of light and colours did not
go beyond a description of the experimental facts that he would not
allow hypotheses more than a secondary role, namely that of *explaining* what has already been *discovered*.

Let us look into some examples. Newton wrote in his reply to
Pardies' first letter:

I do not take it amiss that the Rev. Father calls my theory an hypothesis,
inasmuch as he was not acquainted with it [*siquidem ipsi nondum constet*].

---

[21] Cf. E. Mach, *Principles of Physical Optics*, pp. 276–7.
[22] Cf. *Phil. Trans.*, No. 85, 15 July 1672, (English text) pp. 5004–5, (Latin translation) pp.
5006–7; *Correspondence*, I, pp. 208–11.

But my design was quite different, for it seems to contain only certain properties of light, which, now discovered, I think easy to be proved, and which if I had not considered them as true, I would rather have them rejected as vain and empty speculation, than acknowledged even as an hypothesis.[23]

This passage is of special importance because it contains Newton's first public reaction to the application of the word 'hypothesis' to his theory. It not only expresses Newton's belief that his doctrine was but a statement of some experimental facts, but also his attitude to hypotheses in general which we find developed in his later writings without fundamental alterations. The key to this attitude is clearly in the contrast which Newton draws between a hypothesis, a conjecture which can be at most probable, and truth, a property of propositions which are directly derived from experiments.[24] This was made even more clear in the following passage from his answer to Pardies' second letter; it is one of the longest and most important statements that Newton made on hypotheses.

In order to reply to these [Pardies' suggestions that Newton's experiments could be explained by the hypotheses of either Hooke or Grimaldi, without assuming the unequal refrangibilities of rays], it is to be noted that the Doctrine which I have expounded about Refraction and colours consists only in certain *Properties of Light* without considering *Hypotheses* by which these *Properties* should be explained. For the best and safest method of philosophizing seems to be, first diligently to investigate the properties of things and establish them by experiment, and then to proceed more slowly to hypotheses to explain them. For hypotheses ought to be fitted [*accomodari*] merely to explain the properties of things and not attempt to predetermine [*determinare*] them except in so far as they can be an aid to experiments [*nisi quatenus experimenta subministrare possint*]. If any one offers conjectures about the truth of things from the mere possibility of hypotheses, I do not see how anything certain can be determined in any science; for it is always possible to contrive [*excogitare*]

---

[23] *Isaac Newton's Papers & Letters*, p. 92; cf. *Correspondence*, I, p. 142.
[24] In the letter to Oldenburg of 6 July 1672, Newton wrote: 'You know the proper Method for inquiring after the properties of things is to deduce them from Experiments. And I told you that the Theory wch I propounded was evinced to me, *not by inferring tis thus because not otherwise*, that is not by deducing it onely from a confutation of contrary suppositions, but *by deriving it from Experiments concluding positively & directly*' (*Correspondence*, I, p. 209).

hypotheses, one after another, which are found to lead to new difficulties [*quæ novas difficultates suppeditare videbuntur*]. Wherefore I judged that one should abstain from considering hypotheses as from a fallacious argument [*tanquam improprio argumentandi Loco*], and that the force of their opposition must be removed, that one may arrive at a maturer and more general explanation.[25]

Apart from the assertion that the propounded doctrine did not involve any hypotheses, the following points emerge from the preceding passage which expounds in fact almost the whole of Newton's methodological position. It is first clear that he does not wish to confine the task of scientific inquiry to ascertaining the properties of things by observation and experiment; he allows for the task of explanation by means of hypotheses; he allows, that is, that one may attempt to explain the discovered properties by hypotheses. Nor does he wish to restrict the use of hypotheses to a stage in which they are merely fitted to explain the experimental facts and, so to speak, made to their measure. For he recognized that hypotheses may help to supply (*subministrare*) experiments that have not yet been performed. From this it appears that, according to Newton, hypotheses are not allowed in only one case: namely, when they are unjustifiably made to predetermine the properties of things. In other words hypotheses should not be maintained *against* what has been (or will be) experimentally established. Thus understood, Newton's advice presupposed two things. First, it presupposed his own interpretation of his experiments on dispersion, and in particular the *experimentum crucis*; this interpretation was to the effect that they *proved* the expounded doctrine of colours. Second, it presupposed the particular attitude taken by his critics towards his doctrine, especially their failure to grasp the full significance of the experimental results. That particular attitude itself presupposed (or was at least connected with) the Cartesian doctrine of hypotheses and, together with this, the Cartesian manner of attempting physical

---

[25] *Correspondence*, I, p. 164. Except for the first sentence and a few alterations the translation given here from the Latin original is that by Florian Cajori in the Appendix to his edition of Newton's *Principia*, p. 673. Compare the English translation in *Isaac Newton's Papers & Letters*, p. 106.

explanation. The latter, as we have seen, consisted in conjecturing hypotheses in terms of motion about the mode of production of natural phenomena, and then showing how they agreed with the observed facts. But then, Newton objects at the end of the passage, how can anything *certain* be determined in science? The Cartesian procedure appears to him to be an improper way of discovering truth; for the initial hypotheses will always remain in the realm of possibility even though their consequences may be shown to be true. Against this 'improper' method Newton therefore proposes what he believes to be the proper method, namely that in which the properties of things are derived with certainty from the observed facts of nature.

It is now clear that in order to declare the innocence of his theory of colours, Newton believed it was sufficient to divest it of the hypothesis regarding the corporeal nature of light. If we are still in any doubt about this, we need only look at his replies to Huygens. There we find him asserting that it was beside his purpose to explain colours hypothetically and that he

never intended to show wherein consists the nature and difference of colours, but onely to show that *de facto* they are originall & immutable qualities of the rays wch exhibit them, & to leave it to others to explicate by Mechanicall Hypotheses the nature & difference of those qualities; wch I take to be no very difficult matter.[26]

Less than three months afterwards, Newton presented a new formulation of his doctrine in which he described the proposition 'The Sun's light consists of rays differing by indefinite degrees of refrangibility' as a 'matter of fact'.[27] If we were now to agree that only statements expressly stating the corporeal (or wave) character of light are to be called 'hypotheses', then we should accept Newton's claim that his doctrine was indeed free from hypothetical elements. Such an agreement, however, would not be useful. It would turn us into victims of the illusion that by formulating his theory in terms of *rays* instead of *corpuscles*, Newton successfully made it neutral to

---

[26] Newton to Oldenburg, 3 April 1673, *Correspondence*, I, p. 264.
[27] Newton to Oldenburg, 23 June 1673, ibid., p. 293.

both corpuscular and wave interpretations. As will be seen presently, Newton's rays were always those of the corpuscular theory and were explicitly denied the meaning they would have in a wave theory. They were, in particular, conceived as discrete entities that could, in principle, be isolated from one another. In consequence, Newton's doctrine of the original heterogeneity of light as expressed in terms of such rays was denied, *a priori*, a wave interpretation. Huygens was therefore right in the reservations he expressed against Newton's doctrine: he was right in regarding it as a hypothesis that was greatly confirmed by the experiments without thereby reaching the factual status assigned to it by Newton.

Newton produced the following as a neutral definition of rays of light in his second reply to Pardies:

By light [*Lumen*] therefore I understand, any being or power of a being, whether a substance or any power, action, or quality of it, which proceeding directly from a lucid body, is apt to excite vision. And by the rays of light I understand its least or indefinitely small parts [*et per radios Luminis intelligo minimas vel quaslibet indefinitè parvasejus partes*], which are independent of each other; such as are all those rays which lucid bodies emit in right lines, either successively or all together. For the collateral as well as the successive parts of light are independent; since some of the parts may be intercepted without the others, and be separately reflected or refracted towards different sides.[28]

We find the same definition, in slightly different and clearer terms, in the beginning of the *Opticks* (1704). There, too, it is immediately preceded by the declaration that the design of the author 'is not to explain the Properties of Light by Hypotheses, but to propose and prove them by Reason and Experiments'.[29] It is as follows:

By the Rays of Light I understand its least Parts, and those as well Successive in the same Lines, as Contemporary in several Lines. For it is manifest that Light consists of Parts, both Successive and Contemporary; because in the same place you may stop that which comes one moment, and let pass

---

[28] *Isaac Newton's Papers & Letters*, pp. 106–7; cf. *Correspondence*, I, p. 164.
[29] Newton, *Opticks*, p. 1.

that which comes presently after ; and in the same time you may stop it in any one place, and let it pass in any other. For that part of Light which is stopp'd cannot be the same with that which is let pass. The least Light or part of Light, which may be stopp'd alone without the rest of the Light, or propagated alone, or do or suffer any thing alone, which the rest of the Light doth not or suffers not, I call a Ray of Light.[30]

Newton is here thinking of something like the following situation. A beam of light is let through a small hole. By making the hole narrower and narrower, more and more of the light arriving simultaneously at the hole is intercepted and only some of it is allowed to pass. At the same time a cogwheel (say) rotates in front of the hole. The light which arrives at a moment when the hole is closed is reflected backwards, whereas that which has just escaped proceeds forward. So far there seems to be no reason why Newton should not interpret this experiment to indicate that light has 'parts', without thereby necessarily prejudicing his conception of light towards one theory or another. But what does he mean by the 'least' or 'indefinitely small' parts of the light? For what is being made indefinitely smaller and smaller is, in the first operation, a region of space and, in the second, an interval of time. Newton is obviously making the assumption that this double process by which the beam is being chopped both spacially and temporally may be imagined to come to an end before the hole is completely closed. Thus by making the hole narrow enough only those rays coming successively in the same line will be let through. And, further, by making the interval during which the hole is open sufficiently small, only one of these rays will escape. It was this assumption which made Newton's critics suspect that his rays (of which the passage from the reply to Pardies said that they are entities independent of one another and which the doctrine endowed with various original colours and degrees of refrangibility) were simply the corpuscles, in spite of Newton's refusal to attach the proper label upon them.

The preceding interpretation of Newton's rays is confirmed by the

---

[30] Ibid., pp. 1-2.

explicit statement in the *Opticks* that they were not the straight lines of geometrical (and wave) optics,[31] and further by the fact that in his free speculations on the nature of light in the Queries at the end of that book, Newton never permits that light may consist of waves. At most, he allows that vibrations in an ethereal medium may play a complementary part—for example by putting the rays into 'fits'— but they never constitute light. As to the rays themselves, they understandably turned out in the Queries to be the corpuscles. We are led to discuss the nature of these next.

4. In what sense were colours asserted by Newton to be original *qualities* of the rays? Huygens, it will be remembered, remarked that even if the rays were from their origin endowed with colours, it would still be necessary to explain in mechanical terms wherein the diversity of colours consists. Newton on the other hand, realizing the uncertain and unstable character of hypotheses, had preferred to speak of light in general terms:

And for the same reason I chose to speake of colours according to the information of our senses, as if they were qualities of light without us. Whereas by that Hypothesis [of the corporeity of light which Hooke and the others attributed to him] I must have considered them rather as modes of sensation excited in the mind by various motions figures or sizes of the corpuscles of light making various Mechanicall impressions on the Organ of Sense. . . .[32]

Although Newton does not actually assert here that colours are in fact qualities of light existing outside our minds, his words must have sounded sinfully un-Cartesian to Huygens: how could anyone speak of colours as if they were qualities without us? Had not

---

[31] Newton, *Opticks*, p. 2: 'Mathematicians usually consider the Rays of Light to be Lines reaching from the luminous Body to the Body illuminated, and the refraction of those Rays to be the bending or breaking of those lines in their passing out of one Medium into another. And thus may Rays and Refractions be considered, *if Light be propagated in an instant.*' Newton remarked, however, that the last condition (my italics) had been proved false by Roemer. Cf. ibid., p. 2 and p. 277.

[32] *Correspondence*, I, p. 174 (Answer to Hooke's Considerations).

qualities been completely exorcised from matter once and for all? Had they not been very strictly confined to the mind? And, further, what does it mean that colours are to be understood as qualities of the rays 'according to the information of our Senses'? For our senses inform us that colours are qualities of things, not of light. But this is precisely what Newton's doctrine denies; since natural bodies appear in the dark in any colour which they happen to be illuminated with, it is concluded that these bodies are only variously disposed to reflect one sort of rays rather than another. Are colours then qualities of the rays in the same sense in which they *appear* to be qualities of natural objects?

Some writers have answered this question for Newton in the affirmative. M. Roberts and E. R. Thomas, for example, have explained Newton's preceding passage in the following words: 'That is, he has spoken of colours as if they were qualities of light independent of any observer—that is to say, he has regarded light as a primary property, not a secondary property which must be explained as a sensation produced by the action of the primary matter on the sense organs.'[33] And, as if trying to convince the reader that Newton had himself made up for his blunder, the same authors have added: 'His own work had, of course, made it possible to replace a description of colour by a numerical statement of refrangibility.'[34]

Émile Meyerson has also expressed the opinion, to which he was presumably led by similar statements to the one just quoted from Newton, that there is an obvious affinity between Newton's light corpuscles and the qualitative atoms of the seventeenth-century philosopher, Claude G. Berigard.[35] Meyerson[36] describes Berigard's doctrine as a legitimate development of peripatetic philosophy though it was also atomistic; it hypostatized qualities and conceived

---

[33] M. Roberts and E. R. Thomas, *Newton and the Origin of Colours*, p. 97, note 1.
[34] Ibid.
[35] Émile Meyerson, *Identity and Reality*, translated by Kate Loewenberg, London and New York, 1930, p. 334: 'The "caloric", semi-material fluid, bearer of a quality, belongs to the same family as phlogiston, and the relation of Newton's luminous corpuscles to the qualitative atoms of Berigard is equally obvious.'
[36] Cf. ibid., p. 329.

them as atoms. To explain the infinite diversity of the perceptible world, Berigard supposed an infinite variety of elements which he understood as spherical corpuscles each representing an elementary quality. These corpuscles were thus atom-qualities or corporeal qualities (*qualitates corporatae*) which conferred upon bodies their properties as for example (we may add) Newton's rays conferred colour upon them.

Now although Newton's position as he expressed it in 1672 was certainly ambiguous, Meyerson's interpretation of it is no doubt exaggerated.[37] A clearer statement of what Newton in fact wanted to maintain is to be found in the *Opticks*. There he makes it clear that the rays are not bearers of colours (coloriferous), as Berigard's corpuscles would be, but rather makers or producers of colour (colorific) by their action on the sensorium. Properly speaking, colours are, in the rays, *dispositions* to excite this or that sensation.[38] Newton's assertion that the rays are endowed with colour should

---

[37] Gaston Bachelard has passed this judgement on Meyerson's opinion. Cf. *L'activité rationaliste de la physique contemporaine*, Paris, 1951, p. 33. See also Léon Bloch, *La philosophie de Newton*, Paris, 1908, pp. 359–60.

[38] Newton, *Opticks*, Bk. I, Pt. I, Definition, pp. 124–5: 'The homogeneal Light and Rays which appear red, or rather make Objects appear so, I call Rubrifick or Red-making; those which make Objects appear yellow, green, blue, and violet, I call Yellow-making, Green-making, Blue-making, Violet-making, and so of the rest. . . . For the rays to speak properly are not coloured. In them there is nothing else than a certain Power and Disposition to stir up a Sensation of this or that Colour. For as Sound in a Bell or musical String, or other sounding Body, is nothing but a trembling Motion, and in the Air nothing but that Motion propagated from the Object, and in the Sensorium 'tis a Sense of that Motion under the Form of Sound; so Colours in the Object are nothing but a Disposition to reflect this or that sort of Rays more copiously than the rest; in the Rays they are nothing but their Dispositions to propagate this or that Motion into the Sensorium, and in the Sensorium they are Sensations of those Motions under the Forms of Colours.' See also ibid., Bk. I, Pt. II, Prop. V, Theor. IV, Exper. 12, p. 146. In the Waste Sheets, of probably 1676–7 (see above, p. 244, n. 32), Newton wrote: 'Understand therefore these expressions [i.e. substance, qualities, etc.] to be used here in respect of the Peripatetick Philosophy. For I do not my self esteem colours the qualities of light, or of anything else without us, but modes of sensation excited in our minds by light. Yet because they are generally attributed to things without us, to comply in some measure with this notion, I have in other places of these letters, attributed them to the rays rather than to bodies, calling the rays from their effect on the sense, red, yellow, &c. whereas they might be more properly called rubriform, flaviform, &c.' (*Correspondence*, I, p. 106). In the Note-Book of 1661–5 he already spoke of the rays as colour-making rather than colour-bearing, and he was already associating colours with mechanical properties of the light globules (cf. A. R. Hall, *Cambridge Historical Journal*, IX, 1948, p. 247; above, p. 247, and n. 41). Similarly, in the 1665–93 Note-Book, Newton wrote: 'I call those blew or red rays &c, wch make ye Phantome of such colours' (quoted by A. R. Hall, *Annals of Science*, XI, 1955, p. 28).

not, therefore, be taken literally; when he makes this assertion he is speaking roughly, not properly: 'And if at any time I speak of Light and Rays as coloured or endued with Colours, I would be understood to speak not philosophically and properly, but grossly, and accordingly to such Conceptions as vulgar People in seeing all these Experiments would be apt to frame.'[39]

At the same time it would not be a correct description of Newton's position to say that he was completely satisfied with replacing colours by their measurable degrees of refrangibility; that colours were sufficiently *explained* by substituting the difference in refrangibility for the difference in colour. He himself explicitly denied this in one of his replies to Huygens:

But I would not be understood, as if their difference [i.e. the difference of colours] consisted in the different refrangibility of those rays. For that different refrangibility conduces to their production no otherwise than by separating the rays whose qualities they are.[40]

That is to say, the difference in refrangibility only explains the *position* of the colour on the spectrum; it helps in the production of colours only by sorting out the rays which already possess those qualities, or rather the dispositions to excite the corresponding sensations in the sensorium. But this means that Huygens' demand for a physical explanation of colours still remains unsatisfied. Had Newton's intention in regarding colours as qualities of the rays been clarified to him, he would have still asked what those *dispositions* of the rays to produce the various colours consisted in. This question would have been the same as the question in which, as we have seen, he was asking for a theory in which a definite physical (or mechanical) property would represent the various colours and, if possible, would explain that 'accident', their degrees of refrangibility.[41]

Newton, however, always refrained from offering a direct and definite answer to Huygens' question. Just as he wished to speak

---

[39] *Opticks*, p. 124.
[40] *Correspondence*, I, pp. 264–5.
[41] See above, p. 271.

of colours in 1672 'according to the information of our senses', he also preferred to speak in the *Opticks* (1704) 'not philosophically and properly' but as the 'vulgar' would speak. He knew that speaking philosophically would inevitably be in terms of a particular view about the nature of light. But as this would explicitly tie up his theory with a *contestable hypothesis*, he preferred to expound the allegedly *experimentally proved* properties of light in dispositional terms. Nevertheless, he indulged in an attempt to answer Huygens' question—by way of a 'query'—at the end of the *Opticks*, thus admitting at least the validity of Huygens' demand, and making explicit the physical nature of his rays:

Nothing more is requisite for producing all the variety of Colours, and degrees of Refrangibility, than that the Rays of Light be Bodies of different Sizes, the least of which may take violet the weakest and darkest of the Colours, and be more easily diverted by refracting Surfaces from the right Course; and the rest as they are bigger and bigger, may make the stronger and more lucid Colours, blue, green, yellow, and red, and be more and more difficultly diverted.[42]

Physiologically, the light corpuscles produce the various colour sensations by exciting in the optic nerves vibrations of various bignesses according to their different sizes. The smallest and most refrangible corpuscles (for violet) excite the shortest vibrations; the largest and least refrangible (for red) excite the largest vibrations; and the corpuscles of intermediate sizes and degrees of refrangibility excite vibrations of intermediate bignesses to produce the inter-mediate colours.[43]

5. Having seen that Newton's conception of rays was, from the beginning, that of the corpuscular theory, we can now understand his interpretation of the *experimentum crucis*. It had been generally assumed before Newton that colours are created as a result of light

---

[42] Newton, *Opticks*, Query 29, p. 372.
[43] Cf. ibid., Queries 12 and 13, pp. 345–6. These two Queries are formulated in terms of rays, not of corpuscles; but they have to be understood in the light of Query 29 which was added in the second edition of the *Opticks* (1706).

coming into contact with reflecting or refracting bodies. Thus Descartes explained the generation of colours by the rotatory motion which the ethereal globules acquire when, under certain conditions, they obliquely strike a refracting surface. Grimaldi, for whom light was a fluid substance, made colours depend on the undulatory motions which are excited in the fluid when it meets with obstacles. Hooke attributed colours to the disturbing influence of refraction on the pulses which are originally simple and uniform. Newton argued that if colours were modifications of light due to the action of the reflecting or refracting bodies, then these bodies should produce the same effects whether the light falling on them has been already disturbed or not. But this consequence was proved false by the *experimentum crucis*: the rays emerging from the first prism preserved their respective colours and degrees of refrangibility upon their refraction through the second prism. Therefore, Newton concluded, colours must be unalterable properties of the rays and, consequently, white light must be an aggregate of rays that already possess those properties.

This conclusion, however, does not follow from the experiment. What the experiment does prove is that the light emerging from the first prism behaves differently from the light coming directly from the sun. But it does not prove that the properties of the refracted light exist primarily and unaltered in white light. It might well be that those properties are manufactured out of white light by the first prism but, once generated, they are not alterable by further refractions. The problem of how to represent white light thus remains open; it cannot be solved solely on the basis of Newton's experiment. In Gouy's interpretation, for example, white light is conceived as a succession of isolated pulses which do not possess the regularities characteristic of monochromatic light. Anticipating Gouy's interpretation, Hooke suggested that the composition of white light might be the kind of composition in which the component elements neutralize one another. But this suggestion Newton found 'unintelligible'; indeed, to him, any interpretation of light in wave terms was 'impossible'.

Newton was perhaps justified in his view of Hooke's suggestion by the backward state of the wave theory of his time. This only explains, however, his rejection of Hooke's idea; it does not account for his inclination towards a corpuscular interpretation of light in which the colours are original properties of the rays. Why, then, did Newton prefer such an interpretation?

Newton was an atomist who believed that matter is composed of hard and permanent particles which are endowed with various properties from the beginning of creation. The properties of natural things, on this conception, depend on the connate properties of the ultimate particles which go into their constitution. Natural changes consist, not in the creation or annulment of properties, but in the separating and combining of the particles.[44] These operations merely result in making manifest or concealing the original properties. From this point of view the dispersion of light by a prism could be readily explained: the prism simply *separates* the mixed corpuscles and, as a result, colours appear. Since the colours and their degrees of refrangibility are not altered by further refractions, no further separations seem to be possible and it is inferred that the colours on the spectrum and their degrees of refrangibility are connate properties of the corpuscles. The function of the prism thus consists in sorting out the corpuscles with respect to two of their original properties, colour and refrangibility.

---

[44] Newton, *Opticks*, Query 31, p. 400: '... it seems probable to me, that God in the beginning form'd Matter in solid, massy, hard, impenetrable, moveable Particles, of such Sizes and Figures, and with such other Properties, and in such Proportion to Space, as most conduced to the End for which he form'd them ... While the Particles continue entire, they may compose Bodies of one and the same Nature and Texture in all Ages: But should they wear away, or break in pieces, the Nature of Things depending on them, would be changed ... And therefore, that Nature may be lasting, the Changes of corporeal Things are to be placed only in the various Separations and new Associations and Motions of these permanent Particles.' Newton's belief in atomism has been generally recognized for a long time. His early attraction to this doctrine—as is shown by his first Note-Book of 1661-5—has been pointed out, first by A. R. Hall (*Cambridge Historical Journal*, IX, 1948, pp. 243-4) and more recently by R. S. Westfall (*British Journal for the History of Science*, I, 1962, pp. 171ff). What is not generally realized is that Newton's specific doctrine of white light cannot be deduced from the *experimentum crucis* without adding his belief in atomism as a necessary premise. It should be noted, moreover, that not only refrangibility and colour, but also reflexibility, polarity and (as we shall see in Ch. XIII) the 'fits' of easy reflection and of easy transmission are viewed as *original* properties of the rays. Thus Newton's explanations of all light phenomena exhibit one invariable pattern which is perfectly in keeping with the kind of atomism outlined in the passage just quoted from Query 31.

The history of science has shown that this is not the only possible interpretation of the experiment, but it should be remembered that, in Newton's time, atomism provided the picture in terms of which the clearest representation of white light could be most easily conceived without running into serious difficulties.

Chapter Twelve

THE TWO LEVELS OF EXPLANATION:
NEWTON'S THEORY OF REFRACTION

In this and the following chapter two more examples from Newton's researches in optics are discussed in turn. These are his theory of refraction and his theory of the colours of thin transparent plates. It will be shown that, apart from determining the experimental laws governing the phenomena in question, Newton proposed two kinds of explanation occurring on two different levels which he carefully distinguished from one another. According to him, the explanations of the first level are *theories* propounding only certain properties of light which have been deduced from the phenomena; that is, they are positive discoveries which are to be accepted irrespective of whatever opinion one might have concerning the nature of light; these theories therefore have the same value as that which he attributed to his doctrine of the original heterogeneity of light. Such is his explanation of the experimental law of refraction by the existence of a force acting perpendicularly at the refracting surface, and his explanation of the colours of thin bodies in terms of certain states or dispositions of the rays which he called their 'fits' of easy reflection and easy transmission.

The explanations proposed on the second, higher, level constitute merely an attempt to show how the refracting force and the fits themselves, whose existence is taken for granted, could possibly be accounted for. They are avowedly tentative explanations, or *hypotheses*, whose truth or falsity is regarded as irrelevant to the truth of the theories asserted on the first, lower, level.

Newton's radical distinction between these two levels of explanation is interesting in that it clearly reveals the logical status which he conferred upon his theories, and it helps towards an understanding of his view that scientific theories should be deduced from experiments rather than assumed or conjectured. It also provides an instructive illustration of his simultaneous appreciation of what role or roles hypotheses might possibly have in scientific procedure. Our main purpose will be to examine Newton's explanations of refraction and of the colours of thin plates, and to see whether the theories proposed on the first level were in fact obtained by a method such as that advocated by him, that is, whether they were purely experimental discoveries.

1. The law of refraction first appears at the beginning of Newton's *Opticks* as one of the eight Axioms or propositions giving 'the sum of what hath hitherto been treated of in Opticks. For what hath been generally agreed on I content my self to assume under the notion of Principles, in order to what I have farther to write'.[1] The form of the law as given in that place is the following: '*The Sine of Incidence is either accurately or very nearly in a given Ratio to the Sine of Refraction.*'[2] One might think that the cautious qualification 'either accurately or very nearly' is merely due to the fact that the law is stated here without reference to the unequal refrangibilities of differently coloured rays. But Newton later formulates the law in such a way as to take that fact into account: Proposition VI of Bk. I, Pt. I in fact reads '*The Sine of Incidence of every Ray considered apart, is to its Sine of Refraction in a given Ratio*'.[3] Nevertheless, and although he describes an 'experimental Proof'[4] of the sine law in this qualified and correct form, he is still not willing to accept the truth of the law 'as far as appears by Experiment', but wants to 'demonstrate' that it is 'accurately true'.[5] In other words, he wants to provide a demonstration in which the law is *rigorously* deduced from some assumption, as distinguished from a table of experimental readings which, alone, do not warrant the establishment of a *mathematical* ratio. This demonstration he bases on the following 'supposition': '*That Bodies refract Light by acting upon its Rays in Lines perpendicular to their Surfaces.*'[6] Obviously, this assertion involves an *explanation* of refraction in terms of a perpendicular force.

---

[1] Newton, *Opticks*, Bk. I, Pt. I, pp. 19–20.   [2] Ibid., Bk. I, Pt. I, Axiom V, p. 5.
[3] Ibid., p. 75.   [4] Ibid., pp. 76–79.   [5] Ibid., p. 79.   [6] Ibid.

There is a definite and interesting relation between the Newtonian explanation of refraction and Descartes' proof of the sine law; a relation which Newton was aware of but which has not been adequately defined.[7] Let us first quote what Newton himself had to say about the Cartesian proof in the *Optical Lectures* of 1669–71 :[8]

The Ancients determined Refractions by the Means of the Angles, which the Incident and refracted Rays made with the Perpendicular of the refracting Plane, as if those Angles had a given Ratio . . . the Ancients supposed, that the Angle of Incidence . . ., the Angle of Refraction . . ., and the refracted Angle . . . are always in a certain given Ratio, or they rather believed it was a sufficiently accurate Hypothesis, when the Rays did not much divaricate from the Perpendicular . . . . But this estimating of the Refractions was found not to be sufficiently accurate, to be made a Fundamental of Dioptricks. And *Cartes* was the first, that thought of another Rule, whereby it might be more exactly determined, by making the Sines of the said Angles to be in a giving Ratio . . . The Truth whereof the Author had demonstrated not inelegantly, provided he had left no room to doubt of the Physical Causes, which he assumed.[9]

The last sentence in this passage in fact foreshadows the kind of relationship that was going to hold between the Cartesian proof and Newton's explanation as he published it, first in the *Principia* (about eighteen years after this passage had been written), and later in the *Opticks*. Newton, in strong contrast to Fermat, has no objections against Descartes' demonstration; indeed he finds it 'not inelegant'. Quite naturally, however, he displays a reserved attitude towards the 'physical Causes', that is, the mechanical considerations that were

---

[7] E. W. Whittaker has the following remark on the relation between Descartes' and Newton's demonstrations of the law of refraction: 'Newton's proof of the law of refraction, in *Opticks*, i, prop. 6, does not differ greatly in principle from Descartes' proof' (*A History of The Theories of Æther and Electricity*, the classical theories, Edinburgh, 1951, p. 20, note 1).

[8] See above, p. 237, n. 14.

[9] Newton, *Optical Lectures*, London, 1728, Sec. II, pp. 46–47. It is seen from this passage that in 1669–71 (see above, p. 37, n. 14) Newton attributed the discovery of the sine law to Descartes. He also attributed the demonstration of the law to Descartes in 1675 (see letter to Oldenburg, 7 December 1676, in *Correspondence of Isaac Newton*, ed. Turnbull, I, p. 371). When the *Principia* was published (in 1687) he had adopted the view, first expressed in print by Isaac Vossius in 1662, that Descartes had got the law from Snell and merely exhibited it in a different form. Cf., on Vossius, J. F. Scott, *The Scientific Work of René Descartes*, pp. 36–37. See below, p. 304.

involved in it. What Newton showed later was that if refraction is given a dynamical interpretation, then the mathematical assumptions in Descartes' proof and the conclusion to which they lead (i.e. the law giving the sines in inverse ratio to the velocities) become perfectly acceptable.

For let us look at refraction from a dynamical point of view and assume that the ray of light is something or other which obeys Newton's second law of motion. The fact that the ray changes direction in passing from one medium into another will mean, according to that law, that there is a force acting at the surface of separation. Further, since a ray falling perpendicularly on a refracting surface passes through in the same line without suffering any deflection, we will conclude that the force acts only in the direction of the normal to the surface—it has no components in any other direction. Therefore, when the ray falls obliquely, only the perpendicular component of its velocity will be accelerated or retarded according as the perpendicular force is directed towards or away from the refracting medium; and thus Descartes' assumption asserting the conservation of the parallel component of the ray's velocity is preserved. Whether the refracting force is directed towards the medium of refraction or away from it will be indicated by the manner in which the ray is deflected: if the deflection is towards the normal, the perpendicular force must be acting towards the refracting body and consequently the velocity must have been increased; and if the deflection is away from the normal, the force must be acting in the opposite sense, and the velocity must have been diminished. Finally, since we observe that light passing from a rare medium like air into a denser medium such as water or glass is deflected towards the normal, we will conclude that the velocity of light in water or glass is greater than in air, also in accordance with Descartes' proof. Thus the proposition that the velocity of light is greater in denser media is seen to be an *inevitable* result of the supposition that refraction is caused by a *perpendicular* force.

But, so far, we have not established that the velocity in the refracting medium will be to the velocity of incidence in a *constant* ratio. Had Newton simply *assumed* that proposition (as Descartes did), his demonstration of the sine law would have run as follows: Having considered that the refracting force acts only in the direction of the normal to the surface, the following proposition is to be asserted, namely, that

(1) $$v_i \sin i = v_r \sin r,$$

where i, r are the angles of incidence and of refraction; and v_i, v_r the velocities of incidence and of refraction respectively.
Further, *assuming* that

(2) $$v_r = n\, v_i$$

(where n is a constant), it follows:

(3) $$\frac{\sin i}{\sin r} = \frac{v_r}{v_i} = n.$$

It is seen that this demonstration exactly coincides with the Cartesian deduction of the same conclusion (3),[10] with the one difference that here the word *force* is explicitly introduced in virtue of the initial supposition that the ray follows the laws of dynamics. This would not have constituted a significant advance on the Cartesian treatment. But the important merit of Newton's demonstration, as will be seen presently, is that, by providing an *independent* expression E for the refracting force, that is, an expression which does not presuppose equation (2) above, it established for the first time a connection between equations (1) and (2) which were in the Cartesian theory entirely unrelated. On the basis of E and assumption (1), Newton deduces the sine law in the form 'sin $i = n$ sin r'; and this result, again in conjunction with (1), allows him to obtain Descartes' law (3) as well as equation (2). One more advantage gained by introducing the expression E is that it further allows Newton to calculate the refractive power of a given medium,[11] such a calculation being impossible to perform in the Cartesian theory.

Before giving a detailed account of Newton's demonstration in the *Opticks*, it will be instructive to look at the character of that demonstration as it was first published in the *Principia*. At the end of Bk. I, in Section XIV (dealing with '*The motion of very small bodies*

---

[10] See above, p. 111.   [11] See Newton, *Opticks*, Bk. II, Pt. III, Prop. X, pp. 270-6.

when agitated by centripetal forces tending to the several parts of any great body'),[12] Newton considers (in Prop. XCIV) an imaginary situation in which a moving particle passes through a region of space that is terminated by two parallel planes. When the particle arrives at the first plane, it is assumed to be 'attracted or impelled'[13] by a perpendicular force whose action is the same at equal distances from either plane (referring always to the same one). The influence of this force does not extend outside the terminated region and the particle is not disturbed at any point of its journey by any other force. Since the acting force is, like that of gravity, a constant one, the particle will describe a parabola composed of the unaltered parallel component of the incident velocity and the accelerated or retarded motion in the perpendicular direction. The particle will emerge through the second plane tangentially to the element of the parabola at that plane. Newton demonstrates that the direction of emergence out of the terminated region will make with the normal an angle whose sine is to the sine of incidence on the first plane in a given ratio.

The next Proposition (Prop. XCV) further proves that, the same things being supposed, *the velocity of the particle after emergence will be to the velocity of incidence* (before the particle enters the region) *as the sine of incidence to the sine of emergence.* But since the preceding Proposition asserts that the sines will be in a given ratio, it follows that the ratio of the velocities will also be constant.

Suppose now (as is considered in Prop. XCVI) that the velocity of emergence is less than the velocity of incidence; this means that the perpendicular force is directed from the second plane towards the first and that the parabola described by the particle is convex towards the second. As soon as the particle enters the region, the perpendicular component of its motion will be gradually retarded until (if at all) the farther plane is reached and the particle emerges with a velocity composed of the original parallel velocity and what is left of the original normal component. Newton shows that if the angle of incidence is continually increased, the particle will be finally

---

[12] Newton, *Principia*, p. 226.　　　　[13] Ibid.

reflected at an angle equal to the angle of incidence. This will happen when the perpendicular component of the incident velocity is so small that it will be completely lost before the particle escapes from the region of the force's influence. At the point where the normal velocity is lost, the particle will momentarily move in a direction parallel to either plane but will immediately be forced to move symmetrically backwards to the first plane, through which it will emerge with a velocity composed of the original parallel component and a perpendicular component that is equal in magnitude but opposite in direction to the corresponding component in incidence.

Nothing is said about light in particular in any of Newton's demonstrations of the preceding Propositions; but, as he himself remarks in the Scholium to Prop. XCVI,

These attractions bear a great resemblance to the reflections and re-fractions of light made in a given ratio of the secants, as was discovered by *Snell*; and consequently in a given ratio of the sines, as was exhibited by Descartes.[14]

Thus, although Propositions XCIV-XCVI do not refer to any particular real situation, we may assume that they have been de-veloped with the purpose that they may be of help in the investiga-tion of the properties of light.

How then are these Propositions applied in the explanation of optical refraction?

In the *Principia*, the terminated region is supposed to lie in the midst of one and the same medium, and the distance between the two parallel planes may be of any finite magnitude. If now we assume that the interface separating two dissimilar media (e.g. air

---

[14] Newton, *Principia*, p. 229. If the angles of incidence and of refraction are taken to be the angles made with the normal to the refracting surface, then the given ratio must be between the cosecants (not the secants) of those angles. If therefore Snell's law, as here expressed by Newton, is to be correct at all, the angles whose *secants* are in a constant ratio must be measured by the inclinations of the incident and refracted rays to the interface, that is by the angles $(90°—i)$ and $(90°—r)$, where i and r are the angles made with the normal. See Snell's formula-tion of the law above, p. 100, n. 12.

and glass) lies somewhere between two *very near* parallel planes, and that a ray of light passing through the terminating planes suffers the same action as the particle (and no other action), we will come to the same conclusions about the ray as we have about the particle. Hence, in the *Opticks*, Newton lays down the following proposition concerning the motion of a ray of light falling on any refracting surface:

If any Motion or moving thing whatsoever be incident with any Velocity on any broad and thin space terminated on both sides by two parallel Planes, and in its Passage through that space be urged perpendicularly towards the farther Plane by any force which at given distances from the Plane is of given Quantities; the perpendicular velocity of that Motion or Thing, at its emerging out of that space, shall be always equal to the square Root of the sum of the square of the perpendicular velocity of that Motion or Thing at its Incidence on that space; and of the square of the perpendicular velocity which that Motion or Thing would have at its Emergence, if at its Incidence its perpendicular velocity was infinitely little.[15]

This obviously reduces optical refraction to the imaginary situation examined in *Principia*; except that the thinness of the terminated space is here assumed owing to the fact that no *sensible* curvature of the rays' path is normally observed near refracting surfaces.

The equation stated in the above passage (expression *E* referred to before) is the familiar law for constant acceleration:

$$v = \sqrt{u^2 + 2gs},$$

where v is the final velocity (i.e. the *perpendicular* velocity at the emergence from the farther plane), u the initial velocity (i.e. the *perpendicular* velocity at the incidence on the first plane), s the distance covered (between the two planes), and g is a constant.

Putting $s = 1$ and equating $2g$ with some constant f, we get the following by squaring the two sides of the equation:

(1) $$v^2 = u^2 + f.$$

---

[15] Newton, *Opticks*, Bk. I, Pt. I, pp. 79–80. As Newton immediately adds, if the motion is retarded, then the difference of the squares should be taken instead of their sum.

To determine f, Newton considers a case in which u is nearly equal to zero. For such a case, when the incident ray is almost parallel to the refracting surface, we have:

$$(2) \qquad\qquad v^2 = f,$$

where v is the perpendicular velocity of the refracted ray.

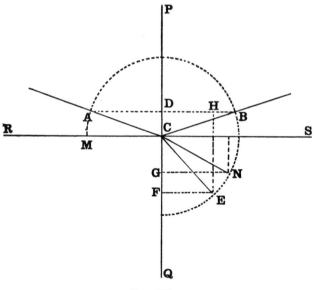

FIG. XII. 1

Suppose then[16] that a ray falling along the line MC (Fig. XII. 1) impinges on the refracting surface RS at C making an angle with the normal PQ almost equal to 90°, and is refracted into the line CN. QCN is thus a *given* angle.

The refracted ray along CN will have a horizontal velocity equal to the velocity of incidence, the latter being unaffected (or very little affected) in this case by the action of the refracting force. It will also have a perpendicular velocity v given by:

$$(3) \qquad\qquad v = \frac{MC}{GN} \times CG$$

---

[16] Newton, *Opticks*, Bk. I, Pt. I, pp. 80–81.

(on the assumption that MC represents the velocity of incidence and that this is equal to the horizontal component, along GN, of the refracted velocity along CN).

Therefore, from (3) and (2), we obtain:

(4) $$v^2 = \frac{MC^2}{GN^2} \times CG^2 = f.$$

Now if AC is any other incident ray, the problem is to find the line CE into which it will be refracted.

The perpendicular velocity of the refracted ray will be equal to

$$\frac{AD}{FE} \times CF$$

(again on the assumption that the parallel component AD of the incident velocity AC is unchanged by refraction). Putting $u = CD$ in (1), it follows from (1) and (4) that the perpendicular velocity

$$\frac{AD}{FE} \times CF = \sqrt{CD^2 + \frac{MC^2}{NG^2} \times CG^2},$$

i.e.

$$\frac{AD^2}{FE^2} \times CF^2 = CD^2 + \frac{MC^2}{NG^2} \times CG^2.$$

Adding the equals AD^2 and $(MC^2 - CD^2)$, we obtain:

$$AD^2 + \frac{AD^2}{EF^2} \times CF^2 = MC^2 - CD^2 + CD^2 + \frac{MC^2}{NG^2} \times CG^2,$$

so that

$$\frac{(EF^2 \times AD^2) + (AD^2 \times CF^2)}{EF^2} = \frac{(NG^2 \times MC^2) + (MC^2 \times CG^2)}{NG^2},$$

that is,

$$\frac{AD^2}{EF^2}(EF^2 + CF^2) = \frac{MC^2}{NG^2}(NG^2 + CG^2).$$

But since

$$EF^2 + CF^2 = NG^2 + CG^2,$$

therefore,

$$\frac{AD^2}{EF^2} = \frac{MC^2}{NG^2} = \frac{CN^2}{NG^2}$$

(CN being equal to MC).

307

But, since *AD* represents sin *i*, *EF* represents sin *r*, and $\dfrac{CN}{NG}$ is the cosecant of the *given* angle *NCG*, it follows that

$$\frac{\sin i}{\sin r} = \frac{CN}{NG} = n, \text{ a constant.}$$

Or,

$$\sin i = n \sin r,$$

the result obtained by Newton. He does not derive in the *Opticks* a relation of the sines in terms of the velocities. But such a relation can be deduced by combining the last result with the assumption which we have been making throughout, namely, that

$$v_i \sin i = v_r \sin r,$$

and we finally get:

$$\frac{\sin i}{\sin r} = \frac{v_r}{v_i} = n,$$

as in Descartes' proof.

2. At the end of the preceding demonstration Newton writes:

And this Demonstration being general, without determining what Light is, or by what kind of Force it is refracted, or assuming any thing farther than that the refracting Body acts upon the Rays in Lines perpendicular to its Surface; I take it to be a very convincing Argument of the full truth of this Proposition.[17]

This comment has to be understood in conjunction with the declaration at the beginning of Book I (where the demonstration in question occurs):

My Design in this Book is not to explain the Properties of Light by Hypotheses, but to propose and prove them by Reason and Experiment.[18]

We may thus gather that, for Newton, the 'supposition' or 'assumption' on which his demonstration (or, as we have regarded it,

[17] Newton, *Opticks*, Bk. I, Pt. I, pp. 81–82. The expression 'this proposition' refers to the sine law in the form 'sin *i* = *n* sin *r*'.
[18] Ibid., p. 1.

explanation) rests, is not a hypothesis. It is not a hypothesis because it does not determine what the *nature* of light is or by what *kind* of force it is refracted. The phenomenon of refraction indicates to him the existence of *a* perpendicular force acting at the refracting surface. From the force as assumption he deduces the sine law whose 'full truth' is thereby established. This ends, as far as refraction is concerned, the double process of 'analysis' and 'synthesis' of which Newton speaks in the *Opticks* as the proper method of investigating natural phenomena.[19] But to determine *a priori*, and without sufficient evidence from experiments, what the nature of the refracting force is—*that* would be a hypothesis.[20]

Although Newton was convinced that his demonstration of the law of refraction did not need a hypothesis to explain it, he in fact indulged in what he himself regarded as hypothetical explanations of refraction. These were of two opposite types, the one being in terms of an *impulsive* force due to the action of an ethereal medium and the other involving the idea of action at a distance.

An explanation of refraction of the former type was first proposed by Newton in a paper presented to the Royal Society on 9 December 1675 containing 'An Hypothesis Explaining the Properties of Light'.[21] In this hypothesis Newton imagines an ethereal medium similar to air, but far rarer, subtler and much more elastic than air. He still does not determine what light itself is, believing that if he abstains from pronouncing on this point, his hypothesis 'will become so generall & comprehensive of other Hypotheses as to leave little room for new ones to be invented'.[22] In any case, light is asserted to be 'neither this Æther not its vibrating motion, but something of a different kind propagated from lucid bodies'.[23] It might be an aggregate of peripatetic qualities, or a multitude of very small and swift corpuscles, or the motion of some other medium permeating the main body of ether—'let every man here take his fancy'.[24] But

---

[19] Cf. Newton, *Opticks*, Query 31, p. 404–5.
[20] It should be noted in this connection that the demonstration of the sine law in the *Principia* (like that in the *Opticks*) is meant to be generally valid, that is, regardless of whether the acting force is impulsive or attractive. See above, p. 303.
[21] Cf. Birch, *History*, III, pp. 248–60; *Correspondence*, I, pp. 362–86.
[22] *Correspondence*, I, p. 363.   [23] Ibid., p. 370.   [24] Ibid.

whatever light may be, one thing is requisite in order to explain its properties: It should be capable of interacting with the ether. Light acts on the ether by warming it, or by exciting in it swift vibrations (to explain, as will be seen below, the 'fits' of easy transmission and of easy reflection). Refraction is a case in which the ether acts upon light.

To explain this Newton supposes the density of ether within opaque and transparent bodies to be inversely proportional to the density of the body, and that it is greatest in free space. He then imagines at the surface separating any two dissimilar media a very thin region (call it the refracting region) in which the ether grows rarer from one side to the other—its density being the same along lines parallel to the surface of separation, but uneven along the perpendiculars to that surface. Beyond the region of refraction the ether is uniformly distributed on both sides. It is further supposed that the denser the ether is, the stronger is its action on light. When, therefore, an oblique ray penetrates the refracting region in its passage from one medium into another, it will be continually *impelled* perpendicularly from the side where the ether is denser (and, therefore, where the body is rarer) and be forced to recede to the other side, and thus its passage within the refracting region will be curved:

Now if ye motion of the ray be supposed in this passage to be increased or diminished in a certaine proportion according to the difference of the densities of the æthereall Mediums, & the addition or detraction of the motion be reckoned in the perpendicular from ye refracting Superficies, as it ought to be, the Sines of incidence & refraction will be proportionall according to what Des Cartes has demonstrated.[25]

---

[25] *Correspondence*, I, p. 371. The picture is that of atmospheric refraction; Newton imagines at the surfaces of refracting bodies a thin ethereal atmosphere which refracts light by the continuous variation in density of its layers. He made use of the same picture in 1675 (see *Correspondence*, I, pp. 383–84) in his hypothetical explanation of the diffraction phenomena (such as the bending of light around the sharp edge of a shadow-forming body) that had been noted and so named by Grimaldi (see Grimaldi, *Physico-mathesis*, Prop. I). By calling these phenomena by the name *inflection* (see Newton, *Principia*, p. 230; *Opticks*, Bk. III), Newton was using a term that had been introduced into optics by Hooke to denote a situation similar to that of atmospheric refraction, as when a beam of light passes through a solution of varying

These considerations were proposed by Newton about eleven years before his own demonstration of the sine law appeared in the *Principia* and, presumably, some time before this demonstration was formulated. But although they were briefly referred to again, after the publication of the *Principia*, by way of a question in the second English edition of the *Opticks* (1718), [26] the Queries of the *Opticks* simultaneously suggested the possibility of explaining refraction by supposing the refracting body to act upon the light ray at a distance.[27] We read in Query 29:

---

density (Hooke, *Micrographia*, p. 220: 'This *inflection* (if I may so call it) I imagine to be nothing else, but a *multiplicate refraction*, caused by the unequal *density* of the constituent parts of the *medium*, whereby the motion, action or progress of the Ray of light is hindered from proceeding in a straight line, and *inflected* or *deflected* by a curve.' See also Hooke's *Posthumous Works*, p. 81). Hooke had also applied that term, in discourses to the Royal Society, to the phenomena of diffraction which he discovered independently of Grimaldi though after the publication, in 1665, of Grimaldi's *Physico-mathesis* (see Birch, op. cit., III, pp. 54, 63, 69, 194–5, 268–9; also Hooke's *Posthumous Works*, 186–90). It was mainly Hooke's discourses (the first of which was delivered on 19 June 1672) that drew Newton's attention seriously to the investigation of diffraction (see Newton's statement in *Correspondence*, I, p. 383). Not only did Newton borrow the term *inflection* from Hooke, but he also inherited from him the particular interpretation of the phenomena denoted by that term, that is, diffraction was only a special kind of refraction. Thus Newton explained in the 1675 Hypothesis referred to above that the bending of light observed in the formation of shadow by a sharp-edged body was due to the varying density of the ethereal atmosphere surrounding the body. Since it was assumed that the ether grew rarer towards the body, the light should be inflected *into* the geometrical shadow. This explanation may therefore be taken, as far as it goes, to account for the inner fringes that appeared *inside* the shadow. But it fails to explain both the appearance of the outside fringes and the fact that the shadow was bigger than it should been have, had the light proceeded in straight lines. In the *Opticks* Newton proposed an alternative explanation of diffraction (or inflection) by the action – at a distance – of the opaque body upon the light rays (see *Opticks*, Queries 1–3). Here (in the experiments related in Bk. III) we find him concerned with the outer fringes and, consequently, the action of the body should be understood as repulsive, not attractive. But, then, what about the *inner* fringes? Mach has understandably found it strange that they are neither mentioned nor illustrated in the *Opticks* (see Mach, *Principles*, p. 143). Newton's investigations of diffraction phenomena were not completed to his own satisfaction (see *Opticks*, pp. 338–9). It would seem that, in order to explain the six coloured fringes (three inside the shadow and three outside it), which are indicated in the figure accompanying his 1675 discourse (see *Correspondence*, I, p. 384), he would have to assume that while some of the rays are attracted towards the body to form the inner fringes, others are repelled by it and thereby give rise to the outside fringes. To allow for this assumption, he would have to appeal to something like his theory of fits which he formulated to explain another class of what is now recognized as interference phenomena. Newton's undoubted merit in this field lies in the detailed experimental analysis and remarkably accurate measurements which he made.

[26] Cf. Newton, *Opticks*, Queries 19 and 20, pp. 349–50.

[27] The Queries exhibited this dual character of hypothetical explanations (sometimes mechanical, sometimes attractionist) with regard to most properties of light and to gravity.

Are not the Rays of Light very small Bodies emitted from shining Substances? . . . Pellucid Substances act upon the Rays of Light at a distance in refracting, reflecting, and inflecting them, and the Rays mutually agitate the Parts of those Substances at a distance for heating them; and this Action and Re-action at a distance very much resembles an attractive Force between Bodies. If Refraction be perform'd by Attraction of the Rays, the Sines of Incidence must be to the Sines of Refraction in a given Proportion, as we shew'd in our Principles of Philosophy: And this Rule is true by Experience.[28]

On this interpretation the particles of the refracting body are supposed to be endowed with attractive powers. The resultant force of all the attractions from the particles at the surface of the body will be perpendicular to that surface. And, as is always necessary to assume, the influence of the force must not extend beyond the very near neighbourhood of the refracting surface. When the light particle falls into the region of influence of the refracting force, it is attracted perpendicularly towards the surface. But once the particle is totally immersed in the refracting body, the attractions will be equal on all sides and thus neutralize one another. The particle will then proceed in a straight line with its velocity at emergence from the refracting region.

That refraction is caused by an ethereal medium acting on the ray, or by an attractive power residing in the particles of the refracting body, these are *hypotheses* advanced by Newton without consideration of their truth or falsity. The fact that they appear in the *Opticks* as Queries and not as Propositions means that they do not form part of the *asserted* doctrine of light and of refraction. The propositions themselves, and in particular those of Bk. I, have a different status: they stand on a lower level where they are firmly supported by 'Reason and Experiment'.[29] So far as the sine law is concerned this means that the experiments show the law to be 'either accurately or very nearly' true, while a demonstration based on the assumption of a

---

[28] Newton, *Opticks*, Query 29, pp. 370-1. This Query was originally published in the first Latin edition of *Opticks* (1706) as Query 21. It later appeared in the second English edition of 1718 (and in all subsequent English editions) as Query 29, and a final paragraph was added.
[29] See above, p. 308.

perpendicular force proves it to be 'accurately' true. [30] For Newton, this demonstration constitutes 'a very convincing Argument of the full truth'[31] of the law, not only because it leads to a conclusion that is found to agree with the experiments, but also, and more important, because in his view it is so general as to leave open all questions regarding the nature of the refracting force.

But why should there be a *perpendicular* force at all?—Because, as we have seen, this must be supposed if the ray is to behave in accordance with the laws of dynamics. What Newton, therefore, takes for granted throughout is a dynamical interpretation of refraction. This, of course, was not the only possible interpretation in his time, and it was not general enough to allow for all views regarding the nature of light. For it presupposed a conception of the ray as something that could be acted upon by a force and it thereby excluded the wave conception that was advanced, for example, by Huygens.

Newton had reason to believe that his was the true interpretation. For if he was asked to justify it against Huygens' geometrical principle (which equally leads to the sine ratio though it is not compatible with the assumption of a perpendicular force since they give contradictory results about the velocity of light in different media), he would simply cite his objections against the wave theory itself which embodied Huygens' principle. One of these objections was that it could not explain rectilinear propagation. As has been remarked before, he was justified in this objection.[32]

Newton was mistaken, however, in believing that the difficulties facing the wave theory in his time were insurmountable.[33] All these difficulties have been overcome after Young and Fresnel had enriched the wave hypothesis with new ideas. Moreover, his own dynamical theory of refraction had to be abandoned after Foucault's 1850 experiment had decided against it and in favour of the wave explanation.

An examination of the role of this experiment will help to make clear the logical status of Newton's explanation of refraction.

During the eighteenth century Newton was generally understood

---

[30] See above, p. 299. [31] See above, p. 308. [32] See above, pp. 282ff. [33] See above, p. 282.

as having propounded a corpuscular view of light. This understanding was justified not only by his repeated denunciations of the wave hypothesis, but also because it was the only natural interpretation that could be attached to his conception of light rays, in spite of his deliberate abstaining from explicitly declaring their real nature. It was particularly obvious from his explanation of refraction that he was relying on a corpuscular view, since this explanation (especially as it appeared in *Principia*) clearly presupposed the assumption that the rays obeyed the laws of *particle* dynamics.[34] The rigorous deduction of the sine law on the basis of that assumption provided further evidence of the fruitfulness of Newtonian mechanics, showing as it did that the idea of attraction could be successfully extended to the domain of optics. It was therefore natural for the adherents of Newton's system to subscribe to his corpuscular view of light; and Huygens' theory was almost completely forgotten until it was revived and improved by Young and Fresnel in the beginning of the nineteenth century. It was then shown that all kinds of optical phenomena known at the time could be adequately explained by wave considerations making use of Huygens' principle of secondary wavelets, Young's principle of superposition and the idea of the transversality of light waves introduced by Fresnel to account for the phenomena of polarization. Newton's corpuscular theory was thus pushed aside, having failed to offer any explanations of diffraction, interference or polarization phenomena that could be compared with the convincing mathematical representations of Fresnel. But some Newtonians were not convinced, and, perhaps, understandably; the impressive success of one theory does not by itself constitute a refutation of a rival theory. After all, the corpuscular theory could reasonably claim to have succeeded just as well in explaining at least some of the properties of light (reflection and ordinary refraction); and only a few years before Fresnel entered the scene, Laplace was trying to formulate an explanation of double refraction in terms of forces acting at a distance and in accordance with Maupertuis'

---

[34] See the title of the section in *Principia*, which contained the demonstration of the sine law—quoted on pp. 308f.

principle of least action.[35] There was thus no compelling reason to abandon the corpuscular view until Foucault performed his experiment in 1850.

The purpose of the experiment was to decide between the corpuscular and wave theories by testing their respective explanations of refraction. As we have seen, the corpuscular theory demanded, *as a necessary result of its initial dynamical point of view*, that light should travel faster in denser media; whereas Huygens' principle required that the contrary should be the case. When Foucault finally succeeded in devising an apparatus by which he could compare the velocities of light in air and in water, he found that the light travelled more slowly in the latter, denser medium.

When this result was announced it was generally accepted as a definitive refutation of Newton's explanation of refraction; and Foucault's experiment was regarded as a *crucial experiment* deciding against the corpuscular view and supporting the wave hypothesis. The impact of this experiment on the minds of nineteenth century physicists was such that attempts to develop the corpuscular hypothesis were generally abandoned until Einstein advanced his photon theory in 1905. But, shortly before Einstein published his theory, Pierre Duhem had formulated his well-known view that crucial experiments are not possible in physics and, therefore, that Foucault's 1850 experiment cannot properly have the role which had been assigned to it.[36] This view has recently gained many followers, especially after quantum optics established itself as a universally accepted theory. It is now sometimes argued that since the corpuscles have returned in the form of photons, Foucault's experiment was not crucial after all. But this argument should not be confused with Duhem's own views.

Duhem asserted two theses which should be distinguished from one another. First, he pointed out that a physical experiment cannot,

---

[35] Cf. P. S. de Laplace, 'Sur la loi de la réfraction extraordinaire de la lumière dans les cristaux diaphanes,' in *Journal de physique*, LXVIII (1809), pp. 107–11.

[36] Cf. P. Duhem, *La théorie physique, son objet, sa structure*, 2nd ed., Paris, 1914, Pt. II, Ch. VI, Secs. II & III. The first edition appeared in 1906, but the chapters of the book had been published successively (in the *Revue de philosophie*) in 1904–5; see Preface to the second edition.

logically, refute an isolated hypothesis, but only a whole system of hypotheses taken together. For in deducing a certain prediction from a given hypothesis, in setting up the experimental apparatus, and in interpreting the experimental result, the physicist has to make use of a whole body of propositions which he takes for granted. If, therefore, the result is not in agreement with his prediction, this only tells him that there is something wrong *somewhere* in his system, but the experiment itself does not help him to locate the error. Applying these considerations to Foucault's experiment Duhem concluded that it had not logically refuted the corpuscular *hypothesis* (that the rays of light are small bodies emitted by luminous objects), but had only shown that the corpuscular *theory* (embodying all the propositions that had been admitted by Newton and his successors) was not compatible with the facts.

Duhem's second thesis was that physical experiments cannot, as Bacon had believed, be assigned a role similar to that of the indirect proof in mathematics. That is to say, when an experiment shows that a conclusion obtained in a given theory is not in agreement with the facts, it does not thereby *establish* the truth of any other theory which leads to the contradictory conclusion. It is in this *positive* sense that Duhem asserted the impossibility of crucial experiments in physics. In his view, no experiment is able to prove in this indirect way any given theory or hypothesis, because (contrary to what Bacon had assumed) there can never be a complete enumeration of all possible theories or hypotheses that might successfully fit the facts. If some of Foucault's contemporaries assigned a positive role to his experiment, that was only because they had not conceived the possibility of a different conception of light other than those of Newton and Huygens. Before the end of the century, however, Maxwell had already developed such a different conception in his electromagnetic theory.

These two theses of Duhem should, of course, be granted.[37] But

---

[37] With regard to the first thesis, it is not quite clear whether Duhem would go so far as to maintain that the argument in an experimental refutation necessarily involves the whole of knowledge as the ultimate background of any particular theory. If he does, then, Popper has pointed out, his criticism 'overlooks the fact that if we take each of the two theories (between which the crucial experiment is to decide) *together* with all this background knowledge, as

he also advanced the view that had the nineteenth century physicists wished to preserve the corpuscular *hypothesis*, 'they would no doubt have succeeded in founding on this supposition an optical system that would have been in agreement with Foucault's experiment'.[38] In this he has been followed by more recent writers. Alexander Wood, for example, suggested[39] that the adherents of the corpuscular view could have assumed that when the ray (the corpuscle) passes from a rare into a dense medium, *the perpendicular component of its velocity remains constant* while the component parallel to the refracting surface is diminished by (say) action of a frictional kind. Since this assumption leads to the conclusion that the actual velocity in the dense medium of refraction will be less than in the rare medium of incidence, the corpuscular hypothesis would thus have been brought into agreement with the facts and Foucault's experiment would have ceased to be crucial. Since the appearance of Wood's book his suggestion has been endorsed by Florian Cajori in the Appendix which he wrote to his edition of Newton's *Principia*.[40]

This kind of suggestion might appear to be justified by the power of mathematics to adapt itself to new situations and its capacity to assimilate new facts. Its advocates, however, seem to presuppose that this process of assimilation which has indeed continued throughout the history of science can always be carried out—even in an arbitrary and *ad hoc* fashion. This presupposition is mistaken. It is

---

indeed we must, then we decide between two systems which differ *only* over the two theories which are at stake. It further overlooks the fact that we do not assert the refutation of the theory as such, but of the theory *together* with that background knowledge; parts of which, if other crucial experiments can be designed, may indeed one day be rejected as responsible for the failure. (Thus we may even characterize a *theory under investigation* as that part of a vast system for which we have, if vaguely, an alternative in mind, and for which we try to design crucial tests.)' (Karl R. Popper, *Conjectures and Refutations*, p. 112.)

[38] *Théorie physique*, ed. cit., p. 284.

[39] Cf. A. Wood, *In Pursuit of Truth*, London, 1927, pp. 47–48.

[40] Cf. Newton, *Principia*, Appendix, note 26, pp. 651–2. It may be remarked that Wood's particular suggestion happens to lead to at least one result, namely a *false* law of refraction, which would have been sufficient to dismiss it at once. For his assumption concerning the perpendicular velocity can be expressed thus:

$$v_i \cos i = v_r \cos r.$$

Combining this with the definition of the refractive index n,

$$v_r = n \, v_i,$$

a cosine law of refraction is obtained:

$$\cos i = n \cos r.$$

not, for example, easy to see how the nineteenth century physicists could have effected the desired modification of the corpuscular theory without doing violence to the accepted Newtonian physics.[41] For we have seen[42] that the result refuted by Foucault's experiment is a *necessary* consequence of Newton's second law of motion. It is true, as the current argument runs, that a synthesis has been made (first by L. de Broglie) between particle dynamics and wave optics. But this has been achieved within the framework of a *new* mechanics making use of the relativistic relation $E = mc^2$ and Planck's quantum of action, that is, two concepts completely foreign to Newtonian mechanics.[43]

---

[41] See G. Bachelard's remarks in *L'activité rationaliste*, pp. 46–8.

[42] See above, p. 301.

[43] A. I. Sabra, 'A note on a suggested modification of Newton's corpuscular theory of light to reconcile it with Foucault's experiment of 1850,' *British Journal for the Philosophy of Science,* V (1954), pp. 149–51.

Chapter Thirteen

THE TWO LEVELS OF EXPLANATION: NEWTON'S THEORY OF THE COLOURS OF THIN PLATES

Newton rejected the explanation of reflection by the impinging of light rays on the solid parts of bodies.[1] Three of his reasons for abandoning this idea may be mentioned here. First, he pointed out that when light passed from a dense medium like glass into a rarer medium like air, some of it was reflected at the separating surface, even when the adjacent air was drawn away. This, clearly, could not be ascribed to the impinging of the light rays on something outside the glass.[2] Second, it was known that at a sufficiently great angle of incidence, the whole of the incident light was reflected back into the glass. He naturally found it difficult to imagine that at a certain inclination the light should find enough pores to let it through while these pores should not be available at another inclination, especially considering that when the passage was from air into glass some of the light was always transmitted regardless of the degree of incidence. Third, he noticed that if the coloured rays emerging from a prism were successively cast at the same angle of incidence on a second prism, the latter could be so inclined as to totally reflect the blue rays while rays of other colours (red, for example) were copiously transmitted.[3] 'Now if the Reflexion be caused by

---

[1] Newton, *Opticks*, Bk. II, Pt. III, Prop. VIII: 'The Cause of Reflexion is not the impinging of Light on the solid or impervious parts of Bodies, as is commonly believed.' Cf. pp. 262–9.

[2] This objection was first publicly stated by Newton in 1675; see *Correspondence*, I, p. 360: 'I remember, in some discourse with Mr Hook, I happened to say, yt I thought light was reflected, not by ye parts of glas, water, air, or other sensible bodies. . . .' (7 December 1675). Regarding Newton's possible meeting with Hooke in October 1675, see Birch, *History*, III, p. 228, and *Correspondence*, I, p. 417, editor's note (3).

[3] See *Opticks*, Bk. I, Pt. I, Expers. 9 and 10, pp. 54–63. These experiments led Newton to postulate a new original property of the rays, their 'reflexibility' (cf. ibid., Definition III and Bk. I, Pt. I, Prop III). He measured that property by the critical angle for a certain colour and a given pair of media; he found that it was smallest for the most refrangible rays (blue) and greatest for the least refrangible rays (red).

the parts of Air or Glass, I would ask, why at the same Obliquity of Incidence the blue should wholly impinge on those parts, so as to be all reflected, and yet the red find Pores enough to be in a great measure transmitted.'[4]

The simple picture of mechanical collision was thus found to be incapable of accounting for some of the relevant facts. Newton succeeded in providing an adequate explanation of total reflection by ascribing it to the action of the same force which he supposed to be responsible for refraction:[5] Whether or not a beam of a certain colour (passing from a dense into a rare medium) will be *wholly* reflected can be determined from knowledge of the degree of incidence; the beam will be totally reflected when the angle of incidence is greater than the critical angle for that colour. But the problem of *partial reflection* remains, so far, unsolved: how does one and the same force that is acting at the separating surface refract some of the incident light and reflect the rest when rays of the same colour are passing from the rare into the dense medium at any angle of incidence?[6]

Newton's answer to this question is to be found in his theory of fits.

But although that theory was called upon to solve the problem of partial reflection, it was primarily designed to explain the colours of thin transparent bodies; and it is this explanation which will concern us in the present chapter. The preceding remarks may be taken to illustrate the interesting fact that a seemingly simple phenomenon does not sometimes receive an adequate explanation in a given theory until after a relatively more sophisticated hypothesis has been formulated to account for more complicated phenomena that fall within the scope of the theory. A striking example from the history of optics is the apparent rectilinear propagation of light (i.e. the formation of shadows in the ordinary way) which was not satisfactorily explained from the wave-theory point of view until after Young and Fresnel had developed a complicated system

---

[4] See *Opticks*, Bk. II., p. 264.
[5] See above, pp. 303f.
[6] See formulation of the problem in Kamāl al-Dīn and in Huygens, above, pp. 220f.

designed mainly to account for diffraction and interference phenomena.

Newton was not the first to make a careful study of the colours of thin transparent bodies. And although the principal ones among those phenomena have traditionally borne the name 'Newton's rings,' it will be seen that Robert Hooke had gone a long way in investigating them, thus preparing the road on which Newton followed.

1. About the time that Hooke's *Micrographia* appeared,[7] Robert Boyle published his *Experiments and Considerations touching Colours* as the first part of an experimental history of colours.[8] Both books contained observations on the colours exhibited by thin bodies; but the degree to which these observations were developed in Boyle's treatise was far below that reached in Hooke's book, and the aims of the two authors were different.

Boyle cited his observations as negative instances refuting the view, held by the 'Chymists', that the generation of colours was due to the presence of the sulphurious or the saline or the mercurial element in bodies.[9] He remarked that when essential oils or spirit of wine, for example, are shaken until they form bubbles, 'those bubbles will (if attentively consider'd) appear adorn'd with various and lovely colours'. He drew attention to the fact that both colourless bodies (water) and coloured ones (turpentine) exhibit, when made thin, colours which they do not normally possess. The colours which appear on the surface of bubbles, he further remarked, are 'not always the same Parts of them, but vary'd according to the Incidence of the Light, and the Position of the Eye'.[10] Thin-blown glass exhibited the same rainbow colours 'which were exceedingly vivid'. (Expressing this fact in general terms, Hooke—and after him

---

[7] The date of publication of Hooke's *Micrographia* is 1665, but the *imprimatur* (of Brouncker president of the Royal Society) is dated 23 November 1664.

[8] The full title ran as follows: *Experiments and Considerations touching Colours. First occasionally Written, among some other Essays, to a Friend; and now suffer'd to come abroad as the Beginning of an Experimental History of Colours*, London, 1664.

[9] Cf. Boyle, op. cit., pp. 85 and 242–5.

[10] Newton gave Hooke the credit for having first made this important observation. See below, pp. 329ff.

Newton—asserted that the intensity of the colours depends on the refractive power of the body—see below.) Finally, holding a feather before his eye against the sun when it was near the horizon, 'me thought there appear'd to me a Variety of little Rain-bows, with differing and very vivid Colours, of which none was constantly to be seen in the Feather'. This last observation (being one of diffraction) does not in fact belong to the same class as the preceding ones.[11] It did not matter to Boyle to make the differentiation (if he noticed it, that is) since all he was concerned to conclude from these experiments was that the appearance of colours could not be uniquely determined by the constant presence of any or all of the constituting elements mentioned before.

Both Boyle and Hooke were greatly influenced by Bacon's teaching; both were ardent seekers after new and extraordinary phenomena. But whereas Boyle confined his work on colours to fulfilling the first part of the Baconian programme, that is the recording of instances,[12] Hooke was primarily interested in finding an explanation of the phenomena of colours in general and those of thin bodies in particular. In fact the whole of Hooke's theory of colours was devised ultimately to account for the latter phenomena.

Hooke made most of his experiments concerning the colours of thin bodies on plates of muscovy glass (mica), liquids of various kinds pressed between two plates of ordinary glass, liquids and glass blown into bubbles, and the surfaces of metals.[13] By splitting the

---

[11] Cf. Newton, *Opticks*, Bk. III, Pt. I, Obs. 2, p. 322.

[12] This is already indicated by the title of Boyle's book. The rule under which the book was written is manifest from the following sentence of Bacon's which figures as a motto on the title-page: 'Non fingendum, aut excogitandum, sed inveniendum, quid Natura faciat, aut ferat.' It comes from *Novum Organum*, II, 10, where it occurs in the following context: 'Primo enim paranda est *Historia Naturalis et Experimentalis*, sufficiens et bona; quod fundamentum rei est: neque enim fingendum, aut excogitandum, sed inveniendum, quid natura faciat aut ferat—For first of all we must prepare a Natural and Experimental History, sufficient and good; and this is the foundation of all; for we are not to imagine or suppose, but to discover, what nature does or may be made to do' (B, I, p. 236 and IV, p. 127). The motto 'Non fingendum, aut excogitandum, . . .' is not unlike Newton's 'Hypotheses non fingo'. Newton also opposed *discovery* (or finding what really exists) on the one hand to what is merely *imagined* or *supposed* or *feigned* on the other, and claimed for his theories the status of the former. It is interesting to note, however, that works claiming to follow the Baconian method may vary as much as the difference between Boyle's *Considerations* and Newton's *Opticks*.   [13] Cf. Hooke, *Micrographia*, pp. 48–53.

muscovy glass with a needle into thin flakes he observed that there were several white specks or flaws in some parts of the flake, while other parts appeared tinged with all the rainbow colours. These colours appeared under the microscope to be arranged in rings, circular or irregular according to the shape of the spot which they terminated. The order of the colours was, from the middle, as follows: blue, purple, scarlet, yellow, green; and this series repeated itself around every spot from six to nine times. He called these systems of colours *rings*, *lines* and *irises*—counting all the gradations between the ends of each series for one. (They were later called 'Newton's Rings'!)

He also observed that the rings varied in brightness and breadth with the distance from the centre—being dimmer and narrower as they were farther from it. The spot in the middle was 'for the most part all of one colour'. Another thing which he found 'very observable' was that the colours could be made to change places by pressing the glass where they originally appeared.

Plates which were equally thick in all parts were seen under the microscope to be all over tinged with one determinate colour depending on the plate's thickness. It was necessary for the appearance of any colour at all that the thickness of the plate should lie within two definite limits; when the plates were too thick or too thin the colours disappeared. Two inclined plates (in the shape of a wedge) with some transparent medium interposed between them exhibited the rainbow colours arranged according to the various thicknesses of the medium.

Hooke then observed that when a transparent medium (air or liquid) was pressed between two glass plates, the same coloured rings were visible. He found that the colours could be changed by increasing or relaxing the pressure. The colours were the more vivid the greater the difference was in the refractive power between the glass and the interposed medium.

When the plates of muscovy glass, for example, were formed in the shape of a double-convex lens, the order of the colours was—from the middle: red, yellow, green, blue. This order was reversed when the plates had the shape of a double-concave lens.

He proposed to explain the preceding observations by the following consideration which expressed, in his own terms, the *material cause* of the phenomena in question:

it is manifest . . . that the material cause of the *apparition* of these several Colours, is some *Lamina* or Plate of a transparent or pellucid body of a thickness very determinate and proportioned according to the greater or less refractive power of the *pellucid* body.[14]

The same phenomena were produced by liquid and glass bubbles. Hooke found the appearance of the rings on the surfaces of the glass bubbles surprising 'having never heard or seen any thing of it before'.[15] (As we have seen, the same observation was made by Boyle.)

In accordance with the explanation just stated, he ascribed the appearance of colours on a polished surface of hardened steel when softened by heat to the fact that heating causes the metal to form on its surface a lamina of the requisite thinness and transparency for the production of the colours. It was not, therefore, necessary for producing the colours that the lamina should be terminated on both sides by one and the same medium.

To Hooke, the interest of all these observations consisted in that they supplied him with a *crucial experiment* against the Cartesian theory of colours; and, therefore, they called for a new theory. For if, as Descartes had maintained, the production of colours by refraction was due to the rotary motion which the ethereal globules acquired at their impact on a refracting surface, it would follow—as Descartes himself had pointed out[16]—that no colours would appear if the light passed through two *parallel* surfaces. For the rotation acquired at the first surface should be counteracted by the second (the refraction there being in the opposite sense); and since the two surfaces are parallel, the effect of the second surface should be to restore the globules to exactly their original state before meeting the

---

[14] Cf. Hooke, *Micrographia*, p. 50.
[15] Ibid.
[16] Cf. Descartes, *Meteors*, D, VI, pp. 330–1.

first. But, as Hooke observed, colours did appear by viewing the source of light through laminae of *uniform* thickness:

This Experiment therefore will prove such a one as our *thrice excellent Verulam* calls *Experimentum Crucis*, serving as a Guide or Land-mark, by which to direct our course in the search after the true cause of Colours. Affording us this particular negative Information, that for the production of Colours there is not necessary either a great refraction, as in the Prisme; nor Secondly, a determination of Light and shadow, such as is both in the Prisme and Glass-ball. Now that we may see likewise what affirmative and positive Instruction it yields, it will be necessary, to examine it a little more particularly and strictly; which that we may the better do, it will be requisite to premise somewhat in general concerning the nature of Light and Refraction.[17]

Then follows in the *Micrographia* the theory of light and colours which has been discussed on two previous occasions. The fundamentals of that theory, we have seen, are: that light consists in a succession of pulses generated in the ether by a vibrating movement of the luminous body; that colours are produced when the pulses (or wave-fronts) are made, by refraction, oblique to the direction of propagation; that blue is the effect on the retina of an oblique pulse whose weaker part or side precedes and whose stronger part follows; that red is the effect of an oblique pulse whose stronger side precedes and whose weaker side follows; and that all the intermediate colours on the spectrum are either different degrees of blue or red, or the intermixture of these degrees in various proportions.

How does Hooke explain the colours of thin plates on these principles?

He imagines[18] a pulse impinging in an oblique direction on the first surface of a lamina having the minimum requisite thickness. Some of the light is reflected by that surface while the rest is transmitted into the lamina. When the refracted light reaches the second surface, some of it will be reflected back to the first where it will be again transmitted out of the lamina. Two pulses will thus proceed from the first surface: a pulse that has been reflected at that surface, followed by another which has undergone one reflection at the

---

[17] Hooke, *Micrographia*, p. 54.   [18] Cf. ibid., pp. 65–66.

second surface and two refractions at the first. Owing to these two refractions and the time spent in traversing the thickness of the lamina twice, the succeeding pulse will—according to Hooke—be weaker than the one preceding it. But as the distance between the surfaces of the lamina is very small, the impression of these two pulses on the eye will be rather that of *one* pulse whose stronger part precedes and whose weaker part follows. This generates the sensation of yellow.

If a slightly thicker lamina is used, the distance between the leading strong pulse and the succeeding weak pulse will naturally increase, and, therefore, the action of the latter on the retina will be more retarded than in the first case. This gives the sensation of red.

When the thickness of the lamina is such that the weak part of a first original (i.e. incident) pulse lies exactly in the middle between its preceding strong part and the strong part of a second pulse (i.e. the part that has just been reflected at the first surface), a sensation of purple is generated. Purple is therefore an impression on the retina of a succession of strong and weak pulses, alternating at equal intervals.

By further increasing the thickness of the plate, the weak part of a first pulse will be nearer to the strong part of the immediately following pulse with which it will tend to be associated in the eye, rather than with the strong part that precedes. The order of the pulses with respect to their weakness and strength will thus be reversed; the impression on the eye will be that of a pulse whose weaker part precedes and whose stronger part follows. This gives the sensation of blue.

As the weaker pulse gradually approaches the stronger pulse behind, the blue turns into green, until such thickness of the plate is reached that the weaker pulse coincides with the following stronger pulse. With further increase in the thickness of the plate, the order of the pulses and, consequently, the order of the colours, will be repeated:

if these surfaces [of the plate] ... are further remov'd asunder, the weaker pulse will yet lagg behind [the stronger] much further, and not onely be *coincident* with the second [pulse] . . . but lagg behind that also, and that so much the more, by how much the thicker the Plate be; so that by degrees it will be *coincident* with the third . . . backward also, and by degrees, as the Plate grows thicker with a fourth, and so onward to a fifth, sixth, seventh, or eighth; so that if there be a thin transparent body, that from the greatest thinness requisite to produce colours, does, in the manner of a Wedge, by degrees grow to the greatest thickness that a Plate can be of, to exhibit a colour by the reflection of Light from such a body, there shall be generated several consecutions of colours, whose order from the thin end towards the thick, shall be Yellow, Red, Purple, Blue, Green; . . . and these so often repeated, as the weaker pulse does lose paces with its Primary, or first pulse, and is *coincident* with a second, third, fourth, fifth, sixth, &c. pulse behind the first.[19]

It is clear from the preceding account that Hooke ascribed the colours of thin plates to the manner in which the pulses reflected at the first surface are associated with those reflected at the second. Fundamentally the same idea underlies the wave-theory explanation of these colours. It is not therefore surprising that the originator of that explanation, Thomas Young, has said that had he not independently satisfied himself regarding these phenomena, Hooke's explanations 'might have led me earlier to a similar opinion'.[20] Following another remark by Young in which he pointed out potentialities in Hooke's theory, David Brewster wrote:

had Hooke adopted Newton's views of the different refrangibility of light, and applied them to his own theory of the coincidence of pulses, he would have left his rival behind in this branch of discovery.[21]

It should be noted in this connection that Newton asserted—in the discourse on colours produced before the Royal Society on 9 December 1675—that Hooke *had* in fact abandoned his dualistic view of colours:

---

[19] Ibid., pp. 66–67.

[20] Cf. Thomas Young, 'On the theory of light and colours'. Paper read before the Royal Society on 12 November 1801. Printed in his *A Course of Lectures on Natural Philosophy*, II, London, 1807, pp. 613–31. The quotation is from p. 627.

[21] Brewster, *Memoirs*, I, p. 160.

I was glad to understand, as I apprehended, from Mr Hooks discourse at my last being at one of your Assemblies, that he had changed his former notion of all colours being compounded of only two Originall ones, made by the two sides of an oblique pulse, & accomodated his Hypothesis to this my suggestion of colours, like sounds, being various, according to the various bignesse of the Pulses.[22]

And yet, there is no evidence that Hooke ever applied Newton's ideas to his own explanation of the colours of thin plates. Had he done so, Young has remarked, 'he could not but have discovered a striking coincidence with the measures laid down by Newton from experiment'.[23]

In any case, Hooke himself felt at the time of writing the *Micrographia* that he had not carried his investigations as far as he would have liked. We read towards the end of his account of the colours of thin plates:

One thing which seems of the greatest concern in this *Hypothesis*, is to determine the greatest or least thickness requisite for these effects, yet so exceeding thin are these coloured Plates, and so imperfect our Microscope, that I have not been hitherto successful.[24]

That important task was first undertaken by Newton.

Hooke's experiments on the colours of thin plates were instructive and his explanation of those colours was far from being absurd. He was the first to make a serious investigation of those phenomena,

---

[22] *Correspondence*, I, pp. 362–3. Newton's allusion is (probably) to the remarks made by Hooke during a meeting of the Royal Society (11 March 1674/5) which are rendered in Birch's *History* (III, p. 194) in the following words: 'as there are produced in sounds several harmonies by proportionate vibrations, so there are produced in light several curious and pleasant colours, by the proportionate and harmonious motions of vibrations intermingled; and, those of the one are sensated by the ear, so those of the other are by the eye.' Thus, as far as dispersion was concerned, Hooke finally conceded all the modifications that his theory would have to undergo in order to accommodate Newton's experiments from the wave-theory point of view. We do not know whether he wished to maintain in 1675 the possibility (first suggested by him in 1672) of representing white light by the coalescence or super-position of the waves associated with the different colours. But there was nothing in these experiments that would have forced him to give up the possibility of that representation. See above, pp. 259ff, 278ff.
[23] Thomas Young, op. cit., II, p. 628.
[24] *Micrographia*, p. 67.

and his study of them was quite elaborate. Many of his experiments were later repeated by Newton, and the problems which he left unsolved constituted the starting point of Newton's own researches in this field.

Newton himself expressed his debt to Hooke on more than one occasion. Thus, in a letter to Oldenburg dated 21 December 1675,[25] Newton credited Hooke with having preceded him in observing and attempting an explanation of the colours of thin plates. And we are further informed by Brewster that there accompanied that letter an unpublished paper, entitled 'Observations', in which Newton said that Hooke in the *Micrographia* 'had delivered many very excellent things concerning the colours of thin plates, and other natural bodies, which he [Newton] had not scrupled to make use of as far as they were for his purpose'.[26]

Later in a letter to Hooke of 5 February 1675/6, Newton wrote:

What Des-cartes did [that is, in optics] was a good step. You have added much several ways, & especially in taking ye colours of thin plates into philosophical consideration. If I have seen further it is by standing on ye sholders of Giants. But I make no question but you have divers very considerable experiments beside those you have published, & some it's very probable the same wth some of those in my late papers.[27] Two at least there are wch I know you have observed, ye dilatation of ye coloured rings by the obliquation of ye eye, & ye apparition of a black spot at ye contact of two convex glasses & at the top of a water bubble; and it's probable there may be more, besides others wch I have not made: so yt I have reason to defer as much, or more, in this respect to you as you would to me . . . .[28]

---

[25] Cf. Isaaci Newtoni *Opera*, IV (London, 1782), pp. 378–81; also printed (in part) in Birch, *History*, III, pp. 278–9; *Correspondence*, I, p. 406.

[26] Brewster, *Memoirs*, I, p. 139, note 2.

[27] These are the papers read before the Royal Society (in Newton's absence) on the following dates: 9 December 1675 (Birch, *History*, III, pp. 248–60), 16 December 1675 (ibid., pp. 262–9), 30 December 1675 (ibid., p. 270), 20 January 1676 (ibid., pp. 272–8), 3 February 1676 (ibid., pp. 280–95), 10 February 1676 (ibid., pp. 296–305). These comprised the 'Hypothesis Explaining the Properties of Light' and the 'Discourse on Observations' which accompanied Newton's letter to Oldenburg, 7 December 1675; see *Correspondence*, I, pp. 362–92.

[28] *Correspondence*, I, p. 416. Turnbull remarks that 'No reply to this letter has been found' (p. 417, note 4). The passage was quoted with some variations by David Brewster, *Memoirs*, I, pp. 142–3.

Newton here credits Hooke with two important observations which, as Brewster has noted, are not to be found in the *Micrographia*. Both observations are, however, described in a paper by Hooke which he read in a meeting of the Royal Society on 19 June 1672.[29] He reported that by pressing 'two thinn pieces of glass'[30] (not two convex glasses) together, a red spot first appeared in the middle. When he increased the pressure, he saw several rainbow colours encompassing the middle point,

& continuing to presse the same closer and closer, at last all the colours would disappear out of the middle of the circles or Rainebows, and the middle would appear white; and if yet I continued to press the said plates together, *the white would in severall places thereof turne into black*.[31]

Concerning the second observation, Hooke wrote:

all the said Rings, or Rainbows would vary their places by varying the position of the eye by wch they were observed; and not only their position but their colours . . . .[32]

This observation was later made the subject of a detailed quantitative examination by Newton. In the *Opticks*, Bk. II, Pt. I, Obs. 7, he determined the rule according to which the diameters of the rings vary with the obliqueness of the line of vision.

In the same discourse of 19 June 1672, Hooke mentioned another observation (also lacking in the *Micrographia*) which later received attention from Newton. Hooke wrote:

Moreover that [part of the plate] wch gives one colour by reflection [that is, when viewed from the side of the source of illumination], gives another by trajection [when viewed from the other side].[33]

---

[29] Cf. Birch, op. cit., III, pp. 52–54; *Correspondence*, I, pp. 195–7.

[30] *Correspondence*, I, p. 196.

[31] Ibid.; italics added. A previous paper of Hooke's on the colours of soap bubbles (read to the Royal Society on 28 March 1672), does not contain the observation of a black spot on top of the bubble (Cf. Birch, op. cit., III, p. 29).

[32] *Correspondence*, I, p. 197.

[33] Ibid. Compare with Newton, *Opticks*, Bk. II, Pt. I Obs. 9, pp. 206–7, where it is remarked that the order of the colours appearing by reflection is the reverse of that of the colours appearing by trajection or refraction.

This observation revealed one of the fundamental features to be accounted for by Newton's theory of fits.

The sentences quoted above from Newton's letters to Oldenburg and to Hooke greatly contrast with the fact that when Newton published his *Opticks* in 1704, after Hooke had died, he failed even to mention Hooke's name in connection with the subject of the colours of thin plates. Newton's account of those colours simply begins with the following:

It has been observed *by others*, that transparent Substances, as Glass, Water, Air, &c. when made very thin by being blown into Bubbles, or otherwise formed into Plates, *do exhibit various Colours according to their various thinness*, altho' at a greater thickness they appear very clear and colourless.[34]

The attribution of this observation to 'others' is surprisingly inadequate; for the discovery of what is asserted here, namely that the appearance of a particular colour depends on a particular thickness of the plate, was due to Hooke and to no one else.

2. Newton's interest in the colours of thin plates dates from a period before 1672. This we gather from his reference to those phenomena in his Answer to Hooke's Considerations which was written in that year.[35] There he pointed out what had been observed by Hooke in the *Micrographia*—namely, that those colours depended for their production on the thickness of the plate; and further proposed the *hypothetical* explanation (to which he did not then nor ever afterwards commit himself) which was developed in his later publications. The first detailed account of those colours by Newton was contained in the discourses produced before the Royal Society in 1675-6.[36] We learn from two letters of Newton to Oldenburg (dated, respectively, 13 November 1675 and 30 November 1675)

---

[34] Newton, *Opticks*, Bk. II, Pt. I, p. 193; my italics.
[35] Cf. *Philosophical Transactions*, No. 88, 18 November 1672, pp. 5088-9; *Correspondence*, I, pp. 179-80.
[36] See above, p. 329, n. 27.

that he had written those discourses—apart from a 'little scrible', namely his 'hypothesis' for explaining the properties of light—before he sent his first letter on colours to the Royal Society in 1672.[37] But this should be understood as only generally true; his interest in diffraction phenomena which he also considered in the same discourses did not in fact begin until at least the middle of 1672.[38] In any case it is clear from his Answer to Hooke that he had in 1672 the leading idea in his explanation of the colours of thin plates which is contained in the revised discourses of 1675. The part of these discourses dealing with those phenomena was later reproduced, with additions, in the *Opticks* (1704). The following analysis will be based on the final account given in the *Opticks*.[39] My aim will be to give a description of some of the fundamental experiments with a view to understanding Newton's explanations which will be the subject of the subsequent discussion.

As we have seen, the experimental problem before which Hooke's investigations came to a stop was the determination of the requisite thickness in a plate of a given medium for the production of a particular colour. Hooke ascribed his failure to solve this problem to the defectiveness of his microscope. Newton, however, overcame the difficulties involved, not by using a better microscope, but because he struck upon a happy experimental device—a lens combination which allowed him to make the necessary measurements and calculations. His study was also facilitated and made more effective by using monochromatic light. This had not been attempted by Hooke.

In open air, Newton placed upon a double-convex lens of a large radius of curvature another plano-convex lens with its plane side downwards.[40] There was thus between the lenses a thin film of air, bounded by a plane and a spherical surface, which increased in

---

[37] Cf. Brewster, *Memoirs*, I, pp. 131–2; *Correspondence*, I, p. 358 and note 8, p. 359.

[38] See above, p. 310, n. 25; also A. R. Hall, 'Further optical experiments of Isaac Newton', *Annals of Science*, XI (1955), pp. 27–43; 'Newton's First Book (I)', *Archives internationales d'histoire des sciences*, XIII (1960), p. 48; Richard S. Westfall, 'Newton's reply to Hooke and the theory of colors', *Isis* LIV (1963), pp. 88–89 and notes 27 and 30.

[39] It would be interesting to study in detail the development of Newton's ideas on this subject, but this will not be attempted here.

[40] Cf. Newton, *Opticks*, Bk. II, Pt. I, Obs. 4.

thickness from the point of contact. Viewing the lenses by reflection and slowly pressing them together he saw coloured circles successively emerging from the central point and being formed in multicoloured rings; the last to appear was a black spot in the middle. He observed that while the diameters of the rings were gradually becoming larger (by increasing the pressure), the breadths of their orbits decreased. When he slowly released the pressure the circles gradually approached the central spot and disappeared in it one after the other. This indicated (what had been concluded by Hooke) that the particular colours were related to the air thicknesses at which they appeared.

Illuminating the lenses in a dark room with the prismatic colours one after the other he noticed that the rings were more distinct and greater in number than when white light was used.[41] The largest rings were formed by red light, the smallest by violet, and the other colours occupied places between these two extremities according to their order in the spectrum.[42] In each of these cases the phenomenon simply consisted in a series of alternating dark and bright rings, with the dark spot in the middle, and the bright rings all exhibiting the same colour with which the lenses were illuminated. It was therefore apparent that the multi-coloured rings—when white light was used—were the combined effect of an indefinite number of monochromatic rings. Newton remarked that when the lenses were viewed from the side of the illuminating source the bright rings indicated the places where the light was reflected, while the dark ones indicated the places where it was transmitted. Thus, the light was transmitted at the point of contact—hence the black spot, reflected at the place where the first bright ring appeared, transmitted again, and so forth. When viewed by transmission, the rings were exactly complementary to those perceived by reflection and the central point appeared bright.

Viewing the rings perpendicularly, or almost perpendicularly, by reflection (other specifications need not be mentioned here but are

---

[41] Cf. ibid., Bk. II, Pt. I, Obs. 12 and 13.
[42] Cf. ibid., Bk. II, Pt. I. Obs. 14.

carefully stated by Newton), he measured their diameters with a pair of compasses. He found that while the squares of the diameters of the dark rings were in the arithmetic progression of the even numbers, 0, 2, 4, etc. (0 being the central point), the squares of the diameters of the bright rings were in the progression of the odd numbers 1, 3, 5, etc.[43] From this he concluded that the air thicknesses (or intervals) corresponding to the dark rings were in the progression of the first sequence, whereas the intervals corresponding to the bright rings were in the progression of the second.

Having determined the refractive index of the double-convex lens and its focal length he calculated its radius of curvature. Using this result he computed the air interval corresponding to a sufficiently large ring (of, say, the colour in the confines of yellow and orange) whose diameter he had already measured. Finally he calculated the air interval corresponding to the first (innermost) bright ring produced by this colour. That was the $\dfrac{1}{178000}$th part of an inch. It follows that the thicknesses corresponding to the remaining bright rings of the same colour were given by $\dfrac{3}{178000}$, $\dfrac{5}{178000}$, etc.; and the thickness of the dark rings, by 0, $\dfrac{2}{178000}$, $\dfrac{4}{178000}$, etc. [44]

He also observed that the absolute values of the intervals corresponding to rings of a certain colour changed with the density (and, consequently, with the refractive index) of the enclosed medium. Thus, by allowing a drop of water to creep slowly into the air gap, the rings gradually contracted until the gap was full.[45]

Newton's investigations of these phenomena did not come to an end with the determination of the experimental laws governing them. Such, for example, was the law expressing the thicknesses in terms of the angles of incidence and of refraction,[46] or the law stating that the rings made, successively, by the limits of the seven colours—

[43] Cf. Newton, *Opticks*, Bk. II, Pt. I, Obs. 5.
[44] Cf. ibid., Bk. II, Pt. I, Obs. 6.
[45] Cf. ibid., Bk. II, Pt. I, Obs. 10.
[46] Cf. ibid., Bk. II, Pt. I, Obs. 7, p. 205.

red, orange, yellow, green, blue, indigo and violet—were to one another as the cube roots of the squares of a certain musical sequence.[47] He offered an explanation: his theory of fits.

His observations showed that the appearance of the rings was independent of the degree of convexity of the lenses,[48] and the nature of the material used. From this he concluded (like Hooke) that their appearance solely depended on the film surfaces and the distances between those surfaces.[49] By wetting either face of a muscovy-glass plate the colours grew faint. This indicated to him that both faces played a role in the production of the phenomenon. What, then, are the roles he assigned to the surfaces? The following is Newton's answer:

Every Ray of Light in its passage through any refracting Surface is put into a certain transient Constitution or State, which in the progress of the Ray returns at equal Intervals, and disposes the Ray at every return to be easily transmitted through the next refracting Surface, and between the returns to be easily reflected by it.[50]

That is to say, *assuming* that the ray has been transmitted by a refracting surface, it will always[51] be alternately disposed to be easily transmitted or easily reflected by a second surface—at equal intervals between every two dispositions. Newton called the returns of the state which disposed the ray to be easily transmitted its *fits of easy transmission*, and the returns of the state disposing it to be easily reflected its *fits of easy reflection*. He defined the 'interval of fits' as 'the space [which the ray traverses] . . . between every return and the next return', that is the *distance* covered by the ray between any two consecutive and similar fits.[52]

---

[47] Cf. ibid., Bk. II, Pt. I, Obs. 14, p. 212; Bk. II, Pt. III, Prop. XVI, p. 284.
[48] Cf. ibid., Bk. II, Pt. I, Obs. 6, pp. 200 and 202.
[49] Cf. ibid., Bk. II, Pt. III, p. 279.
[50] Ibid., Bk. II, Pt. III, Prop. XII.
[51] Cf. ibid., Bk. II, Pt. III, p. 279: 'this alternation seems to be propagated from every refracting Surface to all distances without end or limitation.'
[52] Ibid., Bk. II, Pt. III, p. 281.

Suppose then that a ray of a particular colour has been transmitted by the surface of the air film considered in the two-lens experiment. The ray will, after a whole interval, be in a fit of easy transmission. Therefore, if the thickness of the film happens to be equal to the interval of fits, the ray will be transmitted by the second surface. And the same will take place where the thicknesses are 2, 3, 4, . . . times a whole interval. This, according to Newton, explains the dark rings: they appear where the light is transmitted by the second surface.

But if the ray meets the second surface after a half interval, then, being in a fit of easy reflection, it will be reflected back towards the first. When it arrives there it will have completed a whole interval and will consequently be transmitted to the eye. The same of course will happen where the thicknesses are 2, 3, 4, . . . times a half-interval. This explains the bright rings: they are visible where the light is reflected by the second surface.

It is clear that the interval of fits (for a given colour and a given medium) can be calculated from the thickness at the first (innermost) bright ring produced by that colour and medium. For, by supposition, the interval must be twice that thickness. For example, having determined that thickness—for the colour in the confines of yellow and orange and for the medium air—to be the $\frac{1}{178000}$th part of an inch, the interval of fits of easy reflection must be $\frac{2}{178000} = \frac{1}{89000}$ parts of an inch.[53] The same number also measures the interval of fits of easy transmission for the same colour and medium.

It should be noticed that the interval of fits depends not only on the colour but also on the nature of the medium used. This is evident from the fact—observed by Newton—that the rings produced by a given colour were displaced when the air gap was filled with water, thus changing their diameters and the corresponding thicknesses.

So far we have neglected the *reflection* which takes place at the *first* surface of the transparent plate. According to Newton, the light

---

[53] Cf. Newton, *Opticks*, Bk. II, Pt. III, Prop. XVIII.

reflected at that surface does not play any part in the production of the rings. But the problem of partial reflection, which applies to both thick and thin transparent bodies, still remains. Newton's answer to that problem is contained in the following formulation of his theory of fits:

The reason why the Surfaces of all thick transparent Bodies reflect part of the Light incident on them, and refract the rest, is, that some Rays at their Incidence are in Fits of easy Reflexion, and others in Fits of easy Transmission.[54]

And hence Light is in Fits of easy Reflexion and easy Transmission, before its Incidence on transparent Bodies. And probably it is put into such fits at its first emission from luminous Bodies, and continues in them during all its progress.[55]

According to this formulation the fits are ('probably') with the rays from their origin. When the rays meet a refracting surface, those in a fit of easy reflection will be reflected and those in a fit of easy transmission will be refracted. Neglecting the former rays and following the subsequent course of the latter, the phenomena of thin transparent bodies are explained by Newton in the manner described before.

We are now in a position to discuss the logical status which Newton assigned to his theory. The following is what he wrote regarding the fits in a sequel to Prop. XII of Bk. II, Pt. III:

---

[54] Ibid., Bk. II, Pt. III, Prop. XIII, p. 281. That Newton did not mean to restrict this proposition to the partial reflection made at the surfaces of *considerably* thick bodies is clear from the sentence immediately following the statement of the proposition, and in which he mentions as an example the light 'reflected by thin Plates of Air and Glass' (ibid., p. 281) and also from the following comment: 'In this Proposition I suppose the transparent Bodies to be thick; because if the thickness of the Body be much less than the Interval [exactly, less than half the interval] of the Fits of easy Reflexion and Transmission of the Rays, the Body loseth its reflecting power. For if the Rays, which at their entering into the Body are put into Fits of easy Transmission, arrive at the farthest Surface of the Body before they be out of those Fits, they must be transmitted. And this is the reason why Bubbles of Water lose their reflecting power when they grow very thin; and why all opake Bodies, when reduced into very small parts, because transparent' (ibid., p. 282). Therefore, by the phrase *thick transparent bodies* Newton meant here bodies thicker than half an interval.

[55] Ibid., p. 282.

What kind of action or disposition this is: Whether it consists in a circulating or a vibrating motion of the Ray, or of the Medium, or something else, I do not here enquire. Those that are averse from assenting to any new Discoveries, but such as they can explain by an Hypothesis, may for the present suppose. . . .[56]

The proposed supposition or hypothesis will be quoted later. Having stated it, Newton concluded his sequel thus:

But whether this Hypothesis be true or false I do not here consider. I content my self with the bare Discovery, that the Rays of Light are by some cause or other alternately disposed to be reflected or refracted for many vicissitudes.[57]

From this the following points clearly emerge:

First, so long as he did not assert what kind of action or disposition the fits were, or assign their cause, the theory of fits was not—for Newton—a hypothesis; it constituted a positive contribution, a 'bare Discovery' borne out solely by the experiments. It therefore had to be accepted regardless of all hypotheses about the nature of light.

Second, only an attempt to explain the fits themselves, that is, an attempt to state their cause or nature or mode of production, would constitute a hypothesis—as long as there were no *new* experiments to support it.

Third, Newton himself proposed such a hypothesis, though rather unwillingly, and definitely without committing himself to it.

In this hypothesis the reader is invited to suppose:

that as Stones by falling upon Water put the Water into an undulating Motion, and all Bodies by percussion excite vibrations in the Air; so the Rays of Light, by impinging on any refracting or reflecting Surface, excite vibrations in the refracting or reflecting Medium or Substance, and by exciting them agitate the solid parts of the refracting or reflecting

---

[56] Newton, *Opticks*, Bk. II, Pt. III, Prop. XIII, p. 280.
[57] Ibid., pp. 280–1.

Body, and by agitating them cause the Body to grow warm or hot; that the vibrations thus excited are propagated in the refracting or reflecting Medium or Substance, much after the manner that vibrations are propagated in the Air for causing Sound, and move faster than the Rays so as to overtake them; and that when any Ray is in that part of the vibration which conspires with its Motion, it easily breaks through a refracting Surface, but when it is in the contrary part of the vibration which impedes its Motion, it is easily reflected; and by consequence, that every Ray is successively disposed to be easily reflected, or easily transmitted, by every vibration which overtakes it.[58]

Since it is here supposed that the rays are put into fits by the waves excited *in the reflecting or refracting medium*, it is clear that this hypothesis does not suit the formulation according to which the fits exist in the rays *before* their incidence on the surfaces of transparent bodies. In Query 18[59] it is alternatively suggested that a medium 'exceedingly more rare and subtile than the Air, and exceedingly more elastick and active' might be responsible for putting the rays into fits of easy reflection and easy transmission. Since this ethereal medium is also supposed to expand 'through all the heavens' the rays would excite vibrations in it at their emission from luminous objects, and would thus be put into fits before they encounter any reflecting or refracting media.

Led by Newton's reference—as for example in the above passage (see also Query 17)—to water waves, I. B. Cohen has remarked[60] that 'the disturbance [in the refracting medium, or in the ether] would be transverse'. But it is equally justifiable to argue—as Cohen also observed—that the analogy with sound waves suggests that the disturbance would be longitudinal. The question can perhaps be decided by considering the function assigned to these vibrations by Newton. For since they are supposed to accelerate the ray or retard

---

[58] Ibid., p. 280. 'And do they [the vibrations excited in the refracting or reflecting medium] not overtake the Rays of Light, and by overtaking them successively do they not put them into the Fits of easy Reflexion and easy Transmission described above? For if the Rays endeavour to recede from the densest part of the Vibration, they may be alternately accelerated and retarded by the Vibrations overtaking them' (ibid., Query 17, p. 348).

[59] Cf. ibid., p. 349.

[60] Cf. I. B. Cohen, Preface to the New York edition of Newton's *Opticks*, 1952, note 32, p. iv.

it according as their motion conspires with the motion of the ray or impedes it, the movement to and fro of the ether particles ought to be in the line of direction of the ray; i.e. it is longitudinal. References to water waves are found in the works of many seventeenth-century writers on optics other than Newton (as we have seen, in those of Hooke, Pardies, Huygens, Ango); but none of them—including Newton—tried to make use of transverse vibrations as such in their theories.

The preceding explanation of the fits is mechanical, that is, conceived in terms of mutual action by contact between the rays and the particles of ether or other bodies. But, as in the case of refraction, the Queries of the *Opticks* propose another kind of explanation—in terms of an attractive power:

Nothing more is requisite for putting the Rays of Light into Fits of easy Reflexion and easy Transmission, than that they be small Bodies which by their attractive Powers, or some other Force, stir up Vibrations in what they act upon, which Vibrations being swifter than the Rays, overtake them successively, and agitate them so as by turns to increase and decrease their Velocities, and thereby put them into those Fits.[61]

Of the colours of thin plates Newton therefore offered an explanation, his theory of fits, which in his view was safely based on the experimental results which he obtained. Of the fits themselves he offered an explanation in terms of ether waves excited either by the impinging of the light rays on the ether particles, or by means of an attractive power with which the rays are endowed. This higher level explanation was proposed by him as a hypothesis whose truth or falsity he did not consider, a query to be examined in the light of new experiments; it did not form part of his doctrine of light and colours. Whether the experiments would pronounce for or against it, the fits would, according to him, remain as real properties of light.

The question must now be asked whether the fits had this status. To answer this question it is not necessary to enumerate inherent

---

[61] Newton, *Opticks*, Query 29, pp. 372–3.

difficulties; it is sufficient to point out what has been recognized since Young and Fresnel—namely, that the phenomena which Newton tried to explain by his theory of fits are wave phenomena. In the wave interpretation the rings are fully explained by the interference of the rays reflected at both surfaces of the film. This explanation is therefore nearer to the one proposed by Hooke than to Newton's; since in Hooke's explanation—as distinguished from Newton's—the light reflected at the first surface plays an essential part in the production of the phenomena in question.

The experiments may have *suggested* to Newton a certain periodicity in light; that was revealed in the regular succession of the dark and bright rings in the progression of the natural numbers. But he had to *interpret* this periodicity in accordance with an already formed conception of the rays as discrete entities or corpuscles. This *a priori* conception prevented him from envisaging the possibility of an undulatory interpretation in which the ray, as something distinguished from the waves, would be redundant. Indeed he found in the fits an argument against an independent wave theory of light. For, he argued, [62] if light consisted of waves in an ethereal medium, it would be necessary to postulate the existence of a second ethereal medium whose swifter waves would put those of the first into fits. 'But how two Æthers can be diffused through all Space, one of which acts upon the other, and by consequence is re-acted upon, without retarding, shattering, dispersing and confounding one anothers Motions, is inconceivable.'[63] In this argument Newton was obviously taking for granted what the nineteenth-century wave theory had no need of in order to explain the colours of thin plates— namely, the fits themselves. This can be explained only by his unshakable belief in their existence.

Newton's theory of fits was forced to pass into oblivion after the impressive success of the wave theory during the last century. Since the invention of wave mechanics, however, interest has been renewed—from a historical point of view—in Newton's speculations concerning the origins of the fits. His attempt to combine corpuscular

---

[62] Cf. ibid., Query 28, p. 364.

and wave conceptions has been described as constituting 'a kind of presentiment of wave mechanics'.[64] It is odd (but perhaps understandable) that Newton should now be praised for a hypothetical explanation which he proposed, condescendingly, only for the sake of those who 'are averse from assenting to any new Discoveries, but such as they can explain by an Hypothesis'.[65] Fortunately, his fertile imagination was not hampered by his low opinion of imaginative products.

---

[64] L. de Broglie, *Ondes, corpuscules, mécanique ondulatoire*, Paris, 1945, p. 50. Whittaker has also remarked that Newton's theory of fits was 'a remarkable anticipation of the twentieth-century quantum-theory explanation: the "fits of easy transmission and easy reflection" correspond to the transition probabilities of the quantum theory.'

[65] *Opticks*, p. 280.

Bibliography

(A: 17th-century and earlier works. B: Later works and studies*.)

A

ACADÉMIE ROYALE DES SCIENCES, *Histoire de l'Académie Royale des Sciences*, I, *depuis son établissement en 1666 jusqu'à 1686* [i.e. to the end of 1685], Paris, 1733

ACADÉMIE ROYALE DES SCIENCES, *Mémoires de l'Académie Royale des Sciences, depuis 1666 jusqu'à 1699*, X, Paris, 1730

ACADÉMIE ROYALE DES SCIENCES, *Histoire de l'Académie Royale des Sciences, année 1707, avec les Mémoires de mathématique et de physique pour la même année, tirés des Registres de cette Académie*, Paris, 1708

ALHAZEN, *see* Ibn al-Haytham

ANGO, PIERRE, *L'optique divisée en trois livres, où l'on démontre d'une manière aisée tout ce qui regarde: 1. La propagation et les proprietez de la lumière, 2. La vision, 3. La figure et la disposition des verres*, Paris, 1682

ARISTOTLE, *De anima; Mechanica; Meteorologica; Problemata; De sensu*

AVERROËS, *In Aristotelis de anima*, Venetiis, 1562. [Facsimile reproduction in *Aristotelis Opera cum Averrois Commentariis*, Frankfurt am Main, 1962, Suppl. II]

AVERROËS, *In libros meteorologicorum expositio media*, in *Aristotelis Opera cum Averrois Commentariis*, Venetiis, 1562, fols. 400ʳ–487ᵛ. [Facsimile reproduction, Frankfurt am Main, 1962, Vol. V]

AVERROËS, *Kitāb al-nafs* [i.e. Epitome of *De anima*, Arabic text], in *Rasāʾil Ibn Rushd*, Hyderabad, 1947

AVICENNA, *De anima*, in *Opera philosophica*, Venetiis, 1508. [Facsimile reproduction, Louvain, 1961]

BACON, FRANCIS, *Novum Organum*, edited with introduction, notes, etc., by Thomas Fowler, 2nd edition corrected and revised, Oxford, 1889

BACON, FRANCIS, *The Works of*, collected and edited by James Spedding, R. L. Ellis and D. D. Heath, 14 vols, London, 1857–74

BACON, ROGER, *The 'Opus Majus' of*, edited with introduction and analytical table by John Henry Bridges, 2 vols, Oxford, 1897; Supplementary Volume, London, 1900

BARTHOLINUS, ERASMUS, *Experimenta crystalli Islandici disdiaclastici, quibus mira & insolita refractio detegitur*, Hafniae, 1669

BEECKMAN, ISAAC, *Journal tenu par Isaac Beeckman de 1604 à 1634*, edited by C. de Waard, 4 vols, La Haye, 1939–53

BIRCH, THOMAS, *The History of the Royal Society of London for improving of Natural Knowledge, from its first Rise, in which the most considerable of those Papers communicated to the Society, which have hitherto not been published, are inserted in their proper order, as a supplement to the Philosophical Transactions*, 4 vols, London, 1756–7

BOYLE, ROBERT, *Experiments and Considerations Touching Colours. First occasionally written, among some other Essays to a Friend; and now suffer'd to come abroad as the Beginning of an Experimental History of Colours*, London, 1664

---

*The two following lists include, in addition to works cited in this book, only a small selection of works that have a bearing on the problems discussed in it.

343

BOYLE, ROBERT, *The Works of the Honourable Robert Boyle, to which is prefixed The Life of the Author* [by Thomas Birch], a new edition, 6 vols, London, 1772

CLAGETT, MARSHALL, *Archimedes in the Middle Ages*, I (The Arabo-Latin Tradition), Madison, Wisconsin, 1964

CLAGETT, MARSHALL, *The Science of Mechanics in the Middle Ages*, Madison, Wisconsin, 1959

COHEN, MORRIS R. and I. E. DRABKIN (editors), *A Source Book in Greek Science*, Cambridge, Mass., 1948

'CONSTANTINI AFRICANI *Liber de oculis*' [which is a translation of Ḥunayn ibn Isḥāq's *Book of the Ten Treatises on the Eye*, q.v.], in *Omnia opera Ysaac*, Lugduni, 1515

DE LA CHAMBRE, *see* La Chambre

DESCARTES, RENÉ, *Correspondance*, edited by C. Adam and G. Milhaud, 7 vols, Paris, 1936–60

DESCARTES, RENÉ, *Lettres de Mr Descartes*, edited by C. Clerselier, 3 vols, Paris, 1657, 1659, 1667

DESCARTES, RENÉ, *Œuvres de*, edited by Charles Adam and Paul Tannery, 12 vols and *Supplément*, Paris, 1897–1913

DESCARTES, RENÉ, *Œuvres inédites de*, precédées d'une introduction sur la méthode, par Foucher de Careil, 2 vols, Paris, 1859–60

DESCARTES, RENÉ, *The Philosophical Works of*, translated by Elizabeth Haldane and G. R. T. Ross, 2 vols, 2nd edition, Cambridge, 1931, 1934

DRABKIN, I. E., *see* Cohen, Morris R.

EDLESTON, J., *Correspondence of Sir Isaac Newton and Professor Cotes, including Letters of Other Eminent Men, now first published from the originals in the Library of Trinity College, Cambridge; together with An Appendix containing other unpublished letters and papers by Newton; with notes, synoptical view of the philoso-* pher's *life and a variety of details illustrative of his history*, London, 1850

EUCLIDE, *L'Optique et la Catoptrique*, oeuvres traduites pour la première fois du grec en français, avec une introduction et des notes par Paul Ver Eecke, nouveau tirage, Paris, 1959. [First published in 1938]

EUCLIDIS *Optica, Opticorum recensio Theonis, Catoptrica, cum Scholiis antiquis*, ed. I. L. Heiberg (Euclidis *Opera omnia*, ed. I. L. Heiberg & H. Menge, VII), Leipzig, 1895

FERMAT, PIERRE DE, *Œuvres de*, edited by Paul Tannery and Charles Henry, 4 vols, Paris, 1891–1912; *Supplément*, edited by C. de Waard, Paris, 1922

'GALENI *Liber de oculis*' [which is a translation of Ḥunayn ibn Isḥāq's *Book of the Ten Treatises on the Eye*, q.v.], in Galeni *Opera omnia*, VII, Venetiis, apud Iuntas, 1609

GALILEI, GALILEO, *Dialogues Concerning Two New Sciences*, translated from the Italian and Latin by Henry Crew and Alfonso de Salvio, New York, 1914

GRIMALDI, FRANCESCO MARIA, *Physicomathesis de lumine, coloribus et iride, aliisque annexis libri II*, Bononiae, 1665

GROSSETESTE, ROBERT, *De luce*, in *Die philosophischen Werke des Robert Grosseteste*, ed. Ludwig Baur, Münster i. W., 1912

HEATH, THOMAS L., *Diophantus of Alexandria*, a study in the history of Greek algebra, 2nd edition, Cambridge, 1910. Republished, New York, 1964

HERIGONE, PIERRE, *Cursus mathematicus*, VI, Paris, 1664 [This volume contains the first published account of Fermat's method of maxima and minima in a *Supplementum* beginning with separate pagination after p. 466]

HERONS VON ALEXANDRIA *Mechanik und Katoptrik*, herausgegeben und übersetzt von L. Nix und W. Schmidt: Heronis Alexandrini *Opera quae supersunt omnia*, vol II, fasc. I, Leipzig, 1900

HOBBES, THOMAS, 'Tractatus opticus, prima editione integrale a cura di Franco Alessio', *Rivista critica di storia della filosofia*, anno XVIII, fasc. II, 1963, pp. 147–288

HOOKE, ROBERT, *Micrographia, or some physiological descriptions of minute bodies made by magnifying glasses, with observations and inquiries thereupon*, London, 1665

HOOKE, ROBERT, *The Posthumous Works of*, edited by Richard Waller, London, 1705

HUNAYN IBN ISḤĀQ, *The Book of the Ten Treatises on the Eye Ascribed to Hunain ibn Isḥāq (809–877 A.D.)*, the earliest existing systematic text-book of ophthalmology. The Arabic text edited from the only two known manuscripts with an English translation and glossary by Max Meyerhof, Cairo, 1928. [See 'Constantini Africani *Liber de oculis*', and 'Galeni *Liber de oculis*', which are medieval translations of this work]

HUYGENS, CHRISTIAN, *Œuvres complètes de*, published by the Société Hollandaise des Sciences, 22 vols, La Haye, 1888–1950

HUYGENS, CHRISTIAN, *Treatise on Light, in which are explained the causes of that which occurs in reflexion, & in refraction, and particularly in the strange refraction of Iceland crystal*, rendered into English by Silvanus P. Thompson, London, 1912

IBN AL-HAYTHAM (ALHAZEN), 'Abhandlung über das Licht von Ibn al-Haitam', *Zeitschrift der Deutschen Morgenländischen Gesellschaft*, XXXVI (1882), pp. 195–237. [Edition of the Arabic text of Ibn al-Haytham's *Discourse on Light* (*Qawl* or *Maqāla fi'l-ḍawʾ*) with German translation by J. Baarmann]

IBN AL-HAYTHAM, Alhazeni *Optica*, in *Opticae thesaurus*, ed. F. Risner, Basel, 1572

KAMĀL AL-DĪN AL-FĀRISĪ, *Tanqīḥ al-manāẓir li-dhawi 'l-abṣār waʾl-baṣāʾir*, 2 vols, Hyderabad, 1347–8 H (1928–30) [A commentary in Arabic on Ibn al-Haytham's *Optics*, i.e. al-*Manāẓir*]

KEPLER, JOHANNES, *Ad Vitellionem paralipomena*, Francofurti, 1604. Re-edited in J. K., *Gesammelte Werke*, II (ed. Franz Hammer), Munich, 1939

KEPLER, JOHANNES, *Dioptrice*, Augustae Videlicorum, 1611. Re-edited in J. K., *Gesammelte Werke*, IV (ed. Max Caspar and Franz Hammer), Munich, 1941

AL-KINDĪ, *De aspectibus*, in *Alkindi, Tideus und pseudo-Euklid, drei optische Werke*, herausgegeben und erklärt von Axel Anthon Björnbo und Seb. Vogl, Leipzig und Berlin, 1912. (Abhandlungen zur Geschichte der mathematischen Wissenschaften mit Einschluss ihrer Anwendungen. Heft XXVI.3)

LA CHAMBRE, MARIN CUREAU DE, *La Lumière à Monseignevr l'Éminentissime Cardinal Mazarin*, par le sieur De la Chambre, conseiller du Roy en ses Conseils et son Médecin ordinaire, Paris, 1657

LEIBNIZ, G. W., *Discours de métaphysique*, édition collationnée avec le texte autographe, présentée et annotée par Henri Lestienne, 2nd edition, Paris, 1952. [First published at Hanover in 1846]

LEIBNIZ, G. W., 'Unicum opticae, catoptricae, & dioptricae principium', in *Acta eruditorum*, Leipzig, 1682, pp. 185–90

MAUROLYCUS, FRANCISCUS, *Photismi de lumine et umbra ad perspectivam et radiorum incidentiam facientes. Diaphanorum partes, seu libri tres, in quorum primo de perspicuis corporibus, in secundo de iride, in tertio de organi visualis structura et conspiciliorum for-mis agitur. problemata ad perspectivam et iridem pertinentia*, Neapoli, 1611

MERSENNE, P. MARIN, *Correspondance du*, publiée par Mme. Paul Tannery, éditée et annotée par Cornelis de Waard avec la collaboration de René Pintard, 9 vols, Paris, 1932–1965

MERSENNE, P. MARIN, *Quaestiones celeberrimae...*, Lutetiae Parisiorum, 1623

NEWTON, ISAAC, *The Correspondence of*, ed. H. W. Turnbull, 3 vols, Cambridge, 1959, 1960, 1961. Vol I (1661–75), vol II (1676–87), vol III (1688–94), continuing

NEWTON, ISAAC, *Lectiones opticae, annis 1669, 1670, et 1671 in scholis publicis habitae; et nunc primum ex MSS. in lucem editae*, Londini, 1729

NEWTON, ISAAC, *Sir Isaac Newton's Mathematical Principles of Natural Philosophy and his system of the World*. Motte's translation (of 1729) revised and supplied with an historical appendix by Florian Cajori, Berkeley, California, 1934. [First edition as *Philosophiae Naturalis Principia Mathematica*, Londini, 1687]

NEWTON, ISAAC, *Isaaci Newtoni Opera quae extant omnia*, commentariis illustrabat Samuel Horsley, 5 vols, London, 1779–85

NEWTON, ISAAC, *Optical Lectures, Read in the Publick Schools of the University of Cambridge, Anno Domini 1669, never before printed. Translated into English out of the Original Latin*, London, 1728

NEWTON, ISAAC, *Opticks: or a Treatise of the Reflections, Refractions, Inflections and Colours of Light*. Reprinted from the 4th edition (London, 1730) with a Foreword by Albert Einstein and an Introduction by E. T. Whittaker, London, 1931

NEWTON, ISAAC, *Opticks*, etc., New York, 1952. Reprint of the London

edition of 1931 with an additional Preface by I. Bernard Cohen and an Analytical Table of Contents by Duane H. D. Roller

NEWTON, ISAAC, *Isaac Newton's Papers and Letters on Natural Philosophy and Related Documents*, edited with a General Introduction by I. Bernard Cohen, assisted by Robert E. Schofield, Cambridge, Mass., 1958

NEWTON, ISAAC, *Unpublished Scientific Papers of Isaac Newton*, edited by A. Rupert Hall and Marie Boas Hall, Cambridge, 1962

PARDIES, IGNACE GASTON, *Discours sur le mouvement local, avec des remarques sur le mouvement de la lumière*, Paris, 1670

PARDIES, IGNACE GASTON, *La statique ou la science des forces mouvantes*, Paris, 1673

PTOLEMAEI *Optica: L'Optique de Claude Ptolémée*, dans la version latine d'après l'arabe de l'émir Eugène de Sicile, édition critique et exégétique par Albert Lejeune, Louvain, 1956

RÉGIS, PIERRE-SILVAIN, *Cours entier de philosophie, ou système général selon les principes de M. Descartes*, etc., 3 vols, Amsterdam, 1691

RIGAUD, STEPHAN JORDAN, *Correspondence of Scientific Men of the Seventeenth Century, including Letters of Barrow, Flamsteed, Wallis, and Newton, printed from the originals in the collection of the Right Honourable the Earl of Macclesfield*, 2 vols, Oxford, 1841

RISNERUS, FRIDERICUS, *Opticae libri quatuor, ex voto P. Rami novissimo per F. Risnerum... in usum et lucem publicam producti*, Cassellis, 1606

RISNERUS, FRIDERICUS, *Risneri Optica cum annotationibus Willebrordi Snellii*, ed. J. A. Vollgraff, Gandavi, 1918

ROBERVAL, G. PERS. DE, *Traité de méchanique*, in Marin Mersenne, *Harmonie universelle*, Paris, 1636 [The pages are not consecutively numbered throughout this volume. Roberval's *Traité*, which comprises 36 pages, begins after the first three treatises in the volume]

ROHAULT, JACQUES, *Traité de physique*, Paris, 1671

ROYAL SOCIETY OF LONDON, *Philosophical Transactions*, London, 1671/2–1675

ROYAL SOCIETY OF LONDON, *The Philosophical Transactions . . . abridged*, with notes . . . by C. Hutton, G. Shaw, R. Pearson, 18 vols, London, 1809

SPRAT, THOMAS, *The History of the Royal Society of London for the Improving of Natural Knowledge*, 4th edition, London, 1734. [First published in 1667]

VITELLONIS *Optica*, in *Opticae thesaurus*, ed. F. Risner, Basel, 1572

VOSSIUS, ISAAC, *De lucis natura et proprietate*, Amstelodami, 1662

VOSSIUS, ISAAC, *Isaaci Vossii Responsum ad objecta J. de Bruyn*, in J. de Bruyn, *Epistola ad clariss. virum D.D. Isaacum Vossium*, Amstelodami, 1663

B

ALQUIÉ, FERDINAND, *La découverte métaphysique de l'homme chez Descartes*, Paris, 1950. [Includes interesting discussions of Descartes' *Le Monde* and his theory of the creation of eternal truths]

BAARMANN, J., *see* Ibn al-Haytham in Bibl. A.

BACHELARD, GASTON, *L'activité rationaliste de la physique contemporaine*, Paris, 1951

BACHELARD, SUSANNE, 'Maupertuis et le principe de la moindre action', *Thalès*, année 1958, pp. 3–36

BADCOCK, A. W., 'Physical optics at the Royal Society, 1660–1800', *British Journal for the History of Science*, I (1962), pp. 99–116

BEARE, JOHN I., *Greek Theories of Elementary Cognition, from Alcmaeon to Aristotle*, Oxford, 1906

BECK, L. J., *The Method of Descartes*, a study of the *Regulae*, Oxford, 1952

BELL, A. E., *Christian Huygens and the Development of Science in the Seventeenth Century*, London, 1947

BLOCH, LÉON, *La philosophie de Newton*, Paris, 1908

BOAS, MARIE, 'The establishment of the mechanical philosophy', *Osiris*, X (1952), pp. 412–541

BOAS, MARIE, 'La méthode scientifique de Robert Boyle', *Revue d'histoire des sciences*, IX (1956), pp. 105–25

BOEGEHOLD, H., 'Einiges aus der Geschichte des Brechungsgesetzes', *Zentral-Zeitung für Optik und Mechanik, Elektro-Technik und verwandte Berufszweige*, XL (1919), pp. 94–97, 103–5, 113–16, 121–41

BOEGEHOLD, H., 'Keplers Gedanken über das Brechungsgesetz und ihre Einwirkung auf Snell und Descartes', *Kepler-Festschrift*, Teil I, ed. Karl Stöckl, Regensburg, 1930, pp. 150–67

BOUASSE, HENRI, *Introduction à l'étude des théories de la mécanique*, Paris, 1895. [Contains an illuminating discussion of Descartes' laws of motion; cf. Ch. X: 'Des lois du choc dans Descartes']

BOYER, CARL B., 'Aristotelian references to the law of reflection', *Isis*, XXXVI (1945–1946), pp. 92–95

BOYER, CARL B., 'Descartes and the radius of the rainbow', *Isis*, XLIII (1952), pp. 95–98

BOYER, CARL B., 'Early estimates of the velocity of light', *Isis*, XXXIII (1941), pp. 24–40

BOYER, CARL B., *The Rainbow, from Myth to Mathematics*, New York & London, 1959

BREWSTER, DAVID, *Memoirs of the Life, Writings and Discoveries of Sir Isaac Newton*, 2 vols, Edinburgh, 1855

BROAD, C. D., *The Philosophy of*

Francis Bacon, an address delivered at Cambridge on the occasion of the Bacon tercentenary, 5 October 1926, Cambridge, 1926

BROGLIE, LOUIS DE, *Matière et lumière*, Paris, 1948. [An instructive account of the development of optics (Ch. III), and a discussion of the problem of representing white light from the points of view of the wave and quantum theories, Ch. V, Sec. 4]

BROGLIE, LOUIS DE, *Ondes, corpuscules, mécanique ondulatoire*, Paris, 1949

BROGLIE, LOUIS DE, 'On the parallelism between the dynamics of a material particle and geometrical optics', in *Selected Papers on Wave Mechanics* by Louis de Broglie and Léon Brillouin, authorized translation by Winifred M. Deans, London and Glasgow, 1929

BRUNET, P., *Étude historique sur le principe de la moindre action* (Actualités scientifiques et industrielles, No. 693), Paris, 1938

BRUNSCHVICG, LÉON, *L'expérience humaine et la causalité physique*, 3rd edition, Paris, 1949

BRUNSCHVICG, LÉON, 'La révolution cartésienne et la notion spinoziste de la substance', *Revue de métaphysique et de morale*, 12e année (1904), pp. 755–98. [Remarks on the place of the doctrine of instantaneous propagation of light in Descartes' physics]

BUCHDAHL, GERD, 'Descartes' anticipation of a "logic of scientific discovery"', *Scientific Change* (ed. A. C. Crombie), London, 1963

BUCHDAHL, GERD, 'The relevance of Descartes' philosophy for modern philosophy of science', *The British Journal for the History of Science*, I (1963), pp. 227–49

BURTT, EDWIN ARTHUR, *The Metaphysical Foundations of Modern Physical Science*, a historical and critical essay, revised edition, London, 1950

CAJORI, F., *History of Physics*, 2nd edition, New York, 1929

CLAGETT, MARSHALL, ed., *Critical Problems in the History of Science* (Proceedings of the Institute for the History of Science at the University of Wisconsin, September 1–11, 1957), Madison, Wisconsin, 1959

COHEN, I. BERNARD, 'The first English version of Newton's *Hypotheses non fingo*', *Isis*, LII (1962), pp. 379–88

COHEN, I. BERNARD, 'The first explanation of interference', *The American Journal of Physics*, VIII (1940), pp. 99–106

COHEN, I. BERNARD, *Franklin and Newton*, an inquiry into speculative Newtonian experimental science and Franklin's work in electricity as an example thereof, Philadelphia, 1956

COHEN, I. BERNARD, 'Newton in the light of recent scholarship', *Isis*, LI (1960), pp. 489–514

COHEN, I. BERNARD, 'Roemer and the first determination of the velocity of light (1676)', *Isis*, XXXI (1940), pp. 327–79

COHEN, I. BERNARD, 'Versions of Isaac Newton's first published paper', *Archives internationales d'histoire des sciences*, XI (1958), pp. 357–75

CROMBIE, A. C., *Augustine to Galileo: I* (*Science in the Middle Ages, 5th–13th Centuries*), II (*Science in the Later Middle Ages and Early Modern Times, 13th–17th Centuries*), 2nd edition, London, 1961. [First edition published in 1952]

CROMBIE, A. C., *Robert Grosseteste and the Origins of Experimental Science, 1100–1700*, 2nd impression, Oxford, 1962. [First published in 1953]

DIJKSTERHUIS, E. J., *The Mechanization of the World Picture*, translated by C. Dikshoorn, Oxford, 1961. [Dutch original first published in 1950]

DIJKSTERHUIS, E. J., 'La méthode et les *Essais* de Descartes', in *Descartes et le cartésianisme hollandais*, études et documents, par E. J. Dijksterhuis *et al.* Paris and Amsterdam, 1950, pp. 21–44

DONDER, TH. DE, and J. PELSENEER, 'La vitesse de propagation de la lumière selon Descartes', *Académie Royale de Belgique, Bulletin de la classe des sciences*, XXIII (1937), pp. 689–92

DUGAS, RENÉ, *De Descartes à Newton par l'école anglaise* (Conférences du Palais de la Découverte, série D, No. 16), Paris, 1953

DUGAS, RENÉ, *Histoire de la mécanique*, Neuchâtel, 1950

DUGAS, RENÉ, *La mécanique au xviie siècle, des antécédents scolastiques à la pensée classique*, Neuchâtel, 1954

DUGAS, RENÉ, 'Sur le cartésianisme de Hugens', *Revue d'histoire des sciences*, VII (1954), pp. 22–23

DUHEM, PIERRE, 'Σώζειν τὰ φαινόμενα. Essai sur la notion de théorie physique de Platon à Galilée', *Annales de philosophie chrétienne*, 4e série, VI (1908), pp. 113–39, 277–302, 352–77, 482–514, 561–92

DUHEM, PIERRE, 'L'optique de Malebranche', *Revue de métaphysique et de morale*, 23e année (1916), pp. 37–91

DUHEM, PIERRE, 'Les théories de l'optique', *Revue des deux mondes*, CXXIII (1894), pp. 94–125

DUHEM, PIERRE, *La théorie physique, son objet, sa structure*, 2nd edition, Paris, 1914

FABRY, C., 'Histoire de la physique' [i.e. in France in the 17th century], in G. Hanotaux, *Histoire de la nation française*, XIV, Paris, 1924

FEDERICI VESCOVINI, GRAZIELLA, 'Le questioni di "perspectiva" di Biagio Pelacani da Parma', *Rinascimento*, XII (1961), pp. 163–243

FREDERICI VESCOVINI, GRAZIELLA, *Studi sulla prospettiva medievale*, Torino, 1965

GILSON, É., *Discours de la méthode, texte et commentaire*, Paris, 1930

GILSON, E., 'Météores cartésiens et météores scolastiques', *Études de philosophie médiévale*, Strasbourg, 1921, pp. 247–86

GOUY, M., 'Sur le mouvement lumineux', *Journal de physique théorique et appliquée*, 2e série, V (1886), pp. 354–62

HALBERTSMA, K. T. A., *A History of the Theory of Colour*, Amsterdam, 1949

HALL, A. RUPERT, *From Galileo to Newton, 1630–1720* (The Rise of Modern Science, 3), London, 1963

HALL, A. RUPERT, 'Further optical experiments of Isaac Newton', *Annals of Science*, II (1955), pp. 27–43

HALL, A. RUPERT, 'Newton's First Book (I)', *Archives internationales d'histoire des sciences*, XIII (1960), pp. 39–61

HALL, A. RUPERT, 'Sir Isaac Newton's Note-book, 1661–65', *The Cambridge Historical Journal*, IX (1948), pp. 239–50

HALL, A. RUPERT, *The Scientific Revolution, 1500–1800*, the formation of the modern scientific attitude, 2nd edition, London, 1962. [First published in 1954]

HIRSCHBERG, J., 'Über das älteste arabische Lehrbuch der Augenheilkunde', *Sitzungsberichte der Königlich Preussischen Akademie der Wissenschaften* (1903, XLIX, 26 November. Sitzung der philosophisch-historischen Classe), pp. 1080–94

HORTEN, M., 'Avicenna's Lehre vom Regenbogen nach seinem Werk *al Schifāʾ*', *Meteorologische Zeitschrift*, XXX (1913), pp. 533–44

HOSKIN, M. A., 'Clarke's notes to Rohault's *Traité de physique*', *The Thomist*, XXIV (1961), pp. 353–63

JENKINS, FRANCIS A., and HARVEY E. WHITE, *Fundamentals of Optics*, 2nd edition, London, 1951

KEELING, S. V., *Descartes*, London, 1934

KNESER, A., *Das Prinzip der kleinsten Wirkung von Leibniz bis zur Zeit der Gegenwart* (Wissenschaftliche Grundfragen, herausgegeben von R. Hönigswald, Nr. IX), Leipzig, 1928

KORTEWEG, D.-J., 'Descartes et les manuscrits de Snellius d'après quelques documents nouveaux', *Revue de métaphysique et de morale*, 4ᵉ année (1896), pp. 489–501

KOYRÉ, ALEXANDRE, *Études galiléennes:* I. *À l'aube de la science classique;* II. *La loi de la chute des corps, Descartes et Galilée;* III. *Galilée et la loi d'inertie* (Actualités scientifiques et industrielles, Nos. 852–4), Paris, 1939

KOYRÉ, ALEXANDRE, 'La gravitation universelle de Képler à Newton', *Archives internationales d'histoire des sciences*, IV (1951), pp. 638–53 (Conférence du Palais de la Découverte, Paris, 1951)

KOYRÉ, ALEXANDRE, 'L'hypothèse et l'expérience chez Newton', *Bulletin de la Société Française de la Philosophie*, L (1956), pp. 59–89

KOYRÉ, ALEXANDRE, 'The origins of modern science, a new interpretation', *Diogenes*, No. 16, Winter 1956, pp. 1–22. [Review of A. C. Crombie, *Robert Grosseteste*]

KOYRÉ, ALEXANDRE, 'Les Queries de l'Optique', *Archives internationales d'histoire des sciences*, XIII (1960), pp. 15–29

KOYRÉ, ALEXANDRE, 'Les Regulae philosophandi', *Archives internationales d'histoire des sciences*, XIII (1960), pp. 3–14

KRAMER, P., 'Descartes und das Brechungsgesetz des Lichtes', *Abhandlungen zur Geschichte der Mathematik*, IV (1882), pp. 233–78

KUHN, THOMAS S., 'Newton's optical papers', in I. Bernard Cohen (ed.), *Isaac Newton's Papers and Letters on Natural Philosophy*, Cambridge, Mass., 1958, pp. 27–45

LALANDE, ANDRÉ, L' *'interprétation' de la nature dans le 'Valerius terminus' de Bacon*, Mâcon, 1901

LALANDE, ANDRÉ, 'Quelques textes de Bacon et de Descartes', *Revue de métaphysique et de morale*, 19ᵉ année (1911), pp. 296–311

LALANDE, ANDRÉ, *Les théories de l'induction et l'expérimentation*, Paris, 1929

LAPLACE, PIERRE SIMON DE, 'Sur la double réfraction de la lumière dans les cristaux diaphanes', *Bulletin de la Société Philomatique*, I (1807), pp. 303–10. Reproduced in *Œuvres complètes de Laplace*, XIV, Paris, 1912, pp. 278–87

LAPLACE, PIERRE SIMON DE, 'Sur la loi de la réfraction extraordinaire de la lumière dans les cristaux diaphanes', *Journal de physique*, LXVIII (1809), pp. 107–11. Reproduced in *Œuvres complètes de Laplace*, XIV, Paris, 1912, pp. 254–8

LAPORTE, JEAN, *Le rationalisme de Descartes*, Paris, 1945

LEJEUNE, ALBERT, 'Archimède et la loi de la réflexion', *Isis*, XXXVIII (1947), pp. 51–53

LEJEUNE, ALBERT, *Euclide et Ptolémée, deux stades de l'optique géométrique grecque*, Louvain, 1948

LEJEUNE, ALBERT, 'Les lois de la réflexion dans l'*Optique* de Ptolémée', *L'antiquité classique*, XV (1946), pp. 241–56

LEJEUNE, ALBERT, *Recherches sur la catoptrique grecque d'après les sources antiques et médiévales*, Bruxelles, 1957

LEJEUNE, ALBERT, 'Les tables de réfractions de Ptolémée', *Annales de la Société Scientifique de Bruxelles*, série I (Sciences mathématiques et physiques), LX (1940), pp. 93–101

LE LIONNAIS, F., 'Descartes et Einstein', *Revue d'histoire des sciences et de leurs applications*, V (1952), pp. 139–54

LOHNE, JOHANNES, 'Zur Geschichte des Brechungsgesetzes', *Sudhoffs Archiv*, Band 47, Heft 2, Juni 1963, pp. 152–72

LOHNE, JOHANNES, 'Newton's "proof" of the sine law and his mathematical principles of colours', *Archive for History of Exact Sciences*, vol I, No. 4 (1961), pp. 389–405

LOHNE, JOHANNES, 'Thomas Harriott (1560–1621), the Tycho Brahe of optics', *Centaurus*, VI (1959), pp. 113–21

MACH, ERNST, *The Principles of Physical Optics*, an historical and philosophical treatment, translated by John S. Anderson and A. F. A. Young, London, 1926. [German original first published in 1921]

MEYERSON, ÉMILE, *Identity and Reality*, translated by Kate Loewenberg, London, 1930

MILHAUD, GASTON, *Descartes savant*, Paris, 1921

MILHAUD, GASTON, *Nouvelles études sur l'histoire de la pensée scientifique*, Paris, 1911. [A chapter on 'Descartes et Newton', pp. 219–35]

MONTUCLA, JEAN-ÉTIENNE, *Histoire des mathématiques*, nouvelle édition, considérablement augmentée, et prolongée jusque vers l'époque actuelle, 4 vols, Paris, 1799–1802 (Vols III and IV edited and published by Jérome de La Lande)

MORE, LOUIS TRENCHARD, *Isaac Newton*, a biography, New York and London, 1934

MOUY, PAUL, *Le dévelopment de la physique cartésienne, 1646–1712*, Paris, 1934

MOUY, PAUL, *Les lois du choc des corps d'après Malebranche*, Paris, 1927

NAPIER, MACVEY, 'Remarks, illustrative of the scope and influence of the philosophical writings of Lord Bacon', *Transactions of the Royal Society of Edinburgh*, VIII (1818), pp. 373–425

NAZĪF, MUṢṬAFĀ, *al-Ḥasan ibn al-Haytham, buḥūthuhu wa-kushūfuhu al-baṣariyya*, 2 vols, Cairo, 1942–43. [A study in Arabic of the *Optics* of Ibn al-Haytham, based on the extant MSS]

PAPANASTASSIOU, CH.-E., *Les théories sur la nature de la lumière de Descartes à nos jours et l'évolution de la théorie physique*, Paris, 1935

PELSENEER, JEAN, 'Gilbert, Bacon, Galileo, Kepler, Harvey et Descartes—leurs relations', *Isis*, XVII (1932), pp. 171–208

PELSENEER, JEAN, *see* Th. de Donder and J. Pelseneer.

POGGENDORFF, J. C., *Geschichte der Physik*, Vorlesungen gehalten an der Universität zu Berlin, Leipzig, 1879

POLYAK, S. L., *The Retina*, Chicago, 1941

POPPER, KARL R., *Conjectures and Refutations, the growth of scientific knowledge*, London, 1963

POPPER, KARL R., *The Logic of Scientific Discovery*, London, 1959

PRIESTLEY, JOSEPH, *The History and Present State of Discoveries Relating to Vision, Light and Colours*, London, 1722

REILLY, CONOR, 'Francis Line, peripatetic (1595–1675)', *Osiris*, XIV (1962), pp. 222–53

ROBERTS, MICHAEL and E. R. THOMAS, *Newton and the Origin of Colours*, a study of the earliest examples of scientific method, London, 1934

ROBINSON, BRYAN, *A Dissertation on the Æther of Sir Isaac Newton*, Dublin, 1743

ROBINSON, BRYAN, *Sir Isaac Newton's Account of the Æther*, Dublin, 1745

RONCHI, VASCO, *Histoire de la lumière*, translated by Juliette Taton, Paris, 1956. [First published as *Storia della luce*, Bologna, 1939]

ROSENBERGER, F., *Newton und seine physikalischen Prinzipien*, Leipzig, 1895

ROSENFELD, L., 'Marcus Marcis Untersuchungen über das Prisma und ihr Verhältnis zu Newton's Farbentheorie', *Isis*, XVII (1932), pp. 325–30

ROSENFELD, L., 'Le premier conflit entre la théorie ondulatoire et la théorie corpusculaire de la lumière,' *Isis*, XI (1928), pp. 111–22

ROSENFELD, L., 'La théorie des couleurs de Newton et ses adversaires', *Isis*, IX (1927), pp. 44–65

ROTH, LEON, *Descartes' Discourse on Method*, Oxford, 1948

ROUSE BALL, W. W., *An Essay on Newton's Principia*, London, 1893

SABRA, A. I., 'Explanation of optical reflection and refraction: Ibn al-Haytham, Descartes, Newton', *Actes du dixième congrès international d'histoire des sciences* (1962), Paris, 1964, I, pp. 551–4

SABRA, A. I., 'Newton and the "bigness" of vibrations', *Isis*, LIV (1963), pp. 267–8

SABRA, A. I., 'A note on a suggested modification of Newton's corpuscular theory of light to reconcile it with Foucault's experiment of 1850', *The British Journal for the Philosophy of Science*, V (1954), pp. 149–51

SAMBURSKY, S., 'Philoponus' interpretation of Aristotle's theory of light', *Osiris*, XIII (1958), pp. 114–126

SARTON, G., 'Discovery of the dispersion of light and of the nature of colour (1672)', *Isis*, XIV (1930), pp. 326–41

SCHRAMM, MATTHIAS, *Ibn al-Haythams Weg zur Physik*, Wiesbaden, 1963

SCHRECKER, P., 'Bibliographie de Descartes savant', *Thalès*, 3ᵉ année (1936, published 1938), pp. 145–54

SCHRECKER, P. 'Notes sur l'évolution du principe de la moindre action', *Isis*, XXXIII (1941), pp. 329–334

SCOTT, J. F., *The Scientific Work of René Descartes (1596–1650)*, London, 1952

SHIRLEY, J. W., 'An early experimental determination of Snell's law', *American Journal of Physics*, XIX (1951), pp. 507–8

SIRVEN, J., *Les années d'apprentissage de Descartes (1596–1628)*, Paris, 1930

SMITH, NORMAN KEMP, *New Studies in the Philosophy of Descartes*, London, 1952

SNOW, A. J., *Matter and Gravity in Newton's Physical Philosophy*, a study in the natural philosophy of Newton's time, London, 1926

STOCK, HYMAN, *The Method of Descartes in the Natural Sciences*, New York, 1931

STRONG, E. W., 'Newton's "mathematical way"', *Journal of the History of Ideas*, XII (1951), pp. 90–110

STRONG, E. W., *Procedure and Metaphysics*, a study in the philosophy of mathematical-physical science in the sixteenth and seventeenth centuries, Berkeley, California, 1936

SUPPES, PATRICK, 'Descartes and the problem of action at a distance', *Journal of the History of Ideas*, XV (1954), pp. 146–152

TANNERY, PAUL, 'Sur la date des principales découvertes de Fermat', *Bulletin des sciences mathématiques et astronomiques*, 2ᵉ série, VII (1883), première partie, pp. 116–128

TANNERY, PAUL, 'Descartes physicien', *Revue de métaphysique et de morale*, 4ᵉ année (1896), pp. 478–88

TANNERY, PAUL, *Mémoires scientifiques*, publiés par J. L. Heiberg et H.-G. Zeuthen, 17 vols, Toulouse-Paris, 1912–1950

THOMAS, E. R., see Michael Roberts and E. R. Thomas

TOULMIN, STEPHEN, *The Philosophy of Science*, London, 1953

TURBAYNE, COLIN MURRAY, *The Myth of Metaphor*, New Haven, Conn., 1962

VAN GEER, P., 'Notice sur la vie et les travaux de Willebrord Snellius', *Archives néerlandaises des sciences exactes et naturelles* (published by: La Société Hollandaise des Sciences), XVIII (1883), pp. 453–68

VAVILOV, S. I., 'Newton and the atomic theory', in *Newton Tercentenary Celebrations*, Cambridge, 1947, pp. 43–55

VOLLGRAFF, J. A., 'Pierre de la Ramée (1515–1572) et Willebrord Snel van Royen (1580–1626)', *Janus*, 18e année (1913), pp. 595–625

VOLLGRAFF, J. A., 'Snellius' notes on the reflections and refractions of rays', *Osiris*, I (1936), pp. 718–25

WAARD, C. DE, *L'expérience baro-métrique, ses antécédents et ses ex-plications*, étude historique, Thouars, 1936

WAARD, C. DE, 'Le manuscrit perdu de Snellius sur la réfraction', *Janus*, 39e année (1935), pp. 51–73

WAHL, JEAN, *Du rôle de l'idée de l'instant dans la philosophie de Descartes*, 2nd edition, Paris, 1953

WESTFALL, RICHARD S., 'The develop-ment of Newton's theory of color', *Isis*, LIII (1962), pp. 339–58

WESTFALL, RICHARD S., 'The founda-tions of Newton's philosophy of nature', *The British Journal for the History of Science*, I (1962), pp. 171–82

WESTFALL, RICHARD S., 'Newton and his critics on the nature of colours', *Archives internationales d'histoire des sciences*, 15e année (1962), pp. 47–58

WESTFALL, RICHARD S., 'Newton's reply to Hooke and the theory of colours', *Isis*, LIV (1963), pp. 82–96

WHEWELL, WILLIAM, *History of the Inductive Sciences, from the Earliest to the Present Times*, new and revised edition, 3 vols, London, 1847

WHEWELL, WILLIAM, *The Philosophy of the Inductive Sciences*, London, 1840

WHITE, HARVEY E., see Francis A. Jenkins and Harvey E. White

WHITTAKER, EDMUND T., *A History of Æther and Electricity*, the classical theories, 2nd edition, Edinburgh, 1951

WILDE, EMIL, *Geschichte der Optik*, 2 vols, Berlin, 1838–43

WINTER, H. J. J., 'The optical researches of Ibn al-Haitham', *Centaurus*, III (1954), pp. 190–210

WOLF, A., *A History of Science, Tech-nology, and Philosophy in the 16th & 17th Centuries*, 2nd edition prepared by Douglas McKie, 1950

Wood, ALEXANDER, *In Pursuit of Truth*, a comparative study in science and religion, London, 1927

WOOD, ROBERT W., *Physical Optics*, 2nd edition, New York, 1911

WÜRSCHMIDT, JOSEPH, 'Die Theorie des Regenbogens und des Halo bei Ibn al-Haitam und bei Dietrich von Freiberg', *Meteorologische Zeitschrift*, XXXI (1914), pp. 484–87

YOUNG, THOMAS, 'On the theory of light and colours', in *A Course of Lectures on Natural Philosophy*, II, London, 1807, pp. 613–31

ZOUBOV, V., 'Une théorie aristotéli-cienne de la lumière du xviie siècle', *Isis*, XXIV (1936), pp. 341–60

Additional Bibliography 1966–1980

As in the original bibliography, only works closely related to the questions discussed in this book are included.

BECHLER, ZEV, 'Newton's search for a mechanistic model of colour dis-persion: a suggested interpreta-tion', *Archive for History of Exact Sciences*, XI (1973), pp. 1 – 37.

BECHLER, ZEV, 'Newton's law of forces which are inversely as the mass: a suggested interpretation of his later efforts to normalise a mechanistic model of optical dis-persion', *Centaurus*, XVIII (1974), pp. 184 – 222.

BECHLER, ZEV, 'Newton's 1672 optical controversies: a study in the grammar of scientific dissent', in Y. Elkana, ed., *The Interaction Between Science and Philosophy*, Atlantic Highlands, N.J., 1974, pp.115 – 42.

BECHLER, ZEV, 'A less agreeable matter: the disagreeable case of Newton and achromatic refraction', *British Journal for the History of Science*, VIII (1975), pp.101 – 26.

BIERNSON, GEORGE, 'Why did Newton see indigo in the spectrum?', *American Journal of Physics*, XL (1972), pp. 526 – 33.

BOS, H. J. M., et al., eds., *Studies on Christiaan Huygens*, Lisse, 1980.

BUCHDAHL, GERD, *Metaphysics and the Philosophy of Science: the classical origins, Descartes to Kant*, Oxford, 1969.

BUCHDAHL, GERD, 'Methodological aspects of Kepler's theory of refraction', *Studies in History and Philosophy of Science*, III (1972), pp.265 – 98.

DESCARTES, RENÉ, *Discourse on Method, Optics, and Meteorology*. Translated by Paul J. Olscamp, Indianapolis, 1965.

GRUNER, S. M., 'Defending Father Lucas, a consideration of the Newton-Lucas dispute on the nature of the spectrum', *Centaurus*, XVII (1973), pp.315 – 29.

HOLTZMARK, TORGER, 'Newton's *Experimentum crucis* reconsidered', *American Journal of Physics*, XXXIX (1970), pp.1229 – 35.

KNUDSEN, OLE and KURT M. PEDERSEN, 'The link between "determination" and conservation of motion in Descartes' dynamics', *Centaurus*, XIII (1968), pp. 183 – 6.

LAYMON, RONALD, 'Newton's advertised precision and his refutation of the received laws of refraction', in Peter K. Machamer and Robert G.

Turnbull, eds., *Studies in Perception: interrelations in the history of philosophy and science*, Columbus, Ohio, 1978, pp.231 – 58.

LAYMON, RONALD, 'Newton's *Experimentum crucis* and the logic of idealization and theory refutation', *Studies in History and Philosophy of Science*, IX (1978), pp.51 – 77.

LOHNE, J. A., 'Isaac Newton: the rise of a scientist 1661 – 1671', *Notes and Records of the Royal Society of London*, XX (1965), pp. 125 – 39.

LOHNE, J. A., 'The increasing corruption of Newton's diagrams', *History of Science*, VI (1967), pp.69–89.

LOHNE, J. A., 'Experimentum Crucis', *Notes and Records of the Royal Society of London*, XXIII (1968), pp.169 – 99.

LOHNE, J. A. and BERNHARD STICKER, *Newtons Theorie der Prismenfarben, mit Übersetzung und Erläuterung der Abhandlung von 1672*, Munich, 1969.

LOHNE, J. A., 'Newton's table of refractive powers: origins, accuracy, and influence', *Sudhoffs Archiv für Geschichte der Medizin und Naturwissenschaften*, LXI (1977), pp.229 – 47.

MAHONEY, M. S., *The Mathematical Career of Pierre de Fermat (1601 – 1665)*, Princeton, 1973.

MAMIANI, MAURIZIO, *Isaac Newton filosofo della natura: le lezioni giovanili di ottica e la genesi del metodo Newtoniano*, Florence, 1976.

NEWTON, ISAAC, *The Mathematical Papers of I.N.*, ed. D. T. Whiteside, I (1664 – 1666), III (1670 – 1673), Cambridge, 1967, 1969. [Sections on optics with notes by the editor.]

NEWTON, ISAAC, *The Unpublished First Version of Isaac Newton's Cambridge Lectures on Optics 1670 – 1672: a facsimile of the autograph, now Cambridge University Library MS. Add. 4002*, with an introduction by

D. T. Whiteside, Cambridge, 1973.

SHAPIRO, ALAN E., 'Kinematic optics: a study of the wave theory of light in the seventeenth century', *Archive for History of Exact Sciences,* XI (1973), pp.134 – 266.

SHAPIRO, ALAN E., 'Light, pressure, and rectilinear propagation: Descartes' celestial optics and Newton's hydrostatics', *Studies in History and Philosophy of Science,* V (1974), pp.239 – 96.

SHAPIRO, ALAN E., 'Newton's definition of a light ray and the diffusion theories of chromatic dispersion', *Isis,* LXVI (1975), pp.194 – 210.

SHAPIRO, ALAN E., 'Newton's "achromatic" dispersion law: theoretical background and experimental evidence', *Archive for History of Exact Sciences,* XXI (1979), pp.91 – 128.

SHAPIRO, ALAN E., 'The evolving structure of Newton's theory of white light and color', *Isis,* LXXI (1980), pp.211 – 35.

STRAKER, S. M., *Kepler's Optical Studies: a study in the development of seventeenth-century natural philosophy,* Ph.D. dissertation, Indiana University, 1971. Available from University Microfilms, Ann Arbor, Michigan.

STUEWER, ROGER H., 'A critical analysis of Newton's work on diffraction', in *Isis,* LXI (1970), pp.188 – 205.

TATON, RENÉ, ed., *Roemer et la vitesse de la lumière,* Paris, 1978.

WESTFALL, RICHARD S., 'Isaac Newton's coloured circles twixt two contiguous glasses', *Archive for History of Exact Sciences,* II (1965), pp.181 – 96.

WESTFALL, RICHARD S., 'Newton defends his first publication: the Newton-Lucas correspondence', *Isis,* LVII (1966), pp.299 – 314.

WESTFALL, RICHARD S., 'Uneasily fitful reflections on fits of easy transmission', *The Texas Quarterly,* X (1967), pp.86 – 102. Also in *The Annus Mirabilis of Sir Isaac Newton, 1666 – 1966,* ed. Robert Palter, Cambridge, MA, and London, 1970, pp.88 – 104.

WESTFALL, RICHARD S., 'Huygens' rings and Newton's rings: periodicity and seventeenth-century optics', *Ratio,* X (1968), pp.64 – 77.

WESTFALL, RICHARD S., 'Newton and the fudge factor', *Science,* CLXXIX (1973), pp.751 – 8.

ZIGGELAAR, AUGUST, *Le physicien Iqnace Gaston Pardies S. J. (1636 – 1673). Acta historica scientiarum naturalium, edidit Bibliotheca Universitatis Hauniensis,* vol. 26, Odense, 1971.

Corrections

p.7, l.14: *for* 212 *read* 209

p.30, l.1: realize

p.31, l.32: straightway

p.38, l.25: realized

p.46, n.4, l.5: *for* 55ff *read* 57f

p.55, n.34, l.22: Arabic text

p.97, l.21: $\frac{n}{m}$

p.100, n.10, l.8: *Lichts*

p.123, l.29: *for* continue *read* constitute

p.148, l.30: This way

p.243, l.22: who ever

p.244, n.22, l.8: *for* with *read* from

p.275, l.17: *for* has *read* had

p.300, n.9, l.2: *for* 37 *read* 237

p.311, n.25, l.4: streight

p.311, n.25, l.19: *for* been have *read* have been

p.314, n.34, l.2: *for* 308f *read* 302f

p.347, col.A, l.40: Suzanne

p.353, col.B, l.2: *Sciences,* 2 vols.

p.355, col.A, l.33: Suzanne

p.360, col.B, l.32: *Principia* (Newton)

p.361, col.A, l.36: *for* 113 – 16 *read* 131 – 2

356

Index

Académie Royale des Sciences, 159, 162n, 171, 183, 185, 198, 205, 206, 207, 222n, 229, 343
adaequare, 144
Adam, C. and G. Milhaud, 49n, 344
Adam, C. and P. Tannery, 49n, 344
Alhazen, *see* Ibn al-Haytham
Alquié, Ferdinand, 347
Anaclastic, 29, 30n
Analogy, concept of, 27–33, 116/7, 186; stick analogy, 32n, 48n, 54–55n
Analysis, 30–1, 33, and synthesis, 31, 309
Analysis ad Refractiones (Fermat), 145–7, 149n, 150n
Ango, Pierre, 195, 196, 197, 218, 340, 343
'Anticipation' of nature or of the mind, 34, 175; *see* 'Interpretation of nature'
a posteriori proof, 22, 44
Apriorism, 26–7, 37n, 44, 167, 168, 183–4, 288, 309, 341; *see* Deductivism
Archimedes, 23, 24, 104, 161, 344, 350
Aristotle, 25n, 46, 47n, 106, 159, 343, 347, 352
Atomism, 26, 231, 242–3, 252, 291–2, 296, 297
Averroës, 46, 343
Avicenna, 46, 72n, 343, 349

Baarmann, J., 201n, 345
Bachelard, Gaston, 292n, 318n, 347
Bachelard, Susanne, 347
Bacon, Francis, 22, 30, 34, 35n, 36, 37, 38n, 41, 64, 170, 171, 172n, 173, 174, 175, 176, 177, 180, 181, 183, 248, 249n, 316, 322, 343, 348, 350, 351
Bacon, Roger, 47, 48, 76, 80, 98, 343
Badcock, A. W., 347
Baillet, Adrien, 184n
Bartholinus, Erasmus, 185, 220, 343

Beare, John I., 46n, 347
Beck, L. J., 33n, 35n, 347
Beeckman, Isaac, 48n, 49n, 72n, 101n, 343
Bell, A. E., 162n, 170, 225n, 347
Berigard, Claude G., 291, 292
Bigness of vibrations, 278, 328
Birch, Thomas, 234n, 251, 254n, 273n, 309n, 319, 328n, 329n, 330n, 343, 344
Björnbo, A., and S. Vogl, 46n, 78n, 345
Bloch, Léon, 292n, 347
Boas, Marie, 346, 347
Boegehold, H., 347
Book of the Ten Treatises on the Eye (Ḥunayn) 55n, 345
Bouasse, H., 26n, 347
Boyer, Carl, 63n, 67n, 69n, 110n, 347
Boyle, Robert, 181, 229, 321, 322, 324, 344, 347
Brengger, J. G., 100n
Brewster, David, 165n, 231n, 236n, 327, 329, 330, 332, 347
Broad, C. D., 347–8
Broglie, Louis de, 318, 342, 348
Brunet, P., 348
Brunschvicg, Léon, 348
Bruyn, J. de, 347
Buchdahl, Gerd, 348
Burtt, E. A., 348

Cajori, Florian, 239n, 286n, 317, 346, 348
Careil, Foucher de, 344
Cassini, G. D., 205, 206
Catoptrica (Heron), 70, 71n, 344
Catoptrica (pseudo-Euclid), 93n, 344
Certainty of scientific knowledge, 42–5, 175–7, 179, 180–3, 243, 275, 285, 287
Chief Systems (Galileo), 61n
Chromatic aberration, 234
Clagett, Marshall, 93n, 344, 348

Clarke, Samuel, 349
Clerselier, Claude, 48n, 49n, 57n, 58n, 84, 87, 113, 115, 117, 125–37, 153, 154, 202, 344
'Coalescence' of vibrations, pulses or waves, 259–61, 280, 281, 282, 328n; see superposition of Waves
Cogitationes privatae (Descartes), 32n, 34n, 38n, 105
Cohen, I. B., 10, 207n, 244n, 246n, 248n, 264n, 274n, 339, 346, 348, 350
Cohen, Morris R., and I. E. Drabkin, 70n, 344
Colbert, J. B., 222n
Colours, as connate properties of light, 241, 287, 290, 296; modification theory of, 241, 247, 249, 279–81, 282, 295–6; colours of natural bodies, 243, 291; prismatic colourc, 65–6, 67–8, 185, 234–42; different refrangibility of, 68, 169, 240, 241, 246, 250n, 266, 268, 269, 270, 271, 274, 279, 285, 287, 293, 299, 327; colours of thin plates, 185, 187, 221, 263, 298, 319–42; wave explanation of, 169–70, 185, 196, 256–9
Composition of motions, see 'Coalescence' of vibrations; Parallelogram method; superposition of Waves
conatus, 200; see Tendency to motion
Constantinus Africanus, 55n, 344
Copernican view, 18, 50, 51n
Copernicus, 159
Cotes, Roger, 169, 344
Crombie, A. C., 47n, 63n, 348, 350
Crossing of light rays, 59–60, 186, 207

Damianus, 71n
De anima (Aristotle), 46n, 55n, 343
De anima (Averroës), 46n, 343
De anima (Avicenna), 46n, 49n, 343
De aspectibus (al-Kindi), 46n, 78n, 345
Deductivism, 34n, 37n, 38, 39, 40; hypothetico-deductivism, 44, 175n, 182–3, 248; see Apriorism; Hypotheses
De luce (Grosseteste), 47n, 344

De lucis natura et proprietate (Vossius), 101n
Democritus, 168, 172n
Demonstration, concepts of, 87, 91, 128–30, 176–7, 179, 181–3, 283, 286, 299, 308, 312–13
Démonstration touchant le Mouvement de la Lumière (Roemer), 205n
De multiplicatione specierum (R. Bacon), 76n, 77n, 78n, 98n
De oculis ('Constantini Africani'), 55n, 344
De oculis ('Galeni'), 46, 344
De partibus inferiori ventre contentis (Descartes), 34n,
Descartes, René, 11, 12, 13, 14, Chapters I–IV, 136, 142, 143, 147, 148, 149, 153, 155, 159, 163, 164, 165, 166, 167, 168, 173, 174, 183, 184, 186, 188, 190n, 192, 194, 195, 198, 202, 203, 207, 208, 209, 210, 211, 212, 238, 239n, 240, 241, 255, 261, 275, 277n, 280, 283, 295, 300, 310, 324, 329, 344, 346, 347, 348, 349, 350, 351, 352, 353
De scientia perspectiva (R. Bacon), 47n, 77n, 78n
De sensu (Aristotle), 46n, 343
Determination, distinguished from force or speed of motion, 80, 82, 84, 107, 108, 117–21, 126–9
Dialogues (Galileo), 58n, 61n, 104, 157n, 190n, 203n, 344
Diaphaneon (Maurolyco), 63n, 345
Dijksterhuis, E. J., 216n, 348; et al., 349
Diophantus, 144n, 344
Dioptric (Descartes), 11, 17, 19, 20, 21, 22n, 23, 27, 28, 29, 32, 33, 44, 48n, 52n, 59, 60, 65, 72, 78, 79n, 80n, 81, 82n, 85, 86, 87, 90, 98, 99, 101, 102, 103, 104n, 106, 107, 110n, 112, 114, 115, 116, 117, 121, 124, 127, 128, 130n, 133, 137, 183, 186, 188, 261
Dioptrice (Kepler), 30n, 345
Dioptrique (Huygens), 198, 199, 221, 222
Discours de la Cause de la Pesanteur & Addition (Huygens), 164, 165, 166n, 167, 168, 170, 214

Discours de métaphysique (Leibniz), 104n, 345
Discours du mouvement local (Pardies), 196n, 346
Discourse on Method (Descartes), 11, 17, 20, 37, 38n, 49n, 50n, 60, 61, 86
Donder, Th. de, and J. Pelseneer, 349
Drabkin, E. I., *see* Cohen, Morris R.
Dugas, R., 26n, 148n, 157n, 349
Duhem, Pierre, 56, 179n, 221, 259n, 283, 315, 316, 349

Eclipses of the moon, argument from, 57–9, 190–1, 203–5
Edleston, J., 165n, 344
Einstein, Albert, 246n, 315, 346, 350
Elasticity, 80–2, 189, 199, 207–9, 210, 215
Ellis, R. L., 180, 343
Empedocles, 46
Enumeration, 33, 173, 177, 179, 229; *see* Induction
Ether, 167, 209, 211–12, 213n, 224, 239, 241, 275, 278, 279, 288, 309, 310, 311n, 312, 324, 325, 339, 341; *see* subtle Matter
Euclid, 69, 344, 350
Euler, L., 157, 158
Excerpta anatomica (Descartes), 34n
Exclusion or rejection, method of, 64, 176–81, 249, 321
Experiment, role of, 26, 33–7, 40–1, 62, 90, 103, 137, 142–3, 171, 173–4, 177, 181, 187, 203, 215, 224, 230, 273, 276, 283, 285, 288, 299, 308, 312, 313–18
Experimenta crystalli Islandici (Bartholinus), 185n, 343
Experimental laws or propositions or results, 168, 182, 221, 233, 251, 262, 264, 274, 275, 286, 298–9, 334, 340
Experimentum crucis, 41, 187, 228, 236n, 239, 241, 245, 249–50, 252, 267–8, 269, 276, 286, 294–5, 296n, 315–16, 317n, 324–5; *see instantiae crucis*
Explanation, concepts of, 17–45, 159–84, 229–30, 308–18, 337–42; *see*

Apriorism; Deductivism; Demonstration; Hypotheses; Mechanism; Theories *vs.* hypotheses

Fabry, C., 349
Federici Vescovini, Graziella, 349
Fermat, Pierre de, 11, 12, 72n, 86, 87, 89, 90, 91, 92, 93, 98, 104n, 112–33 *passim*, Chapter V, 197, 218, 344, 352
Fits of easy reflection and of easy transmission, 221, 296, 298, 310, 320; theory of, 331–42
Fontenelle, Bernard Le Bovier de, 198n
Foucault, Léon, 158, 315, 316, 318, 352
Fourier's theorem, 261, 280, 281n
Fowler, Thomas, 181n, 343
Fresnel, A.-J., 214, 215, 228, 283, 313, 314, 320, 341

Galen, 46, 344
Galileo, 18, 58n, 61n, 104, 137, 157n, 190n, 203n, 344, 348, 349, 350, 351
Gascoigne, 231n
Gassendi, Pierre, 168, 209
géométrisation à outrance, 52n, 82, 210n; *see* Mathematics *Geometry* (Descartes), 17n, 29, 60, 86
Gilson., E., 34n, 60n, 349
Golius, Jacob, 100, 101, 102
Gouy, M., 206, 261, 280, 281, 282, 295, 349
Gravity, 159–69, 172, 199, 311n
Great Instauration (F. Bacon), 175n, 176n, 177n, 178n, 249n
Grimaldi, F. M., 185, 186, 266, 273, 277n, 283, 285, 295, 310n, 344
Grosseteste, Robert, 47, 48, 344, 348, 350

Halbertsma, K. T. A., 349
Hall, A. R., 234n, 237n, 240n, 242n, 246n, 247n, 248, 292n, 296n, 332n, 346, 349
Hall (or Linus), Francis, 231n, 351
Harmonie universelle (Mersenne), 346

Harriott, Thomas, 100n, 351
Heath, T. L., 144, 344
Hérigone, Pierre, 144n, 344
Heron of Alexandria, 70, 71n, 72, 74, 136, 139n, 155, 344
Hirschberg, J., 55n, 349
Hobbes, Thomas, 81n, 82, 345
Hooke, Robert, 9, 10, 14, 169, 181n, 185, 186, 187, 188, 189, 190, 191, 192, 194, 196, 198, 201, 202, 210, 213, 231, 233, 240, 241, 242, 245, 248n, 251–64, 266, 271, 273, 276, 277, 278, 279, 280, 281, 282, 283, 295, 310n, 311n, 319–32 passim, 340, 341, 345, 353
Horsley, Samuel, 346
Horten, M., 349
Hortensius, Martin, 49n
Hoskin, M. A., 10, 349
Ḥunayn ibn Isḥāq, 55n, 344, 345
Huygens, Christian, 11, 12, 13, 14, 59, 60n, 92, 100, 101, 104, 113, 135, 137, 156, 158, Chapters VI, VII, VIII, 231, 233, 236n, 242, 266n, 267, 268–72, 273, 275, 288, 290, 293, 313, 315, 316, 340, 347, 349
Huygens, Constantin, 102
Hypotheses, conjectures, suppositions, 18–24, 40, 43–5, 165, 169–70, 171, 173–4, 175n, 180–3, 184, 186, 187, 203, 206–7, 221–4, 229, 231–2, 236, 238–9, 243, 244, 248, 251, 253, 264–5, 269–71, 273–6, 277, 281n, 284–8, 294, 298–9, 300, 308–9, 311n, 312, 316, 331–2, 337, 340, 342

Ibn al-Haytham (Alhazen), 12, 47, 48, 72, 74, 77, 78, 79, 80, 82, 93, 96, 97, 98, 99, 100, 103, 114, 116, 137n, 201n, 207n, 208n, 210n, 345, 351, 352, 353
Iceland crystal, 185n, 221–9; see double Refraction
Incompressibility of matter, 54–6, 81–2, 189–90, 209
Indirect proof, 41, 179
Induction, 34, 35n, 36, 38, 170–1, 173, 175n–7, 180, 181–2, 229, 231, 269n, 270, 285, 288, 294, 298, 322

Inflexion of light, see diffraction of Light
instantiae crucis, 178, 179, 249; see experimentum crucis
Intelligibilism, 52, 162, 166, 168, 184; see Mechanism
Interference phenomena, 311, 321, 341; see Colours of thin plates
'Interpretation of nature', 34, 36, 37, 175, 177, 180
Intuition, 31, 32, 35n

Jenkins, F. A., and H. E. White, 281n, 349
Journal (Beeckman), 49n, 72n

Kamāl al-Dīn al-Fārisī, 62n, 63n, 220n, 345
Keeling, S. V., 34n, 349
Kepler, Johannes, 12, 30n, 47, 48, 76, 78, 82, 98, 99, 100, 103, 111, 345, 347, 350, 351
al-Kindī, 46, 47n, 78n, 345
Kitāb al-nafs (Averroës), 47n, 343
Kneser, A., 349
Korteweg, D.-J., 100n, 101n, 102, 103n, 350
Koyré, Alexandre, 9, 52n, 82, 160n, 210n, 274n, 350
Kramer, P., 100n, 103n, 350
Kuhn, Thomas S., 10, 246n, 350

La Chambre, Marin Cureau de, 135–149 passim, 153, 157n, 344
Lagrange, J.-L., 158
Lalande, A., 34n, 350
Laplace, P. S. de, 314, 315n, 350
Laporte, J., 26n, 350
Leibniz, G. W., 104, 115, 148, 166n, 173, 174, 197, 345
Lejeune, A., 71n, 93n, 346, 350
Lejewski, C., 10
Le Lionnais, F., 350
Lettres de Descartes, ed. Clerselier, 48n, 49n, 57n, 58n, 86n, 343

Light, corpuscular view of, 210, 232, 233n, 239, 243, 245, 248, 252, 264, 273, 276, 284, 287–8, 289, 290, 294, 296, 312, 314, 315, 316, 317, 341; diffraction (inflexion, inflection) of, 185, 186, 273, 283, 310–11n, 312, 314, 321, 322, 332; diffusion of, 266, 277n; dispersion of, 257, 266–7, 286, 296, 328n; heterogeneity of, see White light; instantaneous propagation of, 46–60, 190, 200, 202, 207, 210, 348; periodicity of, 192, 212, 341; polarity (polarization) of, 226–9, 296, 314; rectilinear propagation of, 54, 56, 186, 210, 213, 214, 216, 277, 279, 283, 284, 313, 320; reflexibility of, 250n, 296n, 319n; compared with sound, 167, 199, 200, 210–11, 214, 221n, 261, 278, 283, 328, 339; finite velocity of, 47, 49n, 167, 185, 186, 191, 200, 202, 203, 205, 210, 211; velocity of light in different media, 94–5, 105–7, 114–16, 117, 194, 195, 196, 197, 201, 218, 315, 317; wave conception of, 54, 56, 60, 186, 191, 198, 199, 220–1n, 231, 233n, 277, 282–4, 313, 315, 341; see Reflection; Refraction; White light
Lohne, J., 100n, 350–1
Lucas, Anthony, 231n
La lumière, etc. (La Chambre), 138–9
lux & lumen, 49n, 188n

Mach, E., 104, 105n, 110n, 111n, 115, 215, 216, 225n, 284, 311, 351
Malebranche, Nicolas de, 259, 283n, 349, 351
Maraldi, Giacomo F., 205n
Marcus Marci, 351
Mathematics, 24n, 25, 29–30, 39–40, 43, 81–2, 128, 130–1, 136–7, 154–5, 248, 317; see géométrisation à outrance
Matter, 26, 51, 53, 55–6, 105–6, 138, 162, 164, 167–8, 296; subtle matter, 49n, 54, 66, 114, 160, 164, 174, 186, 209; see Ether
Maupertuis, P.-L. M. de, 157, 158, 314, 347

Maurolyco, Francesco, 63n, 345
Maxima and minima, 136, 144–5; see Principle of economy
Maxwell, James Clerk, 316
Mechanica (Aristotle), 70, 343
Mechanism, 26, 28–9, 32–3, 41–2, 48, 54, 78–9, 116, 162–70, 174, 184, 202, 220, 221, 229, 230, 233n, 244, 245, 270, 271, 276–7n, 287, 290, 293
Meditations (Descartes), 24, 25n, 26
Mersenne, Marin, 17n, 23, 24n, 25n, 27n, 35n, 36, 44, 50n, 58n, 60n, 61, 72n, 80n, 81, 82, 86, 87, 89n, 91, 101n, 104n, 106n, 112, 113, 121, 164, 346
Metaphysics, 24–7, 42, 136, 156
Meteorologica (Aristotle), 62n, 106n
Meteorologica (Averroës), 62n, 343
Meteorologica (Avicenna), 62n
Meteors (Descartes), 11, 17n, 19, 22n, 27, 60, 61, 62, 64n, 66, 86
Method, 17, 61–2, 231, 274, 285n, 287
Methodus ad disquirendam maximam et minimam (Fermat), 44–5
Meyerhof, M., 55n, 345
Meyerson, Emile, 291, 292, 351
Micrographia (Hooke), 185, 187, 188n, 190, 191n, 194n, 249n, 252, 253, 256n, 257, 258n, 259, 262, 263, 311n, 321, 322n, 324n, 325, 328, 329, 330, 331, 345
Michael Scot, 63n
Milhaud, G., 34n, 35n, 85, 100n, 102n, 104n, 106n, 119, 124, 125, 351
Minimum deviation, 235, 237, 265–6
Le Monde ou Traité de la Lumière (Descartes) 18, 28, 49n, 50, 51n, 52n, 53n, 54n, 60n, 104n, 188n, 347
Montucla, J.-E., 205n, 351
Moray, Robert, 266n
More, L. T., 144n, 351
Morin, J.-B., 20, 21, 22, 23, 49n, 50n, 58n, 188
Motte, Andrew, 346
Mouy, P., 85, 86n, 119, 165n, 183, 198n, 207n, 351
Multiplication of species, 77, 80–1, 216n

Mydorge, Claude, 86, 87n, 102, 103, 104n, 106n, 120, 126, 129, 130

Napier, MacVey, 351
Natural history, role of, 34–7, 171–2, 177, 180, 187, 229, 321, 322n
Nazīf, Muṣṭafā, 47n, 201n, 221n, 351
Newton, Isaac, 11, 12, 13, 14, 15, 16, 31, 56, 57n, 85, 87, 105n, 106, 111, 124, 135, 144n, 156, 158, 160, 161, 165n, 166–170, 181, 185, 198, 209, 214, 221, 226–230, Chapters IX–XIII, 344, 346–53
Note Books (Newton), 234n, 237n, 246n, 247n, 292n, 296n, 332n
Novum Organum (F. Bacon), 31n, 34, 36, 37, 38n, 171, 172n, 174n, 175n, 177n, 179n, 180n, 181n, 249, 322n, 343

Observation and experiment, see Experiment
Occam's razor, 279n
Occult qualities, 161, 162
Oldenburg, Henry, 195n, 234, 264, 267n, 268, 269, 270, 271n, 272, 284, 285n, 287n, 329, 331
Olympiodorus, 71n
On the theory of light and colours (Thomas Young), 327n
Optica (Alhazen), 47n, 72, 78, 93, 98, 99, 114, 207n, 210n, 345, 351
Optica (Euclid), 69, 344
Optica (Ptolemy), 71,93n,201n,346,350
Optica (Risner), 101n, 346
Optica (Witelo), 46n, 47n, 72n, 78n, 98n, 347
Optical Lectures (Newton), 105n, 236n, 237n, 247, 266n, 274n, 300, 346
Opticks (Newton), 11, 13, 31, 56, 169n, 211n, 214n, 226, 227n, 228n, 229, 230, 235n, 237n, 246n, 253n, 288, 290, 292, 293, 294, 296n, 299, 302, 305, 306n, 308n, 309, 311n, 319n, 320, 322, 330, 332, 333, 334n, 336n, 338n, 340, 342, 346, 350

L'optique (Ango), 195n, 196, 197, 201n, 343

Papanastassiou, Ch.-E., 351
Parallelogram method, applied to reflection, 70, 71, 75–6, 78, 82–4; applied to refraction, 96–7, 98–9, 107–10; Fermat's view of, 87–9, 136, 142
Paralipomena ad Vitellionem (Kepler), 47n, 72n, 77, 78, 98n, 345
Pardies, Ignace G., 11, 13, 186, 195, 196, 197, 200, 210, 213, 218, 231, 233, 235n, 242, 264–8, 269, 273, 277, 284, 285, 289, 340, 346
παρισότης, 144n
Pelseneer, J., 351, see Th. de Donder
Peripateticism, 233n, 245, 277n, 291, 292n, 309
Philoponus, 46, 352
Photon, 315
Physico-mathesis de lumine, etc. (Grimaldi), 185, 186n, 310n, 344
Planck's quantum of action, 318
Plempius, 58n
Poincaré, H., 281n
Polyak, S. L., 208n, 351
Popper, K. R., 9, 175n, 221, 317n, 351
Posthumous Works (Hooke), 190, 311n, 345
Price, D. J. de S, 244n
Priestley, Joseph, 351
Principia, 57n, 87, 124, 156, 160, 161, 164, 167, 169n, 214, 239n, 286n, 300, 302, 303, 304, 305, 317, 346
Principle of economy, as a principle of easier and quicker path, 96, 98, 137n; as a principle of easiest course, 139, 140, 142, 148, 155; as a principle of extremum, 156, 157n, 158; as a principle of least action, 157, 315; as a principle of least resistance, 148; as a principle of least time, 137, 148, 149, 150, 153–4, 155, 156, 197, 218–19; as a principle of shortest course, 136n, 138, 139; as a principle of shortest path or distance, 70–1, 136, 138, 139n, 155; as a principle of simplest

course, 139n; as a principle of simplicity, 136–7, 156, 157n
Principle of secondary waves, *see* secondary Waves
Principles (Descartes), 20n, 21n, 24n, 26n, 35n, 36n, 41, 42n, 43, 49n, 52n, 55n, 58n, 120, 162, 188, 209n, 261n, 346
the Print (Newton), *see* Waste Sheets
Probability of scientific knowledge, 43–4, 176, 184n, 275, 276, 285; *see* Certainty; *Vraisemblance*
Problemata (Aristotle), 69, 343
Problemata ad perspectivam et iridem pertinentia (Maurolyco), 63n, 346
Projet du Contenu de la Dioptrique (Huygens), 198–202, 221–2
Ptolemy, 23, 24, 71, 72, 93, 99, 114, 201n, 350

Quaestiones celeberrimae (Mersenne), 72n, 346
Quaestiones quaedam philosophicae (Newton), 56n
quies media, 80

Rainbow, 61–7
Ramus, Petrus, 110n, 346, 353
Rayleigh, J. W. Strutt, Lord, 281n
Recueil d'Observations . . . pour perfectionner l'astronomie, etc., p. 207n
Reflection, explained mechanically, 69–70, 71–82, 219, 221, 319–20; law of reflection, deduced kinematically, 80–85, 87–92, explained by economy principles, 70–1, 138–9, 156n; partial reflection, 220, 320, 337; total reflection, 108–9, 111–12, 113–16, 201, 219–20, 319–20
Réflexions ou projet de réponse à . . . Fermat (Rohault), 87n, 127
Refraction, 186, 199, 222; explained dynamically, 135, 299, 301–2; explained by economy principles, 96, 98, 136–58; explained hypothetically, 308–13, 315; explained mechanically,

27–8, 93–9 105–6, 107–10, 133–5, 192–4, 196–7, 202, 300–1; law of refraction, 189, 196, 197, 200, 201, 236, 274, 275, 296, 310, 311, 312, 313, 314, deduced from mechanical analogies, 107–11, 215, 216–18, deduced from principle of least time, 145–7, deduced from principles of dynamics, 302–8; reciprocality law for refraction, 201; double refraction, 185, 186, 221–6, 314, 315n; refractive power, 302, 322, 323, 324
Régis, Pierre Silvain, 346
Regulae (Descartes), 29, 32n, 33, 34, 38, 347
Reilly, Conor, 351
Rigaud, Stephen Jordan, 346
Rings, called 'Newton's rings', 321, 323, 324, 330, 333, 341; *see* Colours of thin plates
Risner, F., 47n, 72n, 93n, 98, 101n, 345, 346, 347
Rivet, André, 101n
Roberts, Michael, and E. R. Thomas, 231n, 262, 291, 351
Roberval, G. Pers. de, 86, 159, 160, 162n, 346–7
Robinson, Bryan, 351
Roemer, O., 167, 185, 186, 190n, 202, 203, 205, 206, 207, 290n, 348
Rohault, Jacques, 87, 113, 127, 129, 200, 347, 349
Roller, Duane, H. D., 346
Ronchi, Vasco, 98n, 110n, 351
Rosenberger, F., 351
Rosenfeld, L., 231n, 351, 352
Roth, L., 34n, 60n, 352
Rouse Ball, W. W., 237n, 352
Royal Society, 11, 181, 187, 234, 249, 251, 254n, 266n, 268, 274, 309, 327, 328n, 329, 330, 331, 332, 347, 348
Russell, L. J., 10

Sabra, A. I., 318n, 352
Sambursky, S., 46n, 55n, 352
Sarton, G., 352
Schramm, Matthias, 352

Schrecker, P., 352
Schuster, A., 281n
Scott, J. F., 49n, 58n, 67n, 100n, 110n, 300n, 352
Scriven, J., 352
Shirley, J. W., 100n, 352
Shofield, Robert E., 346
Simplicius, 55n
Smith, Norman Kemp, 352
Snell, Willebrord, 72n, 100, 101, 102, 105, 115, 304, 346, 347, 350, 352, 353
Snow, A. J., 352
Sound, see Light; Waves
Species, 48, 77, 81, 216n
Sprat, Thomas, 347
La statique, etc. (Pardies), 196n, 201, 346
Stock, Hyman, 352
Strong, E. W., 352
Suppes, Patrick, 352
Sur la loi de la réfraction extraordinaire, etc. (Laplace), 315n
Sur la seconde inégalité des satellites de Jupiter (Maraldi), 205n
Synthesis ad Refractiones (Fermat), 149, 150–2, 153n, 157

Tannery, P., 104, 144n, 352
Tannery, P., and C. Henry, 125, 344
Tanqīḥ al-manāẓir (Kamāl al-Dīn), 62n, 221n, 345
Tendency to motion, 50, 53, 54, 56, 59–60, 66, 79, 162, 188, 200, 202, 207, 208, 211, 277n, 282
Thales, 22n
Theodoric of Freiberg, 62n, 63n, 353
Theories (and discoveries) vs. hypotheses, 177n, 181, 187, 230 232, 284–5, 298–9, 312–13, 322n, 338, 342
Thomas, E. R., see Roberts, Michael, and E. R. Thomas
Toulmin, Stephen, 352
Traité de méchanique (Roberval), 159, 346, 347
Traité de physique (Rohault), 87, 347
Treatise on Light (Huygens), 11, 59n, 104n, 156, 166, 169, 170, 171, 183, 185, 186, 196, 198, 200, 201, 202, 203,
206–7, 210n, 211n, 212n, 214, 216n, 220n, 222, 223n, 225, 229, 345
Tschirnhauss, E. W. von, 173
Turbayne, C. M., 208n, 352
Turnbull, H. W., 234n, 264n, 329n, 346

Unicum opticae, catoptricae, et dioptricae principium (Leibniz), 148

Van Geer, P., 100n, 101n, 105n, 352
Vatier, Father, 17n, 18n, 22n, 28, 49n
Vavilov, S. I., 352
Velocity of light, see Light
vis centripeta, 166n; see Gravity
Vollgraff, J. A., 100, 101n, 346, 353
Vossius, Isaac, 100, 101, 104n, 300n, 347
Vraisemblance, 162, 182, 183, 203, 206, 225, 229, 269, 270, 275; see Certainty; Probability

Waard, Cornelis de, 49n, 100, 101n, 106n, 343, 344, 346, 353
Wahl, Jean, 353
'Waste Sheets' (Newton), 244–5, 276–7n, 292n
Waves, of light, see wave conception of Light; longitudinal waves, 201, 339–40; secondary waves, 197, 212, 214, 215, 216, 219, 222, 224, 283, 314; superposition of waves, 280, 283, 314, 328n, see 'Coalescence' of vibrations; sound waves, 196, 201, 278, 283, see Light compared with sound; transverse waves for light, 201, 228, 314, 339–40; water waves, 191, 196, 199, 200, 214, 215, 260n, 283, 338; wave-front, 192, 194, 195, 196, 197, 212, 214, 217, 218, 219, 224, 225, 254, 325; wave-length, 278, 283, 328; wave trains, 215, 281n
Westfall, Richard S., 56n, 232n, 233n, 245n, 246n, 250n, 296n, 332, 353

Whewell, William, 353
White, Harvey E., *see* Jenkins, Francis A. and Harvey E. White
Wisdom, J. O., 10
White light, heterogeneity of, representation of, 233–4, 242, 243, 244, 249, 250n, 252, 253, 259, 261, 264, 265, 271–2, 274, 276, 278–9, 280–2, 284, 288, 295, 297, 298, 348
Whittaker, E., 188, 195, 228n, 253, 282, 300n, 342n, 346, 353
Wickins, John, 235n
Wilde, Emil, 353
Winter, H. J. J., 97n, 353

Witelo, 12, 23, 24, 46n, 47, 48, 98, 99, 100, 103, 137n, 347
Wolf, A., 353
Wood, Alexander, 317, 353
Wood, R. W., 280n, 281n, 353
Würschmidt, J., 353

Young, Thomas, 313, 314, 320, 327, 328, 341, 353
Ysaac, 55n, 344

Zoubov, V., 353

10/23

THEORIES OF LIGHT

FROM DESCARTES TO NEWTON

A. I. SABRA

Harvard University

CAMBRIDGE UNIVERSITY PRESS

Cambridge

London   New York   New Rochelle

Melbourne   Sydney

Published by the Press Syndicate of the University of Cambridge
The Pitt Building, Trumpington Street, Cambridge CB2 1RP
32 East 57th Street, New York, NY 10022, USA
296 Beaconsfield Parade, Middle Park, Melbourne 3206, Australia

© Cambridge University Press 1981

First published by Oldbourne Book Co. Ltd 1967
First published by Cambridge University Press 1981

Printed in Canada

Library of Congress Cataloging in Publication Data
Sabra, A. I.
Theories of light, from Descartes to Newton.
Bibliography: p.
Includes index.
1. Light – History. I. Title.
QC401.S3 1981   535'.1   81-6108   AACR2
ISBN 0 521 24094 8 hard covers
ISBN 0 521 28436 8 paperback